Handbook of Cheese Chemistry

Food Chemistry, Function and Analysis

Series editors:
Gary Williamson, *Monash University, Australia*
Alejandro G. Marangoni, *University of Guelph, Canada*
Graham A. Bonwick, *AgriFoodX Limited, UK*
Catherine S. Birch, *AgriFoodX Limited, UK*

Titles in the series:
1: Food Biosensors
2: Sensing Techniques for Food Safety and Quality Control
3: Edible Oil Structuring: Concepts, Methods and Applications
4: Food Irradiation Technologies: Concepts, Applications and Outcomes
5: Non-extractable Polyphenols and Carotenoids: Importance in Human Nutrition and Health
6: Cereal Grain-based Functional Foods: Carbohydrate and Phytochemical Components
7: Steviol Glycosides: Cultivation, Processing, Analysis and Applications in Food
8: Legumes: Nutritional Quality, Processing and Potential Health Benefits
9: Tomato Chemistry, Industrial Processing and Product Development
10: Food Contact Materials Analysis: Mass Spectrometry Techniques
11: Vitamin E: Chemistry and Nutritional Benefits
12: Anthocyanins from Natural Sources: Exploiting Targeted Delivery for Improved Health
13: Carotenoid Esters in Foods: Physical, Chemical and Biological Properties
14: Eggs as Functional Foods and Nutraceuticals for Human Health
15: Rapid Antibody-based Technologies in Food Analysis
16: DNA Techniques to Verify Food Authenticity: Applications in Food Fraud
17: Advanced Gas Chromatography in Food Analysis
18: Handbook of Food Structure Development
19: Mitigating Contamination from Food Processing
20: Biogenic Amines in Food: Analysis, Occurrence and Toxicity
21: Nutrition and Cancer Prevention: From Molecular Mechanisms to Dietary Recommendations
22: Health Claims and Food Labelling
23: Nutraceuticals and Human Health: The Food-to-supplement Paradigm
24: Nutritional Signalling Pathway Activities in Obesity and Diabetes
25: The Chemistry and Bioactive Components of Turmeric
26: Foodomics
27: Food Proteins and Peptides: Emerging Biofunctions, Food and Biomaterial Applications
28: Handbook of Antioxidant Methodology: Approaches to Activity Determination
29: Fats and Associated Compounds: Consumption and Human Health

30: Oral Processing and Consumer Perception: Biophysics, Food Microstructures and Health
31: Development of Trans-free Lipid Systems and their Use in Food Products
32: Advanced Spectroscopic Techniques for Food Quality
33: Berries and Berry Bioactive Compounds in Promoting Health
34: Metabolism of Nutrients by Gut Microbiota
35: Bio- and Nano-sensing Technologies for Food Processing and Packaging
36: Edible Fungi: Chemical Composition, Nutrition and Health Effects
37: Chemistry and Nutritional Effects of Capsicum
38: Pigmented Cereals and Millets: Bioactive Profile and Food Applications
39: Dietary Supplements with Antioxidant Activity: Understanding Mechanisms and Potential Health Benefits
40: Handbook of Cheese Chemistry

How to obtain future titles on publication:
A standing order plan is available for this series. A standing order will bring delivery of each new volume immediately on publication.

For further information please contact:
Book Sales Department, Royal Society of Chemistry, Thomas Graham House, Science Park, Milton Road, Cambridge, CB4 0WF, UK
Telephone: +44 (0)1223 420066, Fax: +44 (0)1223 420247,
Email: booksales@rsc.org
Visit our website at www.rsc.org/books

Handbook of Cheese Chemistry

Edited by

Michael H. Tunick
Drexel University, USA
Email: mht39@drexel.edu

ROYAL SOCIETY
OF **CHEMISTRY**

Food Chemistry, Function and Analysis No. 40

Print ISBN: 978-1-83916-309-8
PDF ISBN: 978-1-83916-990-8
EPUB ISBN: 978-1-83916-991-5
Print ISSN: 2398-0656
Electronic ISSN: 2398-0664

A catalogue record for this book is available from the British Library

The Royal Society of Chemistry is a charity, registered in England and Wales, Number 207890, and a company incorporated in England by Royal Charter (Registered No. RC000524), registered office: Burlington House, Piccadilly, London W1J 0BA, UK, Telephone: +44 (0)20 7437 8656.

Visit our website at www.rsc.org/books
Printed in the United Kingdom by CPI Group (UK) Ltd, Croydon, CR0 4YY, UK

Preface

Cheese is a food that has been prepared and savoured for millennia. Its chemistry has been investigated since the nineteenth century, and a clear picture of the processes that occur when milk is converted to cheese has emerged over the years. This book describes these processes, including recent developments, as follows.

Chapter 1 introduces the reader to the chemistry of cheese. Starter cultures and coagulants are added to milk to start the process, which are covered in Chapter 2. Nonstarter lactic acid bacteria are also responsible for cheese ripening, as described in Chapter 3. Chapter 4 details the entire process of cheese manufacture and Chapter 5 focuses on the intricate chemistry of ripening. Accelerated ripening techniques, which follow in Chapter 6, allow cheesemakers to get their product to market faster and are constantly being explored. The chemistry behind flavour development is also an ongoing area of research and is outlined in Chapter 7. The microstructure of the cheese matrix changes as the product ripens, which is demonstrated in Chapter 8. These changes affect the rheology and texture of cheese, which is explained in Chapter 9. The cheesemaking procedures used by artisans as opposed to large companies lead to noticeably different results as detailed in Chapter 10. Chapter 11 summarizes the data on cheese production, consumption, and nutrition of cheeses made from the milks of cows and other mammals. The trend toward non-dairy cheese production is described in Chapter 12 along with the chemistry underlying these products.

This volume encompasses the chemistry of all aspects of the cheese-making process. The references to the literature, over 1300 in total, will provide further information to the interested reader. The authors and I hope that this book will expand your understanding of this fascinating field.

Michael H. Tunick
Drexel University

Food Chemistry, Function and Analysis No. 40
Handbook of Cheese Chemistry
Edited by Michael H. Tunick

Published by the Royal Society of Chemistry, www.rsc.org

Contents

Food Chemistry, Function and Analysis No. 40
Handbook of Cheese Chemistry
Edited by Michael H. Tunick

Published by the Royal Society of Chemistry, www.rsc.org

CHAPTER 1

Introduction to Cheese Chemistry

MICHAEL H. TUNICK*[a] AND YOUNG W. PARK[b]

[a] Department of Food & Hospitality Management, College of Nursing & Health Professions, Drexel University, Philadelphia, PA 19104, USA;
[b] Georgia Small Ruminant Research & Extension Center, Fort Valley State University, The University System of Georgia, Fort Valley, GA 31030, USA
*Email: mht39@drexel.edu

1.1 The Basics

Cheesemaking is not only a chemical process, but it also requires desired microbial growth and enzymatic conversion of food components, which qualifies it as a fermented food.[1] Enzymes and lactic acid bacteria (LAB) are responsible for a wide range of reactions and metabolic pathways that result in textures, flavours, and aromas that many consumers enjoy, while providing proteins, lipids, carbohydrates, vitamins, and minerals that contribute to a healthy diet (see Chapter 11). Nonstarter lactic acid bacteria (NSLAB) also find their way into cheese and contribute to the final product (see Chapter 3).

Cheese starts with milk of course, though consumers sometimes opt for non-dairy "cheeses" (see Chapter 12). Most cheeses and other dairy products worldwide come from bovine milk, but milks of goats, sheep, buffalo, mare, and yak are important for human diet and subsistence in many parts of the globe (see Chapter 11). Although more people worldwide drink or consume goat milk and its products than those of any other species,[2,3] cows produce the most milk in developed and western countries, and a few species such as goats, sheep, and buffalo supply lesser amounts of milk. In developing countries,

Food Chemistry, Function and Analysis No. 40
Handbook of Cheese Chemistry
Edited by Michael H. Tunick

Published by the Royal Society of Chemistry, www.rsc.org

minor dairy species such as yak, mare, camel, reindeer, llama, donkey, and moose are nutritionally and economically important.[4] The milk and milk products of non-bovine dairy species such as goats serve three types of markets in many parts of the world: (i) home consumption, (ii) specialty gourmet interests, and (iii) medical needs.[3] The compositional differences between the milks of these species lead to contrasts in the resulting cheeses.

The procedures for making cheese differ in details from one variety to the next (see Chapter 4), but most follow a general outline.[5] The process starts by adding a starter culture (a specified mixture of LAB strains) to milk, causing lactic acid to form and the pH to drop (see Chapter 2). The cheesemaker then adds rennet (an enzyme preparation that breaks the κ-casein molecule), and the milk coagulates into a solid curd with liquid whey. The action of the bacteria and enzymes subsequently increases when the milk is heated, and continues throughout the manufacturing process and during aging (see Chapter 5). Much of cheesemaking involves removal of whey from the curd by draining and pressing. The storage conditions, especially temperature, humidity, and the type of wrapping or packaging (if any), will affect some characteristics of the product. Some operations employ accelerated ripening to cut down on refrigeration and inventory expenses (see Chapter 6).

The smallest artisanal cheesemakers use the same basic procedures as the largest commercial producers, with the difference in scale resulting in changes in the final product (see Chapter 10). Most of the varieties of cheese in the world are artisanal in nature and unique, and cheese lovers seek them out.

Scientists are still elucidating some of the chemical and microbial processes that go into cheesemaking, which are quite complicated. For instance, we do not know the exact origins of some compounds responsible for flavour and aroma. However, we do know a great deal about the conversion of liquid milk to solid cheese, and this book will describe the chemistry behind it.

1.2 Quality Factors

Microflora, the activity of milk enzymes, and relatively small variations in cheese composition may influence cheese quality but become dominant only under conditions where the principal determinants (moisture, salt, and pH) are within appropriate limits.[6,7] Considering the continued and growing presence of dairy foods in consumers' diets, it is important for food manufacturers to create high quality products in order to maintain a competitive advantage. Cheese quality is determined by its major attributes, such as appearance, microbiological safety, flavour, texture, functionality, and nutritive value, as described below.[7,8]

1.3 Appearance

Appearance is especially important from the consumers' perspective since the appearance of cheese is usually the only attribute for assessing cheese quality when they make a purchase. Appearance attributes include colour, the presence or absence of mould as appropriate for the variety, and the presence or absence of eyes or other openings. The consumer knows from experience if the appearance of cheese is defective or not, and if the product does not look right, its flavour, texture, and functionality will be suspect. As a result, retailers strive to make the appearance of cheeses acceptable to consumers.[6] Even the package greatly affects consumers' attitudes toward the cheese inside.[9]

1.4 Safety

Aged cheese is a very safe dairy food, whereby the consumer can have confidence on the safety of the product. Most frequently, US and Canadian cheese-related outbreaks of foodborne illness are due to post-pasteurization contamination or use of raw milk, but these cases are quite infrequent.[10] Since consumers cannot assess the microbiological quality of cheeses on their own, they must rely on the manufacturer, retailer, or public health inspectors to ensure the microbial safety. In addition, the consumers are heavily dependent on the nutrition label of the package of the product when they make purchasing decisions.

1.5 Flavour

The flavour of milk, cheese, and other dairy products is mainly due to chemical reactions, including Maillard reactions and proteolysis, and lipolysis of cheese nutrients.[11,12] Cheese flavour is a vital part of quality control and overall consumer appeal.[13,14] A deeper understanding of flavour and its causes in cheese often comprises identifying specific flavour compounds and isolating the specific reaction or metabolic pathways that produce them.[15] Maturation directly affects the flavours of cheeses and involves the interplay of several factors, such as proteolytic and lipolytic enzymes, storage temperature and period, salting, pH of the curd, and humidity.[11,16,17] The specific catabolic compounds generated during ripening processes will influence the characteristic flavours of cheeses (see Chapter 7). The products of these reactions result in a vast array of flavour compounds.[18,19] These reactions take place throughout the cheesemaking process and are concentrated in the ripening phase.[11,20] Most of these reactions occur due to both endogenous (naturally present in the milk) and exogenous (added to the milk) enzymes.[21] Proteases and lipases mediate many of the most important flavour-generating reactions in goat and cow milk cheeses by lipolysis and proteolysis.[11,17] Numerous biochemical and physical changes can occur in cheeses during manufacture, distribution, and storage processes due to ripening and degradation of nutrients in the products.[11] Catabolism of carbohydrates, proteolysis, and lipolysis are the primary processes in cheese ripening, with a variety of chemical, physical, microbiological, textural, and rheological changes taking place, usually under controlled environmental conditions.[11,16,17,22]

1.5.1 Carbohydrates

Lactose is the main carbohydrate in cheese curds. Fresh cheese curd contains approximately 1% lactose, which the starter LAB converts to lactic acid (mainly the L-isomer), usually within 24 h.[6] The L-lactic acid is racemized to DL-lactic acid by NSLAB or catabolised to CO_2 and H_2O in mould-ripened and smear-ripened cheese.[6] *Propionibacterium freudenreichii* subsp. *shermanii* breaks down lactic acid into propionic and acetic acids, CO_2, and H_2O in Swiss-type cheese, where the CO_2 causes the formation of characteristic eyes in such cheese. The pH of most cheeses increases during aging due to the catabolism of lactic acid and, in the case of smear- or mould-ripened varieties, to the production of NH_3 at the surface, which diffuses into the cheese. The pH of Camembert cheese, for instance, increases from around 4.6 to 7.5.[6] The pH of Cheddar cheese does not normally increase, probably because the catabolism of residual lactose (1%) in the curd reduces the pH to a value below the buffering maximum of cheese (pH 5.2). Washing the curd or reducing the lactose content of milk for Cheddar cheese avoids the pH increase post-manufacture. Cheddar has a mild flavour, while high-lactose Cheddar has a harsh, strong flavour.[23]

1.5.2 Lipids

Aging of cheese unavoidably results in lipolysis. However, little lipolysis occurs in most cheese varieties since LAB or NSLAB are weakly lipolytic, as is indigenous milk lipase if raw milk is used.[6] Extensive lipolysis occurs in some hard Italian-type cheeses, *e.g.*, Provolone and Pecorino varieties, for which added pre-gastric esterase is responsible, and in Blue cheese because of the action of *Penicillium roqueforti*.[6] Fatty acids break down into lactones or esters that give characteristic flavours to varieties such as Provolone and Pecorino. *P. roqueforti* converts FA to methyl ketones in blue-veined cheese, and these account for the characteristic peppery taste of these varieties. Some methyl ketones break down into secondary alcohols, causing off-flavours.[6]

1.5.3 Proteins

Proteolysis is probably the most important biochemical event that gives a major impact to flavour and texture of most cheese varieties, especially cheeses that undergo ripening by internal bacteria.[11,16] Chymosin in the rennet initially hydrolyses the casein, and plasmin, an endogenous enzyme resistant to pasteurization, attacks β-casein. High-cook varieties almost completely inactivate chymosin, and in this case plasmin is the principal agent of primary proteolysis.[6] Chymosin and plasmin produce peptides that reduce to smaller peptides and amino acids by proteinases and peptidases of starter LAB and NSLAB. Small peptides contribute positively to the background brothy flavour of cheese, while some peptides are bitter. Furthermore, in proteolysis cheeses, many amino acids give characteristic flavour and serve as substrates for many catabolic reactions, such as decarboxylation, deamination, and transamination catalysed by the enzymes of LAB, NSLAB, and secondary culture.[6]

1.6 Texture, Rheology, and Functionality

Flavour is an important characteristic of cheese, but its physical behaviour is also key. People do not want to consume food that does not feel right in their mouth. The relative amounts of protein, fat, and water are the main factors in the cheese structure. The chemical processes taking place in cheese also affect its structure, including the size of fat globules, their arrangement in the casein matrix, the presence of calcium lactate crystals, and more (see Chapter 8). The physical attributes of cheese come under the categories of texture, rheology, and functionality (see Chapter 9).

1.6.1 Texture

Texture is a sensory property of food, derived from its structure and manifested by the five senses.[23] It is particularly important to cheese lovers, who derive much of their enjoyment from the way the cheese feels as they eat it (see Chapter 10). Sensory panels provide information on texture (and flavour), while instrumental tests such as texture profile analysis (TPA) offer non-biased results. Scientists constantly strive to match the results of human sensory testing with instrumental results.[24–26] Terms found in both sensory testing and TPA include hardness, fracturability, adhesiveness, springiness, and cohesiveness.

1.6.2 Rheology

Scientists define rheology as the study of the flow and deformation of matter.[27] Cheese is an interesting material to test due to its unique structure, and it is not surprising that one of the earliest studies concerning the rheology of food (published in 1937) dealt with cheese.[28] The field has expanded greatly since these tests of compression under constant load. Other tests include creep (sudden imposition of stress), stress relaxation (removal of force during a creep experiment), steady shear (internal portions sliding past each other), and small amplitude oscillatory shear (stress and strain varying harmonically with time).[29] The experiments provide information on mechanical properties related to the strength of the protein bonds.

1.6.3 Functionality

The functionality of cheese includes sliceability, adherence (stickiness), meltability, and stretchability. Sliceability and adherence are of concern mainly to processors who have to cut or shred cheese blocks, while meltability and stretchability are mainly related to the use of cheese as an ingredient.[8] For instance, the primary quality attributes of mozzarella include its meltability and stretchability in foods such as pizza.[30] Other parameters such as wheying off (whey escaping from the cheese after packaging) and oiling off (the oil appearing on a layer of heated cheese) may be important in some varieties and applications.

1.7 This Book

The remaining chapters of this book will more fully describe the chemistry that goes into the preparation and consumption of cheese, beginning with the starter culture and coagulants needed to convert milk into curds and whey (Chapter 2). NSLAB play a part in cheesemaking, which is the subject of Chapter 3. Chapter 4 covers the manufacture of cheese in detail and Chapter 5 describes the ripening that occurs during storage. The cheesemaker may accelerate ripening in some cases, and Chapter 6 deals with that aspect. As mentioned above, cheese does not have to come from cow milk, and Chapter 11 shows comparisons of cheeses from cow, goat, and sheep milk. Chapter 7 provides a review of the generation of flavour compounds, Chapter 8 covers the cheese microstructure, and Chapter 9 provides an overview of the rheology and texture of cheese. Artisanal cheesemakers are responsible for many varieties, as shown in Chapter 10. Chapter 11 includes the production, consumption, and nutrition of cheese, and Chapter 12 reviews the expanding market of non-dairy cheese.

References

1. M. L. Marco, M. E. Sanders, M. Gänzle, M. C. Arrieta, P. D. Cotter, L. De Vuyst and C. Hill, *et al.*, *Nat. Rev. Gastroenterol. Hepatol.*, 2021, **18**, 196.
2. Y. W. Park, *J. Dairy Sci.*, 1990, **73**, 3059.
3. Y. W. Park and G. F. W. Haenlein, in *Handbook of Food Products Manufacturing*, ed. Y. H. Hui, Wiley, New York, NY, 2007, p. 447.
4. Y. W. Park, G. F. W. Haenlein and W. L. Wendorff, *Handbook of Milk of Non-Bovine Mammals*, Wiley-Blackwell, Oxford, UK, 2nd edn, 2017, p. 1.
5. M. H. Tunick, *The Science of Cheese*, Oxford University Press, New York, NY, 2014.
6. P. F. Fox and T. P. Guinee, in *Milk and Dairy Products in Human Nutrition*, Wiley-Blackwell, Oxford, UK, 2013, p. 357.
7. G. Moatsou and Y. W. Park, in *Handbook of Milk of Non-Bovine Mammals*, Wiley-Blackwell, Oxford, UK, 2nd edn, 2017, p. 84.
8. P. F. Fox and T. M. Cogan, in *Cheese: Chemistry, Physics and Microbiology*, ed. P. F. Fox, P. L. H. McSweeney, T. M. Cogan and T. P. Guinee, Elsevier, Oxford, UK, 2004, vol. 1, p. 583.
9. J. M. Murray and C. M. Delahunty, *Food Qual. Prefer.*, 2000, **11**, 419.

10. E. A. Johnson, J. H. Nelson and M. Johnson, *J. Food Prot.*, 1990, **53**, 519.
11. Y. W. Park, *J. Dairy Sci.*, 2001, **84**, E84.
12. M. A. Drake and G. V. Civille, *Compr. Rev. Food Sci.*, 2003, **2**, 33.
13. M. A. Drake, *J. Dairy Sci.*, 2004, **87**, 777.
14. M. E. Carunchia Whetstine and M. A. Drake, in *Handbook of Milk of Non-Bovine Mammals*, ed. Y. W. Park and G. F. W. Haenlein, Blackwell, Oxford, UK, 2008, p. 107.
15. M. Johnson and J. Lucey, *J. Dairy Sci.*, 2006, **89**, 1174.
16. P. F. Fox, *J. Dairy Sci.*, 1989, **72**, 1379.
17. Y. K. Jin and Y. W. Park, *J. Dairy Sci.*, 1995, **78**, 2598.
18. M. Yvon and L. Rijnen, *Int. Dairy J.*, 2001, **11**, 185.
19. L. Marilley and M. Casey, *Int. J. Food Microbiol.*, 2004, **90**, 139.
20. F. José Delgado, J. González-Crespo, R. Cava and R. Ramírez, *LWT–Food Sci. Technol.*, 2012, **48**, 268.
21. Y. Collins, P. L. H. McSweeney and M. Wilkinson, *Int. Dairy J.*, 2003, **13**, 841.
22. N. C. Bertola, A. E. Bevilacqua and N. E. Zaritzky, *J. Dairy Sci.*, 1992, **75**, 3273.
23. A. S. Szczesniak, *Food Qual. Prefer.*, 2002, **13**, 215.
24. M. A. Drake, P. D. Gerard, V. D. Truong and C. R. Daubert, *J. Texture Stud.*, 1999, **30**, 451.
25. E. A. Foegeding, J. Brown, M. Drake and C. R. Daubert, *Int. Dairy J.*, 2003, **13**, 585.
26. E. A. Foegeding and M. A. Drake, *J. Dairy Sci.*, 2007, **90**, 1611.
27. M. H. Tunick, *J. Dairy Sci.*, 2000, **83**, 1892.
28. J. G. Davis, *J. Dairy Res.*, 1937, **8**, 245.
29. M. H. Tunick and E. J. Nolan, in *Physical Chemistry of Food Processes*, ed. I. C. Baianu, Van Nostrand Reinhold, New York, NY, 1992, vol. 1, p. 273.
30. P. S. Kindstedt and P. F. Fox, *Crit. Rev. Food Sci. Nutr.*, 1993, **33**, 167.

CHAPTER 2

Coagulants and Starter Cultures

IAN B. POWELL

Faculty of Science, Engineering and Technology, Swinburne University of Technology, Hawthorn, Victoria 3122, Australia
Email: ianpowell.tycaws@gmail.com

2.1 Introduction

Archaeo-chemical evidence[1] shows that cheese-like foods have been made for at least the past 7000 years. There are many cheese types,[2–5] and most cheesemaking is still based on centuries-old traditional practices. The defining features of the process of turning milk into cheese[6] are (i) coagulation of milk into curd and (ii) draining (often with the assistance of pressing) of the aqueous material (whey) expressed from that curd and, for many cheese types, (iii) maturation (ripening) to achieve their final form.

This chapter addresses two key factors in cheesemaking: agents used to induce coagulation (coagulants) and microbial cultures used to produce acid and flavour (starter cultures), with examples and reference to a selection from the vast recent and historical literature.

2.1.1 The Components of Milk

Milk is a complex mammalian secretion.[7] Its major constituents are water, lactose, fats, proteins, and calcium salts. Lactose (a disaccharide of galactose and glucose) and other soluble components exist in aqueous solution (the 'serum phase' of milk). The fat content of milk (mostly triacyglycerols[8]) exists in globules,[9] and so milk can be regarded as an oil-in-water emulsion. Most of the proteins are broadly classified as caseins (traditionally defined as proteins

Food Chemistry, Function and Analysis No. 40
Handbook of Cheese Chemistry
Edited by Michael H. Tunick

Published by the Royal Society of Chemistry, www.rsc.org

that coagulate at pH 4.6) or as serum proteins (also known as whey proteins).[10,11] In the ruminant species widely used in dairying, around 70–80% of the milk protein content (depending on the species) is casein protein.

2.1.2 Caseins and Casein Micelles

There are four casein types (known as α_{s1}-, α_{s2}-, β-, and κ-casein) in the common dairying species, and these are now well characterized as proteins[10–13] and at the genetic level.[13–15] Not all mammalian species have the typical casein types, but their behaviour is remarkably similar.[11–13,16] The caseins of bovine milk are by far the most thoroughly studied. Genetic variants exist within each type, and individual molecules vary in the extent of post-translational modification (phosphorylation, glycosylation, disulphide bonding).[11,15,17,18]

Caseins form roughly spherical particles (known as casein micelles) which exist in stable colloidal dispersion. Bovine casein micelles, for example, average about 120 nm in diameter[11] at a density of 10^{14}–10^{16} mL^{-1} (about 240 nm apart).[15] Casein micelles are held together predominantly by non-covalent interactions: the proteins interact with each other (hydrophobic association and hydrogen bonding have been invoked as the key influences), and the proteins interact with calcium phosphate 'nanoclusters' ('micellar' or 'colloidal' calcium phosphate, CCP) *via* phosphorylated amino acids, particularly the multiple phosphoserine residues of α_{s1}-, α_{s2}-, and β-caseins. A typical bovine casein micelle is composed of (by dry weight) about 93% protein and 7% 'ash' (mostly CCP), with tens of thousands of casein molecules[19,20] and some hundreds of CCP nanoclusters.[21,22] Casein micelles have an open internal structure (water and its solutes can enter and leave), and the precise relative quantities of the four caseins and CCP within a micelle vary depending on the micelle size, pH, temperature, and calcium and phosphate solution equilibria.

Despite considerable research, the internal details of the casein micelle structure are yet to be fully resolved and competing models continue to be debated.[19,21,23–28] Structural models involving distinct precursor units ('sub-micelles') have largely given way to models involving a network of interactions between proteins and CCP nanoclusters. In the absence of CCP, caseins can be isolated as aspherical heteromeric particles ('primary casein particles')[19] which modelling suggests contain nine casein molecules.[26] Not to be conceptually confused with sub-micelles, these possibly represent a non-random repeating structural motif.

κ-Casein is evidently located on the outer surface of casein micelles, where it interacts with the other caseins of the micelle (*via* its amino-terminal hydrophobic domain; in bovine casein, residues 1–105) and with the external serum phase of the milk (*via* its carboxy-terminal negatively-charged hydrophilic domain, residues 106–169) but not significantly with calcium phosphate. The presence of κ-casein in this location appears to be essential for maintaining micelle stability and for preventing aggregation or merging of adjacent micelles, with micelles kept apart by a combination of electrostatic repulsion and steric interference (both provided by the protruding C-terminal domain of κ-casein). This is sometimes described as the 'hairy micelle' model of the micelle structure, though the conformation of the protrusion has not been definitively elucidated; indeed, it may be free to adopt various conformational states.[29,30]

2.2 Coagulation and Coagulants

Coagulation of casein micelles to form cheese curd is achieved by abolishing the repulsive effect of the negatively-charged κ-casein 'hairs' on the micelle surface and allowing micelles to approach and interact with each other. The micelles then associate into a long-range network which entraps other components of the milk (including fat globules and the milk aqueous phase and its constituents). Curd handling procedures (*e.g.* acidification, cutting, stirring, heating, draining, pressing, salting) play a large part in determining the extent of syneresis (curd shrinkage and concomitant whey expulsion) and the essential microstructural characteristics of the cheese produced.[2,3,5]

2.2.1 Coagulation *Via* Acidification Without Heating

The manufacture of some cheese types (*e.g.* 'cottage' cheese, quark, fromage frais) is based on coagulation by acidification, in some cases aided by a small degree of proteolysis (Section 2.2.3) to improve yield and produce a stronger gel.[31,32] For most cheeses, acidification is achieved using starter cultures (see Section 2.3), but some cheeses are made *via* addition of acidulants such as lactic acid, citric acid, glucono-δ-lactone or food-grade inorganic acids. Such acidulants are sometimes described as 'precipitants'[4] or 'flocculants'[33] rather than coagulants.

Coagulation of milk by acidification is most simply thought of in terms of aggregation of casein micelles at or near their isoelectric point (*ca.* pH 4.6).[25,34] Acidification causes a reduction in the net charge of the casein molecules, notably including the surface-located κ-casein; measurements show a rapid decrease in zeta potential[34] below pH 5.0. However, across the range of temperatures used for curd formation (*ca.* 20–43 °C[31]), the coagulum structure (and so its rheological properties and propensity for syneresis) is affected by both the temperature and the rate of acid production.[31,34–36] Acidification leads to pH-dependent solubilization of CCP (time-dependent, but potentially near-complete by *ca.* pH 5.3) and to internal protein rearrangements associated with micelle shrinkage (by about 14% by pH 5.4–5.5[37,38]). With acidification below *ca.* pH 5.4, micelles increasingly interact with each other. Coagulation occurs as the pH drops below *ca.* pH 4.9–5.0.[37]

2.2.2 Coagulation *Via* Acid and Heat

2.2.2.1 Cheeses from Milk

At its native pH of *ca.* 6.7, bovine milk will not readily coagulate when heated, typically withstanding 140 °C for at least 15–20 min[39,40] before visible coagulation occurs. However, a combination of acid and heat can readily bring about coagulation.[37,41–43]

If milk is heated (above *ca.* 65–70 °C) prior to acidification, β-lactoglobulin (a serum protein) begins to denature and complex with κ-casein, by a combination of non-covalent interactions and disulphide bonding.[42,44] Other serum proteins (α-lactalbumin, and bovine serum albumin to a lesser extent) can also be involved in these complexes. This facilitates acid-induced coagulation by increasing the effective isoelectric point of the casein micelle surface.[34,44,45] It is also likely that bridging links between casein micelles are formed by protein complexes created in the serum phase. After heating, acid-induced coagulation occurs at a higher pH, the coagulum is firmer, and there can be a commercially significant increase in cheese yield due to incorporation of whey protein (and associated water) into the coagulum. These effects are accentuated by higher heat treatments: for example, it has been experimentally estimated that heat treatment of bovine milk at 90 °C for 10 min results in >95% denaturation of β-lactoglobulin (>70% of α-lactalbumin), with *ca.* 65% (50%) associated with the casein micelle and most of the remainder in soluble aggregates.[46] Heating at 85 °C for 20 min results in aggregation beginning at pH 5.8, with coagulation at pH 5.3.[37]

Paneer is an example of cheese made by acid-heat coagulation of milk.[47,48] At the domestic scale, acidification is commonly performed with acidulants like lemon juice, vinegar or yoghurt. Industrially, citric acid is widely used[48] and food regulations (*e.g.* in India[33]) permit various acidulants 'such as lactic acid, citric acid, malic acid, vinegar, glucono-δ-lactone, and sour whey.' Flavour can be affected by the choice of acidulant. Yield, moisture content and texture are affected by the heating conditions (typically 80–90 °C, for up to 15 min), the acidification temperature (usually slightly cooler, *ca.* 70–80 °C) and the coagulation pH (*ca.* pH 5.2–5.5). Curd particles are collected, drained, and lightly pressed. Acid-coagulated Queso Blanco is produced in a similar manner, though salted before pressing.[47]

Goat and sheep milk are typically less stable to heat than cows' milk, and some evidence indicates that their whey proteins undergo a higher level of complexation with the casein micelles.[49,50] However, the differences in complexation and stability are at least partly dependent on pH (with associated changes in soluble ionic calcium) and the particular time-temperature treatment used.[40,51]

2.2.2.2 *Cheeses from Whey*

The whey produced as a by-product of proteolysis-based cheesemaking (Section 2.2.3) typically contains less than 1% (w/v) protein, but a significant proportion of this can be recovered in the form of cheese made by acidifying and then heating the whey (conditions vary: pH 5–6; 80–100 °C) and collecting the curd that forms.[47,52,53] Protein recovery can be enhanced by adding milk to the whey (*e.g.* 5–10% v/v milk); the added casein facilitates curd formation and incorporates whey proteins.[53] Ricotta is the best-known of these cheeses.

2.2.3 Coagulation *Via* Proteolysis

The key event in milk coagulation *via* proteolysis is removal from κ-casein of the projecting negatively-charged C-terminal domain (known as the caseinomacropeptide or, acknowledging its typical glycosylation, glycomacropeptide). The remainder of the κ-casein molecule (known as *para*-κ-casein) remains associated with the micelle. Without the projecting domain and its charge, the largely hydrophobic casein micelle surface is able to approach and interact with adjacent micelles, leading to coagulation. The desired rate and extent will vary depending on the type of cheese being made.[54–57] The mechanisms, kinetics and measurement of coagulant-induced coagulation are complex.[58,59]

The use of proteolytic coagulants (enzyme preparations, crude or purified) from both plant and animal sources has been recorded since ancient times.[60] Perhaps 75% of cheese production today involves proteolytic coagulation,[57] mostly using industrially-produced coagulants, although some artisanal cheesemakers prepare their own coagulant.[54–57] In the academic usage, 'coagulant' is the common generic term, and 'rennet' is reserved for animal-derived coagulants. However, in the informal and industrial usage, 'rennet' is sometimes used generically and in casual descriptors such as 'non-animal rennet'.

The common coagulant enzymes are all aspartyl proteinases (EC[61] 3.4.23; also known as aspartic peptidases, and other similar terminology), endopeptidases which show significant similarities in amino acid sequence and tertiary structure, folding to form a two-lobed structure with homologous N- and C-terminal domains forming a substrate-binding cleft in which two aspartate residues (one from each domain) are involved in hydrolysis of the target peptide bond.[57,62–65]

Assays of general proteolytic activity are not useful measures of coagulant activity, and so a direct measure of coagulation under standard conditions is used, the International Milk Clotting Unit (IMCU).[55,66,67] During cheese manufacture, excessive general proteolytic activity can lead to a lower yield as the peptides released may be lost in the whey rather than retained in the curd.[55,68]

In cheesemaking, proteolytic coagulation usually occurs in combination with acidification.[25,69,70] At the native pH of bovine milk (*ca.* pH 6.7), micelle aggregation requires at least 85% cleavage of κ-casein. Acidification reduces the charge on micelles, decreasing the extent of proteolysis needed to achieve coagulation (equally, proteolysis permits coagulation at a higher pH). Acidification also increases the activity of aspartyl proteinases; the combined effect is a *ca.* 3-fold reduction in coagulation time at pH 6.3[71,72] compared with that at pH 6.7, and it is *ca.* 5-fold at pH 6.0.[71] Acidification also leads to dissolution of CCP during cheesemaking. Dissolution of CCP and proteolysis of casein in the maturing cheese are the major determinants of textural development.[73]

Prior heating of milk causes complexation with whey proteins and can be used to increase cheese yield by incorporating whey proteins and associated water into the curd. Mild heating (*e.g.* pasteurization) can be sufficient to give a commercially significant increase in cheese yield. However, excessive heat treatment impairs proteolytic coagulation as coagulation of micelles is physically blocked by the binding of whey protein aggregates to the micelle surface.[69] Even so, the heat-treated micelles of a milk-grown starter culture are largely incorporated into cheese curd.[74]

For maturing cheeses, some proteolysis is necessary for normal development of flavour and texture, but excessive proteolysis can lead to abnormal textural softening and to the generation of bitter-tasting hydrophobic peptides (*e.g.* the β-casein fragment f193–209).[75] High coagulation activity relative to general proteolytic activity (C/P)[76] is considered desirable. The residual proteolytic activity in a maturing cheese is influenced by the choice of coagulant[77,78] and the cheesemaking procedure, depending on factors such as the extent of temperature-dependent inactivation during cheesemaking[78–81] and whether enzymes preferentially bind to the caseins at the pH of separation of curd and whey.[75,81,82]

2.2.3.1 Animal Rennet and Its Enzymes

Animal-derived coagulants (rennet) have historically been obtained from the stomachs of various animals (as diverse as hares, pigs and chickens),[81,83] but predominantly from the abomasum (variously known as the fourth stomach, true stomach, vell, *etc.*) of ruminants (mostly cattle, sheep, goats).[55,84] In the simplest forms of rennet technology, a piece of abomasum (fresh, salted, dried, *etc.*) can be mixed directly with milk[83] or soaked (*e.g.* in whey or brine) to produce an extract with coagulant activity. For example, Edam cheesemaking in the 1850s used an extract made by soaking calf abomasum strips in sweet whey for 24 h.[85] Dried calf stomachs macerated in warm (40–45 °C) acid whey were traditionally used in making Gruyère, Emmental and other cheeses;[86,87] such preparations combined proteolytic activity with starter activity. Dried, salted and ground kid and lamb rennet pastes are still used for various southern European cheeses,[88] often made from stomachs containing their last feed of milk.[89] These preparations contain significant levels of pre-gastric enzymes (notably lipase and esterase activities) in addition to proteinases, the action of which gives cheeses a characteristically piquant flavour.

Bovine calf rennet is historically the most widely used[84] and has been mass-produced as a stable liquid rennet extract (enabling standardization of activity and widespread distribution) since 1874.[54] In modern manufacture, abomasa are held frozen until enzymes are extracted from the mucosa in a brine-based solution, followed by acid activation and clarification by centrifugation and/or filtration.[54,55,57] Chromatography and/or membrane filtration are sometimes used to improve the enzyme purity or concentration.

The preparations are tested, their activity is standardized, and appropriate stabilizers and preservatives are added as desired. Liquid and dried forms are available.

The predominant proteinases in rennet derived from milk-fed neonates of ruminants are chymosin (80–95%; EC[61] 3.4.23.4; MEROPS[90] A01.006; historically widely known as rennin) and pepsin[91] (5–20%; EC 3.4.23.1; MEROPS A01.001). The minor proteinase gastricsin (EC 3.4.23.3; MEROPS A01.003) is not considered significant. All are aspartyl proteinases. Commercial rennets derived from weaned or adult stomachs are rich (55–95%)[81] in pepsin.

Chymosin, typical of aspartyl proteinases, is synthesized as a pre-proenzyme.[57,76] The N-terminal signal peptide is removed on secretion. The proenzyme (zymogen) is activated by removal of the N-terminal propeptide (44 amino acids in bovine chymosin, though some variability in the cleavage site has been reported[92]), which occurs autoproteolytically under acidic conditions (for example, in rennet manufacture, pH 2 for 1 h, then adjusted to pH 5.5).[54,57] Depending on the pH of activation, the autoproteolysis may be a one- or two-step process.[93] Mature bovine chymosin contains 323 amino acids (35 600 Da) and occurs as two common variants (A and B; there is also a C variant). A and B differ in only one amino acid (at position 254), with A having higher initial milk clotting activity but a tendency to auto-degrade to a less active form.[57,94]

Chymosin's primary site of action on caseins is the Phe_{105}–Met_{106} bond in bovine κ-casein, releasing the caseinomacropeptide and facilitating micelle coagulation. A highly-conserved sequence of at least 14 amino acids in κ-casein binds within and outside the catalytic cleft of chymosin,[16,57,95,96] resulting in a remarkable degree of site-specificity in its action (*i.e.* higher C/P).[76]

Chymosins from different species generally show greater milk-clotting activity when the substrate κ-casein is from the same species as the enzyme.[57,76] Chymosin shows measurable proteolytic activity at multiple cleavage sites in other caseins (particularly α_{s1}- and β-caseins and, to a lesser extent, α_{s2}-casein), though this varies depending on pH, ionic strength, level of casein phosphorylation and the properties of casein variants.[57,97] Enzyme specificity (assessed as k_{cat}/K_m) is orders of magnitude higher for the primary site in κ-casein than for sites in other caseins,[98] but these sites come into play during long-term cheese maturation.

Pepsin has broader substrate specificity than chymosin.[91] The C/P of bovine pepsin is more than 20-fold lower than that of bovine chymosin.[99] Porcine pepsin, used in some Italian artisanal cheesemaking,[100] has been estimated as 80-fold lower.[76] (Note: C/P is indicative only. There is no standard method for its estimation, and its impact varies in different cheese types.) Cheese yield with pepsins or adult rennet is lower (*e.g.* by 0.14–0.44%[55,101,102]) than with calf rennet, and bitterness or lack of flavour development may occur (depending on the extent of enzyme retention and inactivation during cheesemaking).[81]

Animal rennet now accounts for only 10–20% of world coagulant use.[54] Pepsin-rich rennets (*e.g.* adult bovine rennet, porcine or chicken pepsin, or blends of calf rennet with bovine or porcine pepsin) saw greater use during wartime shortages[4] and when world demand for coagulants permanently outstripped the available supply of calf rennet (particularly in the 1960s and 1970s),[81,103] but these have since been replaced by other rennet substitutes (see below).

2.2.3.2 Fermentation-produced Chymosin

Chymosin produced *via* industrial-scale liquid culture of genetically-modified microorganisms[104] is known within the dairy industry as fermentation-produced chymosin (FPC). FPC has been in commercial use for over 30 years and has been widely embraced as consistent, high-performing and cost-effective. It is the purest chymosin available, with no pepsin or other contaminating enzymes. Commercial coagulant sales are now mostly FPC (perhaps 60% worldwide[54] and 90% in the USA[105]), though that figure is lower in countries or market sectors that embrace traditional cheesemaking methods or wish to avoid the use of products derived from genetically modified organisms.

The most commercially successful FPC products today (both the B variant of bovine chymosin) have made use of eukaryotic host organisms, *Kluyveromyces lactis* and *Aspergillus awamori* (also known as *Aspergillus niger* subsp. *awamori*), to express a prochymosin fusion protein which is secreted, undergoes autoproteolysis to mature chymosin, and is identical in amino acid sequence to the bovine chymosin in conventional rennet (though glycosylations vary).[54,106] It is readily purified and standardized.

Other chymosins (*e.g.* sheep, goat, water buffalo, camel) have also been experimentally expressed in microbial systems. Camel chymosin has been commercialized as FPC (analogous to bovine FPC, produced in *A. awamori*).[107] Bovine (ruminant; *Bovidae*) and camel (pseudo-ruminant; *Camelidae*) chymosins share 85% identity in amino acid sequence and have similar three-dimensional structures.[108] Camel FPC has twice the milk-clotting activity on bovine milk (IMCU per mg) as its bovine equivalent,[108] attributed to altered charge adjacent to the substrate-binding cleft (enhancing the binding of κ-casein) and greater flexibility within the cleft.[108,109] Camel chymosin also has higher C/P (depending on the assay method, estimated 1.7–13.3 times higher)[75,76,107,110] and thus has a low tendency to produce bitter peptides or to cause textural breakdown.[111] Cheese yield increases with camel FPC have been claimed (*e.g.* 0.2%[55]).

Using established FPC expression systems, novel sequence variants of both bovine and camel chymosin have been created,[112–114] minimising the cleavage of β-casein and modifying the rate of curd texture evolution caused by cleavage of α_{s1}-casein. Commercial claims suggest potential for faster coagulation, higher cheese yields and cheese with more stable suitability for shredding and slicing.

2.2.3.3 Microbial Coagulants

Many microbial coagulant preparations have been evaluated as substitutes for animal rennet, but few are satisfactory.[56,57,115] Three are widely used (25–30% of current world coagulant use[54]). These are typical aspartyl proteinases, synthesized as preproenzymes, trimmed to proenzymes during secretion, and activated *via* removal of the pro-segment.

Endothiapepsin[115,116] (EC 3.4.23.22; MEROPS A01.017) is produced by the ascomycete fungus *Cryphonectria parasitica* (formerly *Endothia parasitica*; pro-segment, 69 amino acids; mature enzyme, 330 amino acids). Coagulation is *via* cleavage of κ-casein Ser_{104}–Phe_{105}.[117] Endothiapepsin shows relatively high proteolytic activity against β-casein[118] but is sufficiently thermolabile that it gives satisfactory results when a high cook/scald temperature (*e.g.* 55 °C) is used in cheesemaking.[54] However, cheese yield is reduced (*e.g.* by *ca.* 1.2%[101]) compared with calf rennet.

The most widely used fungal coagulants are from the zygomycete fungi *Rhizomucor miehei* and (now less commonly) *Rhizomucor pusillus* (both formerly classified in the genus *Mucor*). The enzymes from the two species are regarded as variants of a single enzyme type known as mucorpepsin[115,119] (EC 3.4.23.23; MEROPS A01.013). The mature enzyme contains 361 amino acids, with a pro-segment of 47 (*R. miehei*) or 44 (*R. pusillus*) amino acids. Mucorpepsins cleave Phe_{105}–Met_{106} of κ-casein[117] and their primary cleavage sites in other caseins are also similar to those of chymosin,[118] but the C/P quotient is lower (relative to chymosin or calf rennet, by a factor of 20–30[76] or 5–10,[99] depending on the test method), often resulting in excessive proteolysis and bitterness in matured cheese.[81,120,121] Cheese yield is reportedly lower (*e.g.* by 0.5–1.1%) than that with calf chymosin.[101,102,121]

These organisms can be grown in fermenter culture, with the enzyme secreted into the growth medium. Crude *R. miehei* enzyme preparations (containing other fungal enzymes) involve filtration (to remove the organism), ultrafiltration (to concentrate the enzyme) and standardization to the desired activity.[120] Much improved commercial preparations are produced by strain selection, by adopting more stringent purification procedures, and by chemical modification which increases heat-lability and so decreases residual activity in cheese.[114,115,120,122]

2.2.3.4 Plant-derived Coagulants

Plant-derived coagulants (also known as plant rennet, vegetable rennet, *etc.*) have been used for millennia. Traditional preparations (fig latex, or derived from safflower and sunflower seeds, thistle flowers and leaves, and many other fruits, herbs, shrubs and trees) are generally simple: fresh or dried plant parts or water-soluble decoctions of plant parts. Their use is mostly limited to regional artisanal cheesemaking, but they continue to attract research interest,[56,57,64,123–126] including evaluation as potential replacements for animal rennet.

The best-known plant-derived coagulants are those obtained as aqueous extracts of flowers of cardoon thistle (*Cynara carduncculus*) and similar species, used on the Iberian Peninsula and around the Mediterranean.[123,127–130] Milk coagulation is *via* cleavage of the Phe_{105}–Met_{106} bond in bovine κ-casein,[127,131] but these coagulants typically bring about strong, broad-specificity protein degradation (sometimes to the point of liquefaction) and give a bitter piquancy to cheeses[56] which may be suited to soft ewe's-milk cheeses but is not acceptable in many cheese types. These coagulants contain various aspartyl proteinases known as cardosin, cyprosin and cynarase, with multiple forms of each reported. Commercial preparations (liquid and dried) are available.

Aspartyl proteinases are very common in plants and are collectively known as phytepsins[128,132] (EC 3.4.23.40; MEROPS A01.020). Phytepsins are about 100 amino acids longer than mammalian aspartyl proteinases due to a 'plant-specific insert' between the N- and C-terminal domains. Another significant difference is that the proenzyme is usually processed into two polypeptides which form a functional heterodimer. Cysteine and serine proteinases are also found in plants and have coagulant activity.[124,126,133]

Latex from fig plants (genus *Ficus*) is a coagulant with a long history but is only in very limited farmhouse use today. Coagulant activity is generally attributed to the cysteine proteinase ficain (also known as ficin) (EC 3.4.22.3; MEROPS C01.006). Serine and aspartyl proteinases have also been identified from *Ficus* plants but are not as well characterized.

Plant-derived proteolytic preparations often have multiple proteinases (with different specificities). It is common to have multiple enzymes in one plant, sometimes tissue-specific, as well as different enzymes in different cultivars or subspecies.[128–130,134] Their nomenclature commonly reflects the plants that they have been isolated from, rather than being based on any systematic understanding of their enzymology or phylogeny.

2.2.4 The Role of Calcium Ions

Bovine milk, for example, contains a total of *ca.* 30 mM calcium in all its forms,[135] of which *ca.* 2–3 mM is ionic Ca^{2+}. Calcium ions are essential to proteolytic coagulation, and $CaCl_2$ (*e.g.* 100–500 mg kg^{-1}, *ca.* 1–5 mM) is sometimes added to milk to improve curd formation.[136,137] The mechanism(s) of this effect are not well established (possibly casein conformational changes, charge neutralization, and/or ionic bridging between micelles), but added $CaCl_2$ increases the 'aggregability' of micelles at a given pH and a given extent of κ-casein cleavage.[138] Adding $CaCl_2$ also causes a drop in pH (which increases the activity of coagulant enzymes), but excessive addition inhibits coagulant (especially chymosin) activity.[138] Adding $CaCl_2$ increases both ionic calcium and CCP, and risks altering the microstructure and texture of the cheese.[136]

Added Ca^{2+} (especially at >10 mM)[139,140] can also mediate heat-induced coagulation of micelles, and this effect is enhanced if the pH is lowered beyond the reduction caused by $CaCl_2$ addition.[139] It is not known whether Ca^{2+} from acid-dissolved CCP might play a similar role in acid-heat coagulation of milk, although prior heat treatments (>80–90 °C) might also lower its immediate availability.[141]

Adding $CaCl_2$ does not assist in acid-induced coagulation without heating: experimental systems suggest that at low concentrations it might slightly raise the coagulation pH, but at higher concentrations it retards coagulation.[142]

2.2.5 Transglutaminase

Transglutaminase (EC 2.3.2.13)[143] is not used as a coagulant in its own right but has been shown to increase the coagulum yield *via* covalent protein cross-linking (within micelles), giving enhanced water-binding. Whey proteins are only incorporated if they have been denatured (*e.g.* by heat).[144] Transglutaminase treatment can interfere with proteolytic coagulation and alter the functional properties and subsequent maturation of the coagulum. Consequently, its practical application to traditional products is limited but it may be more useful in making modified products or cheese analogues.

2.3 Starter Cultures

In making most cheese types, microbial metabolism is responsible for acidification, either as the sole basis of coagulation (see Section 2.2.1) or in conjunction with coagulation induced by specific proteolysis (Section 2.2.3). Failure to achieve appropriate acidification leads to cheese of variable quality and safety. Historically, cheesemaking relied on spontaneous acidification (sometimes aided by allowing milk to 'ripen' before cheesemaking) and on other poorly-controlled practices, including adding sour whey ('backslopping' from a previous batch of cheese) or sour buttermilk.[84,85,145]

It was first shown in the 1870s[146] that bacteria present in the microflora of milk were responsible for this acidification. Traditional practices can be interpreted in terms of providing conditions favouring growth and survival of 'lactic acid bacteria' (LAB) and other acid-tolerant bacteria. In modern practice, cheesemakers use carefully prepared microbial cultures.[147]

Cheesemaking can conceptually be divided into a primary phase (the initial manufacture) and a secondary phase ('maturation' or 'ripening') during which tastes and aromas (cheese flavour) develop alongside changes in texture and appearance. Cultures added to cheese-milk to produce lactic acid in the primary phase are known as 'starter' or 'primary starter' cultures.[148] Starters are usually multifunctional (due to multiple organisms or organisms with multiple capabilities) and also contribute significantly to cheese maturation.

Other cultures (bacteria, yeasts, or moulds), used solely for their effects on maturation,[149,150] can be added to cheese-milk ('secondary', 'adjunct', or 'ripening' cultures) or applied later to the surface of a cheese ('secondary', 'smear', or 'surface ripening' cultures); confusingly, these are also sometimes known as 'secondary starters'. The general definition of a cheese starter therefore embraces any microorganisms added to achieve a desired effect.[147]

The main focus of this chapter is primary starters (herein simply known as starter cultures or starters). These cultures are predominantly (in most cases, exclusively) composed of LAB. Table 2.1 lists the most significant LAB species in this context.

2.3.1 The Microflora of Milk

Milk acquires microorganisms from multiple sources, such as the milking animal and its living environment, the dairy and its workers, and biofilms on equipment surfaces (during milking, storage, transport, *etc.*).[151,152] Milk microflora is diverse and variable,[153–156] with over 2000 microbial taxa reported.[155,157] The microflora also changes (quantitatively and quantitatively) depending on milk handling practices, such as the time and temperature between milking and the point of use. Bacterial counts of $<10^3$ to $>10^5$ CFU per mL (colony-forming units per millilitre) are seen in raw (unpasteurized) milk,[152] but only some of these are LAB. Improved animal health and modernization of animal husbandry and other on-farm practices, equipment sanitation, refrigeration of milk, and heat treatments (such as pasteurization, thermization, or high temperatures during cheesemaking) substantially reduce the total microbial numbers, but do not totally eliminate the milk microflora.

To the extent that milk microflora is shaped by influences typical of a particular location or by local farming, milking, milk handling and cheesemaking practices, it can be said that the microflora is 'autochthonous' (indigenous to a place) and has microbial 'terroir' which reflects the context in which it was produced.[157,158] In some traditional spontaneously-acidified cheesemaking systems, biofilms on wooden utensils have been found to be the major source of LAB,[159,160] and cheesemaking facilities (traditional and modern) can develop a resident 'house microflora'.[151,161]

2.3.1.1 Lactic Acid Bacteria (LAB)

The term 'lactic acid bacteria' has no formal definition, but is usually taken to refer to non-spore-forming *Firmicute* bacteria (Gram-positive bacteria with 'low G + C content' DNA) of the taxonomic order *Lactobacillales* which produce lactic acid as a major end-product of carbohydrate metabolism, lack true catalase activity, and do not normally produce cytochrome.[162,163] Their central metabolism is essentially anaerobic, although many of them are more accurately described as micro-aerophilic. LAB do not have a functional Krebs (tricarboxylic acid) cycle. Some (notably *Lactococcus lactis* and various non-starter lactobacilli) can establish an aerobic electron transport chain (coupling NADH oxidation to proton translocation) if supplied with haem, or

Table 2.1 Common cheese starter LAB.

Species/sub-species	Synonyms, notes[a]	M/T[b]	Lactate[c]	Citrate[d]
Lactococcus lactis[287,288]				
Lc. lactis subsp. *cremoris*	*Streptococcus cremoris*, *Lc. cremoris* subsp. *cremoris*	M	L (A)	−
Lc. lactis subsp. *lactis*	*Streptococcus lactis*	M	L (A)	−
Lc. lactis subsp. *lactis* biovar diacetylactis	*Streptococcus lactis* subsp. *diacetylactis*, *S. diacetylactis*	M	L (A)	+
Streptococcus thermophilus[289]	*Streptococcus salivarius* subsp. *thermophilus.* Note: few strains metabolize galactose	T	L (B)	−
Lactobacillus helveticus[185]	Notes: *Lb. helveticus* has both L- and D-lactate dehydrogenases. Citrate is utilized in mixed-acid fermentation (acetate, succinate)[290]	T	D,L (C)	−
Lactobacillus delbrueckii subsp. *lactis*[185]	*Lb. lactis*. Note: some strains use galactose on lactose depletion[291]	T	D (B)	−
***Leuconostoc* species**[292]				
Ln. mesenteroides subsp. *cremoris*	*Ln. cremoris*	M	D (D)	+
Ln. mesenteroides subsp. *dextranicum*	*Ln. dextranicum*			
Ln. lactis	Notes: leuconostocs are heterofermentative. These secondary organisms are added as part of, or at the same time as, the primary culture			

[a] The classification of many microorganisms has been revised in recent years. For example, *Lc. lactis* subsp. *lactis* and *Lc. lactis* subsp. *cremoris* (the nomenclature used in this chapter) are increasingly regarded as separate species (*Lc. lactis* subsp. *lactis, Lc. cremoris* subsp. *cremoris*).
[b] Mesophile (M) or thermophile (T).
[c] Primary product from metabolism of lactose: L- or D-lactate. The letter in parenthesis indicates the relevant pathway in Figure 2.1.
[d] +/− Citrate metabolized/not metabolized to acetate, diacetyl, acetoin, CO_2, and 2,3-butanediol.

haem and menaquinone.[164,165] This has been proposed to improve the growth yield and viability of some commercially-produced cultures.

The LAB species associated with acid production in milk are evolutionarily adapted to growth in environments rich in nutrients and fermentable carbohydrates. It is likely that their ancestors were oral, intestinal, or associated with decaying plant matter, and entered cheesemaking as part of the milk flora or in coagulant preparations. They acidify their local environment with lactic and other organic acids, giving them a competitive advantage over acid-sensitive microorganisms. Functions not required have been lost (*via* gene inactivation or deletion, with the corresponding reduction in genome size), while genes necessary for properties that enhance their growth, survival or competitiveness (*e.g.* lactose transport, extracellular proteinase activity, bacteriocin production, reduced susceptibility to bacteriophages) have been acquired, often on plasmids and other mobile genetic elements.[166–169] These (and many other cheese-related microorganisms) have a long history of safe occurrence in fermented foods.[170,171]

2.3.1.2 *Bacteriophages*

Viruses that infect bacteria ('bacteriophages', also known as 'phages' or 'bacterial viruses') are present wherever their host bacteria are found. Their presence and effects in cheesemaking were first noted in the 1930s.[172] Phage biology, the many types of phages, and the full complexity of their interactions with starter bacteria[173–176] are beyond the scope of this chapter. In the most virulent form of interaction between a phage and a bacterial host cell, infection is followed by rapid multiplication of the phage within the infected cell, with subsequent lysis of the cell releasing progeny phage particles which then infect more cells. If phage numbers reach high levels, death of a significant number of starter cells can adversely affect the rate of lactic acid production during cheesemaking. Each phage strain has a limited 'host range' (usually infecting only some strains of one bacterial species), and so phage infection of a multi-strain (or multi-species) culture also alters the composition of the culture.

Phages are seldom a problem in small-scale cheesemaking, but high-throughput industrial cheesemaking (multiple vats, multiple vat fills per day, tight production scheduling and high-volume whey handling with potential for generation and spread of phage-carrying aerosols) must embrace design, hygiene and culture selection strategies to minimize the impact of phages.

2.3.2 Starter Systems

The central purpose of any starter system is the production of a starter culture which will perform appropriately and reliably in cheesemaking and, for many cheese types, in subsequent cheese maturation. Starters can be classified in various ways,[177] but one of the most fundamental is in terms of the range of temperatures at which they best grow: 'mesophilic' (*ca.* 25–32 °C) or 'thermophilic' (*ca.* 35–45 °C). The starter must also have suitable metabolic properties and be produced in sufficient volume and with sufficient acid-producing activity. When grown at the cheese factory, this is known as 'bulk starter'. For some cheese types the volume can be considerable, such as 3% by volume of a whey-grown culture for making Parmigiano Reggiano. The following sections provide examples of various starter production systems.

2.3.2.1 Artisanal/Natural Starters

Starters prepared by culturing the microflora of milk (milk starter) or of cheese whey (whey starter) are described as 'artisanal' or 'natural' starters. These are autochthonous bulk starter cultures made by incubating milk or cheese whey under selective conditions to allow growth of appropriate microorganisms and suppress undesirables, with LAB typically reaching 10^8–10^9 CFU per mL. Their microbial composition is undefined, complex and continuously evolving (in part in response to phage infections), both at the species level and at the strain level. Consequently, their composition and acid-producing activity are variable. They are best known in making European traditional regional cheeses (many carrying *Protected Designation of Origin* or a similar status).[148,178] More than half the cheese made in Italy is made using such starters.[5]

Although clearly related to the earlier (and, by modern standards, hazardous) practices of spontaneous acidification and backslopping, current processes are more sophisticated. Milk starters for cheeses such as Gorgonzola, Taleggio and 'Traditional Specialty Guaranteed' Mozzarella[5,148,179,180] are grown in thermized milk (*e.g.* 60–65 °C, 15 min) at 39–50 °C to *ca.* pH 4.0–5.0, with or without an inoculum of some of the previous day's starter (culture backslopping, to increase microbial continuity and reduce incubation time). Such conditions favour the survival and growth of thermophilic, aciduric organisms, such as *Streptococcus thermophilus* and *Lactobacillus delbrueckii* subsp. *lactis*, with lactococci a persistent minor component and enterococci and various other bacteria (including potential spoilage organisms) also frequently present in low numbers.[180]

Whey starters for Grana Padano and Parmigiano Reggiano cheeses[5,179,181–183] are typically prepared by collecting whey after the cook/scald (53–56 °C) step of cheesemaking, and incubating it at 43–45 °C or gradually allowing it to cool to ambient temperature, with a final pH of *ca.* 3.5 after 16–24 h. These cultures are consistently dominated by thermophilic, aciduric lactobacilli (*Lactobacillus helveticus* and *Lb. delbrueckii* subsp. *lactis*) with usually lower levels of *S. thermophilus.*[179,183,184] Other lactobacilli (*e.g. Limosilactobacillus*[185] (*Lactobacillus*) *fermentum*) have also been reported. Older reports indicate the presence of *Lb. delbrueckii* subsp. *bulgaricus*, but this is not supported by more recent analyses.[183] Whey starters in this form were introduced for Grana manufacture in 1888 with the whey at 28–30 °C and the resultant microflora consisting of mesophilic species, but the current temperature regime (strongly selective for thermophilic lactic acid bacteria) was not adopted until *ca.* 1950.[5,86] The contemporary approach to preparing these starters is more sophisticated than suggested by the terms 'artisanal' and 'natural'; 'spontaneously selected whey cultures' has been proposed as a better descriptor.[183]

2.3.2.2 Undefined Mixed-strain Starters (MSS)

Mixed-strain starters (MSS) are the domesticated descendants of artisanal starters (particularly those providing good acid production, associated with high-quality cheese, and free of pathogens and spoilage organisms), which have been collected and propagated (in microbiological terms, sub-cultured). Such cultures have been used since the 1890s. Crucially, laboratory propagation and the realization that starter cultures could be thought of as products in their own right (not just as an intermediate stage in the daily making of cheese) gave rise to commercially-supplied reproducible cultures for butter[186–189] but soon also used for cheese.[84,186,187] Like their artisanal forebears, the microbial content of a MSS is undefined (though often well-characterized) and potentially includes multiple species and multiple strains of each species.

Today, MSS are most commonly maintained by culture suppliers external to the cheese factory. Stocks are preserved (in modern use, frozen or freeze-dried), and the number of growth generations between the stock and the supplied culture is kept to a minimum. Growth conditions are chosen so as to retain diversity and minimize compositional changes within each culture.

Distribution of small-volume liquid cultures (to use as an inoculum for bulk starter) is still practised where distribution chains are short, such as the thermophilic cultures used for regional Italian PDO cheesemaking[178] or for PDO Emmental cheese.[190] If collection and distribution are both localized to a defined area, these cultures may be regarded as regionally autochthonous, such as the 'raw mixed culture' whey starters[190–192] for PDO Emmental, which were collected from 1906 through to the 1980s.[192,193] These MSS cultures contain multiple strains of *Lb. delbrueckii* subsp. *lactis* and *S. thermophilus*. In current use, bulk starter is grown at 38 °C for 6–18 h in low-fat milk which has been heat-treated to minimize contamination (*e.g.* 95–98 °C for 90–120 min, treatment in a pressure cooker, or UHT).[194,195] The biodiversity of these thermophilic MSS is indicated by a study of *Lb. delbrueckii* subsp. *lactis* which revealed 108 distinguishable genotypes (DNA restriction fragment patterns) across 13 cultures, with as many as 19 in a single culture.[192]

Frozen or freeze-dried distribution formats are common today,[178,196–198] as these allow wider distribution, longer shelf-life, and more extensive quality control testing (performance, composition, contamination) before they are released to the cheesemaker. MSS for Gouda and Edam cheeses,[198,199] for example, were derived from backslopped starters used in artisanal farmhouse cheesemaking in the Netherlands in the 1950s and 1960s.[200] These are mesophilic cultures, dominated by *Lactococcus lactis* but also containing citrate-fermenting *Lc. lactis* subsp. *lactis* biovar diacetylactis and/or leuconostocs.[201,202]

Citrate fermentation leads to eye formation (holes) in these cheeses. Using an inoculum supplied frozen, bulk starter is grown in heat-treated (95 °C/30 min or 110 °C/1 min) skim milk, achieving up to 10^9 CFU per mL after 16–20 h at 20 °C in vessels designed to exclude phages.[198]

Mesophilic MSS can be microbiologically complex. For example, pangenome sequencing analysis of three commercial cultures revealed a total of 32 lactococcal 'lineages' (11 *Lc. lactis* subsp. *lactis*, 21 *Lc. lactis* subsp. *cremoris*) and *Leuconostoc* content ranging from 0.5 to 29%,[201] with similar diversity seen in other cultures.[200,203] MSS cultures for making Cheddar (usually containing no citrate fermenters) are similarly diverse at the strain level.[204,205] With such diversity, the same MSS has sometimes remained in use in moderate-scale cheese production for some years despite the presence of phages, although with variable acid production.[205]

2.3.2.3 *Defined-strain Starters (DSS)*

Variable culture composition can lead to variable performance and cheese quality, and the presence of inappropriate organisms can lead to cheese defects or safety concerns. 'Pure' cultures of known species were first used for butter starters in the 1890s[188] and later for cheeses, though the conditions of bulk starter preparation and propagation quickly compromised their purity.

Defined-strain starters are the result of a conceptually simple microbiological approach: isolating desirable bacteria, usually from artisanal starters or MSS, and propagating them as pure cultures under conditions designed to minimize contamination. Early DSS were single strains, but a modern DSS is a selected blend (typically only a few strains) which can be reproduced consistently from laboratory stocks of the purified strains. DSS have been adopted widely for many cheese types. Microbiological simplicity facilitates customization, and the concept has been extended to defined adjunct cultures (of various species, *e.g. Leuconostoc*, *Propionibacterium*, and various lactobacilli) used where additional microbial complexity is desired.

Mesophilic DSS for Cheddar cheese emerged in the 1930s in Australia[206] and New Zealand[172] largely as a response to open-texture defects (gas production) associated with citrate-fermenting bacteria in the available commercially-supplied cultures. Cheddar DSS[207–210] usually contain only *Lc. lactis* subsp. *lactis* and *Lc. lactis* subsp. *cremoris*. Typically, two or three strains are used together, but various numbers have been used.[207,209,211] In addition to the necessary acid production activity and phage resistance, other selection criteria may be applied, such as low-bitterness proteinase specificity, 'de-bittering' peptidase activity or an expected level of autolysis. To avoid over-acidity and bitterness in cheese due to high bacterial cell numbers, the preferred strains are *Lc. lactis* subsp. *cremoris* that produce acid with little growth at cheesemaking cook temperatures and are more sensitive to salt and acid than are *Lc. lactis* subsp. *lactis*, although various blends of strains with specific properties have been used. In the most common form of the system, each strain is supplied separately (frozen) and can be grown together or (preferably) separately as bulk starter. Bulk starter is usually grown in skim milk (reconstituted from powder or fresh) which has been heat-treated (*e.g.* 90–95 °C/1.5 h, or 137 °C/2 s or longer). The starter is grown at *ca.* 28 °C, in a system designed to exclude phages[196,207,209] and typically with pH control (*ca.* pH 6) *via* the addition of aqueous ammonia or sodium hydroxide, and reaches in excess of 10^9 CFU per mL after *ca.* 14 h.

With strains carefully selected for resistance to known phages, strain combinations devised using knowledge of genetic relationships, and good cheese factory design and practices, continual use of the same strain blend is possible at an industrial scale with consistent performance,[209] but precautionary culture rotation is also used.[207] Should a novel phage arise, it can be controlled *via* rotation or with strain replacement before significant problems occur. The DSS approach also offers the possibility of targeted strain improvement (*e.g.* selection of phage-resistant variants, transconjugants, *etc.*)[207,209,212–214] and rational culture design (acid production, flavour impacts, resistance to phages),[207,209,214] provided that it is backed by a sufficient knowledge base and skilled laboratory support. DSS have become the system of choice for high-throughput cheesemaking, though increasingly in the direct-to-vat form.[214]

Most industrial low-moisture Mozzarella cheese ("pizza cheese") is made using DSS of *S. thermophilus* and *Lb. helveticus* (or *Lb. delbrueckii* subsp. *bulgaricus*),[215] whereas DSS of *S. thermophilus* are used in Italy for making high-moisture Mozzarella.[148] Their selection, propagation, and management parallels that of mesophilic DSS.

Commercial 'pure' cultures (initially *Lb. helveticus* and other lactobacilli, and later *S. thermophilus*) were in use prior to 1930 for making Emmental-style cheese in Switzerland and elsewhere.[188,216] Today, regionally autochthonous defined-strain blends (2 or 3 strains of *S. thermophilus*, plus up to 10 strains of *Lb. delbrueckii* subsp. *lactis*) are used for making Emmental,[195] a high-value cheese made at a relatively small scale but backed by specialist laboratory support. These cultures are used in the same way as MSS. Defined-strain adjuncts (locally-isolated *Propionibacterium freudenreichii*) are also available for PDO Emmental,[190,191,193] a concept first applied to generic Emmental over a century ago,[217] as are defined adjuncts of *Lacticaseibacillus* (*Lactobacillus*) *paracasei* or *Lacticaseibacillus rhamnosus*.[190,193,195]

2.3.2.4 Culture Concentrates for Direct Inoculation of the Vat

The past half-century has seen the rise of an alternative strategy for delivering the starter to the cheese vat. Frozen or freeze-dried starter culture concentrates (MSS or DSS, concentrated to 10^{11}–10^{12} CFU per g) are produced by major culture suppliers for direct addition to the cheese vat, removing the need for bulk starter growth at the cheese factory. These are commercially known (trademarks may apply) as Direct-Set, Direct Vat Inoculation (DVI) or Direct Vat Set (DVS) cultures. Direct-to-vat cultures (starters and adjuncts) are widely and increasingly used by many large- and small-scale cheesemakers, for whom outsourcing starter preparation to a specialist supplier aligns with their financial strategy and/or operational practicalities.

Direct-to-vat defined cultures can be blended and standardized (as frozen pellets or freeze-dried powders) before packaging or at the point of use, allowing traditional or novel blends to be prepared with greater precision than can easily be achieved with bulk starter growth. As with bulk starters, phage risks are managed *via* the choice of culture, with culture rotation as needed. Culture concentrates also facilitate the use of adjunct cultures, added directly to the vat as required.

Early direct-to-vat culture concentrates suffered limitations[197] (poor bacterial viability, long lag time, applicability to a limited range of cultures, high cost) and were particularly problematic for cheese types that require relatively fast acid production in the vat. However, many advances in growth, concentration (using centrifugal or membrane technologies), cryopreservation and freeze-drying technologies have been made, the details of which are mostly trade secrets. Culture costs and phage control options have also been improved by innovative culture formulation, such as the inclusion of *S. thermophilus*[218] in Cheddar cultures, although such cultures are not suitable for all users.

At an industrial cheese factory scale, a daily bulk starter requirement of $>10^4$ L of MSS or DSS is now common, grown in milk or a whey-based medium.[148,196,198,207,209] Direct-to-vat culture production is on a correspondingly huge scale, with one company boasting 2×10^5 L of fermenter capacity at a single production site. Large-scale cheesemakers increasingly use generic acidification cultures which can be used in combination with various adjunct cultures if product differentiation is required. Small-scale cheesemakers also use direct-to-vat cultures, though many seek microbial complexity through using raw milk for cheesemaking or by exploring alternative sources of cultures.[219]

2.3.3 The Functions of Starter Cultures

Reliable acidification by a starter is crucial to reproducible cheese quality, enhancing coagulation, CCP dissolution and syneresis, and thereby influencing the curd moisture content, residual lactose, calcium content (CCP and soluble) and, indirectly, salt uptake. Acidification also helps to suppress undesirable pathogens and spoilage microorganisms. Initial cheese textural characteristics are largely dependent on the curd pH and calcium content,[211,215] with the calcium content largely determined by the pH at which curd and whey are separated, and textural maturation is modulated by the effects of pH on CCP dissolution and on coagulant retention and cleavage of α_{s1}-casein.[73] Bitterness can result from excessive retention of the coagulant (Section 2.2.3).

Flavour development in cheese begins with the metabolic capabilities of starter organisms[220–223] trapped within the cheese curd. Secondary and adventitious (non-starter) organisms also grow in (and on) cheese, depending on the cheese type. In many cases it is clear which component of the microflora is responsible for a particular aspect of cheese flavour, but there are also metabolic interactions between different components of the starter and non-starter microflora.[151,224–226] Crucially, during maturation microbial cells and their enzymes are often under stress and functioning under sub-optimal conditions.[227,228] The functions of starter cultures in cheesemaking are best understood in the context of their biochemical and physiological responses to cheesemaking and the cheese environment.

The reader should be aware of the usage of the terms 'lactic acid' and 'lactate' in the cheese and starter literature (and in this chapter). The pH of milk, cheese or within a bacterial cell is typically well above the pK_a of lactic acid (3.86 at 25 °C), and so the dissociated form (lactate) is predominant. 'Lactic acid' is often used when emphasizing acidification and is not intended to imply that the molecule is in its undissociated form. Similar considerations apply to acidic amino acids and other organic acids.

2.3.3.1 *Growth and Acidification*

All cheese starter cultures are dominated by homofermentative (homolactic) species of LAB. Under unrestricted growth conditions in milk, these produce lactic acid (lactate, 2-hydroxypropanoate) as the predominant end-product from lactose metabolism[220,229,230] (Figure 2.1). Heterofermentative (heterolactic) LAB are common as minor components of undefined starter cultures, as secondary (adjunct) cultures and in the adventitious non-starter flora of cheeses.

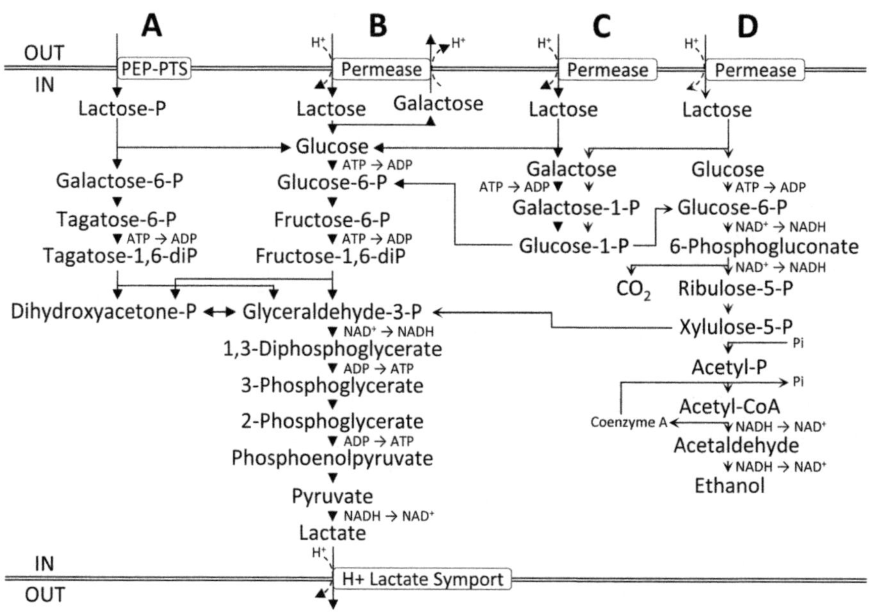

Figure 2.1 Representative pathways for lactose metabolism by LAB.[220,230,286] Homofermentative (closed arrows): (A) PEP-PTS and tagatose pathway; (B) permease with galactose/lactose antiport; (C) permease and Leloir pathway. Heterofermentative (open arrows): (D) permease, Leloir and phosphoketolase pathways. P = phosphate. Not all cofactors are shown.

In milk, glycolytic metabolism of lactose (*ca.* 40–50 g kg^{-1} in ruminant milk) provides for abundant ATP production. In conventional cheesemaking, sustaining the necessary growth and acid production requires a starter culture to (i) be capable of lactose uptake and metabolism and (ii) be able to obtain amino acids *via* degradation of milk proteins. Acidification begins in liquid milk and continues after curd formation, reaching acidities beyond which starters can maintain maximum growth (for lactococci, an external pH of 6.3–6.9; for *S. thermophilus*, pH 6.5–7.5; thermophilic *Lactobacillus* starters are reported to prefer the range of pH 5.5–5.8).[231]

All homofermentative LAB metabolize the glucose moiety of lactose *via* the Embden–Meyerhof–Parnas (EMP) pathway (Figure 2.1), but they differ in the mechanisms used for uptake of lactose and metabolism of its galactose moiety. Most make use of a permease (LacS) which initially uses transmembrane proton-motive force (PMF) to drive lactose uptake. Lactose is then cleaved by β-galactosidase (EC 3.2.1.23). *Lactobacillus helveticus* processes galactose into glucose-6-phosphate *via* the Leloir pathway. Others (most strains of *S. thermophilus*, *Lb. delbrueckii* subsp. *bulgaricus*, *Lb. delbrueckii* subsp. *lactis*) use the transmembrane concentration gradient of galactose generated by lactose hydrolysis to drive further lactose uptake, thereby preserving transmembrane PMF.

Most lactococcal starter strains have a lactose transport system that uses energy from phosphoenolpyruvate (PEP) to simultaneously import and phosphorylate lactose (the PEP-dependent phosphotransferase system, PEP-PTS). Phosphorylated lactose is cleaved by 6-phospho-β-galactosidase (EC 3.2.1.85) into glucose (which feeds into the EMP pathway) and galactose-6-phosphate, which is processed *via* the tagatose pathway. The permease system is also found in lactococci but is of secondary importance. Genes for a lactose PTS have been reported in *Lb. delbrueckii* subsp. *lactis*,[168] where it perhaps functions as a secondary system for use at low lactose levels.

The generation of ATP *via* metabolism of lactose to pyruvate is essentially an anaerobic oxidation process, in which NAD^+ is concomitantly reduced to NADH. NAD^+ is regenerated during reduction of pyruvate to the L- or D-isomer of lactate by L-lactate dehydrogenase (L-LDH, EC 1.1.1.27) and/or D-lactate dehydrogenase (D-LDH, EC 1.1.1.28).

Starter LAB have a nutritional requirement (auxotrophy) for several amino acids (species and strains differ in their needs) which they obtain from casein. The key cell-surface proteinases used by LAB to degrade caseins are known by various names, including cell envelope proteinase (CEP) and lactocepin.[232–234] These are serine proteinases (EC 3.4.21.96; subtilisin-like, MEROPS class S8), synthesized as pre-pro-proteins and located extracellularly, anchored to the cell wall or in the cell membrane. They vary in proteolytic specificity, but all generate oligopeptides that are subsequently taken into the cell *via* oligonucleotide transport systems, and then further metabolized. The lactocepins of lactococci[235,236] are the most thoroughly studied.

In a multi-strain and/or multi-species starter culture, it is not essential that all culture components have the proteolytic ability. Oligopeptides generated by a proteolytic (Prt^+) component can also feed Prt^- components, though growth and acid production may be limited by total peptide availability[237,238] or affected by competition for uptake of the peptides.[239] In cultures of *Lc. lactis*, the lactocepin gene is typically carried on a plasmid; Prt^- variants (due to plasmid loss) arise naturally and persist in association with Prt^+ strains.[240]

2.3.3.2 Establishing the Cheese Environment

Starters are central to establishing the internal environment of a cheese, which influences subsequent chemical and microbiological events. They bring about lowered pH, accumulation of organic acids and other metabolites, reduced levels of fermentable sugars, low oxidation–reduction potential (redox, E_h) and reduced water activity (increased osmolarity, *via* conversion of each mole of lactose to up to four moles of lactic acid, CCP dissolution, proteolysis, *etc.*). These conditions are hurdles to the growth of adventitious flora (pathogens, spoilage organisms and other non-starter organisms), so that each cheese type develops a broadly characteristic non-starter microflora able to survive in that cheese environment. Some LAB produce antibacterial[241] or antifungal[242] peptides with the potential to suppress bacteria such as clostridia or listeria, to induce desirable lysis of other LAB, or to slow mould growth. However, they may also reduce desirable microbial diversity. Their commercial use is limited to specific applications.

Of these factors, E_h is perhaps the least studied.[243] *Lactococcus lactis*, for example, generates negative E_h in milk,[244] initially due to oxygen removal by NADH oxidase activity (NoxE, EC 1.6.3.4, capable of regenerating NAD^+ as an alternative to lactate production *via* LDH) and, at lower pH and after cessation of growth, due to the reducing activities of menaquinones. Similarly, anaerobic conditions and negative E_h are generated in cheeses made with *Lc. lactis*,[245,246] attributed to the same factors and also to the production of reducing molecules (*e.g.* trioses, acetaldehyde). LAB that do not produce menaquinones (*e.g. S. thermophilus*, *Lb. helveticus*, *Lb. delbrueckii*) generally have a much weaker effect on E_h.[247]

2.3.3.3 Survival and Decline in Cheese

LAB vary in their tolerance to stresses during cheesemaking and cheese maturation.[167,227,228,248] Their metabolic responses to carbohydrate starvation and acidification (Sections 2.3.3.4 and 2.3.3.6) lead to catabolism of alternative energy sources and the activation of systems to counter acidification by chemically raising the pH of the internal and external environments or by maintaining a transmembrane potential (PMF) resulting in higher pH within the cell than externally.[227,228,248–252] Initial responses include the urease (EC 3.5.1.5; urea to NH_3 and CO_2) of some *S. thermophilus* strains or citrate metabolism. The arginine deiminase pathway (ADI, found in *Lc. lactis* subsp. *lactis*, in some strains of *Lc. lactis* subsp. *cremoris* and *Lb. delbrueckii* subsp. *lactis*, and in various heterofermentative lactobacilli) converts arginine to ornithine, generating NH_3, CO_2 and ATP. Ammonia (as NH_4^+) raises the pH. Carbon dioxide is poorly soluble at typical cheese curd pH and salt levels,[253] but is not normally generated by ADI in sufficient quantity to cause eye formation or 'blowing' of cheese. Even so, ADI has been implicated in Gruyère cheese defects.[254]

Amino acid decarboxylation (EC class 4.1.1, decarboxylases/carboxy-lyases) generates CO_2 while consuming intracellular H^+, in effect exporting H^+ *via* substrate–product antiport systems. An example is the conversion of glutamate to γ-aminobutyrate, occurring in strains of *Lc. lactis* subsp. *lactis* and *S. thermophilus*. A negative aspect of amino acid decarboxylation is the production of amines (biogenic amines, BA) which can have undesirable physiological effects on the consumer.[255–257] Reports vary, but the likely most significant BA in cheeses are tyramine (from decarboxylation of tyrosine), histamine (from histidine decarboxylation), cadaverine (from lysine decarboxylation) and putrescine (from the agmatine deiminase pathway, analogous to the arginine deiminase system but utilizing agmatine, the decarboxylation product from arginine). Cadaverine and putrescine in particular have unpleasant aromas. Some strains of recognized starter species have been shown to produce BA, leading to recommendations to screen starters. However, when unacceptably high BA levels are found in cheese it is typically attributable to organisms from milk microflora, such as enterococci, non-starter lactobacilli (*e.g. Lentilactobacillus* (*Lactobacillus*) *parabuchneri*) and various *Enterobacteriaceae*, which can also produce sufficient CO_2 from this process to cause cheese defects late in maturation.[258,259]

In maturing cheeses, recoverable starter numbers (estimated as CFU per g) typically decline over time. The details vary depending on species, strain and factors such as cheesemaking cook temperatures, cheese pH and salting levels, but it is clear that this decline is not solely due to simple cell death. *Lc. lactis* is the best-studied example. Lactococcal starter CFU numbers typically decline over a period of days or weeks, but they remain metabolically active (albeit at a minimal rate) for months or years[249,250,260] in various states: (i) live and culturable 'persister' cells, (ii) cells intact but not culturable by any known method, and perhaps also (iii) terminally damaged cells (possibly permeable protoplasts) retaining some function. There is also evidence of lysis of cells (autolysis).[249,260,261] The enzymes released by lysis cannot participate in the ordered pathways typical of intact cells, but potentially have greater access to substrates found in the cheese matrix.

2.3.3.4 *Flavour Production* Via *Pyruvate and/or Lactate*

Lactic acid itself contributes to cheese flavour, especially of fresh curd. However, as lactose levels fall (glycolytic flux falls, and may become growth-limiting) and lactate levels rise, starter LAB may turn to other pathways to process pyruvate, neutralize acid, regenerate NAD^+ and/or produce ATP.

Many of the products and intermediates in these pathways are also contributors to flavour.

Lc. lactis, for example, can enter 'mixed acid' fermentation, generating formic and/or acetic acids in addition to lactic (all are weaker acids than pyruvic acid). Pyruvate can be metabolized to acetyl-CoA *via* either of two reactions (depending on pH and the oxygen level), generating formate or CO_2. At low pH (*e.g.* pH 4.5 and 5.5),[262,263] the reaction catalysed by pyruvate dehydrogenase (EC 1.2.1.104; generating CO_2, consuming NAD^+) is dominant. Acetyl-CoA is metabolized *via* acetaldehyde to ethanol (regenerating NAD^+ at each step) or *via* acetyl phosphate to acetate (generating ATP). The system is subject to complex regulation, including by general catabolite repression, by the NAD^+/NADH ratio, and by allosteric activation and inhibition of enzyme activities.[230,264] Although these pathways are perhaps minor contributors to cheese flavour, transcription of relevant genes has been detected in cheese.[265]

Lc. lactis strains can also condense two pyruvate molecules to α-acetolactate (2-acetoxypropanoate), which is decarboxylated to acetoin (3-hydroxybutanone). Acetoin is excreted or metabolized to 2,3-butanediol (generating NAD^+). Alternatively, at low pH and in the presence of oxygen, α-acetolactate can undergo spontaneous decarboxylation to diacetyl (2,3-butanedione) which can then be enzymatically converted to acetoin.[262,263,266] Of these, diacetyl is the most potent aroma compound (*i.e.* lowest sensory threshold) though it is typically low in abundance. These reactions provide opportunities for NAD^+ regeneration, as well as assisting in the response to falling pH by converting pyruvic acid to CO_2 and neutral molecules.

The latter pathway can be the dominant source of volatile aroma compounds in young cheeses. This is particularly important in Gouda, Edam, and similar cheeses,[198,199] for which these aromas and the formation of gas-filled 'eyes' are characteristic. Some strains of *Lc. lactis* (*Lc. lactis* subsp. *lactis* biovar diacetylactis) are able to metabolize citrate to pyruvate, with acetate and CO_2 as by-products.[263,267,268] This pyruvate can be processed by any of the available pathways (subject to metabolic requirements, such as the NADH/NAD^+ balance), but the practical impact of citrate metabolism on cheese is through enhanced production of diacetyl, acetoin, 2,3-butanediol, acetate, and CO_2. Citrate metabolism also provides a response to acidic conditions: citrate uptake can be driven by lactate efflux, with the net effect of replacing intracellular lactate with citrate, which is then converted to acetate, CO_2 and neutral molecules.

Citrate utilization *via* this pathway is also characteristic of dairy *Leuconostoc* strains.[263,269] These heterofermentative LAB (D in Figure 2.1) grow relatively poorly in milk and are not important acid producers in cheesemaking. They produce mostly lactate from pyruvate (even when the pyruvate is derived from citrate) as long as fermentable sugar (*i.e.* lactose) is also available; acetyl phosphate is diverted to acetate, simultaneously conserving the NADH/NAD^+ balance and generating ATP. In cheese, at low pH and once lactose is exhausted, production of diacetyl, acetoin, 2,3-butanediol and CO_2 becomes significant.

The characteristic 'eyes' and aroma of Emmental and similar cheeses are due to metabolism by propionibacteria, typically *Propionibacterium freudenreichii*.[190,191,270] These propionic acid bacteria (PAB, *Actinobacteria*, Gram-positive bacteria with 'high G + C content') are not LAB. Traditionally part of raw milk microflora, these secondary organisms metabolize the lactate generated by starter bacteria *via* pyruvate to propionate and acetate, CO_2 and ATP, or (coupled with aspartate metabolism) to succinate, acetate, CO_2, NH_3 and ATP. These reactions enhance bacterial survival *via* production of ATP and PMF, conversion of lactic acid to weaker acids, and alkalinization. Aspartate ammonia-lyase (EC 4.3.1.1, aspartase) activity has become an important factor in selection of *P. freudenreichii* adjunct strains; higher activity gives faster growth (with associated generation of flavour compounds, and increased 'umami'[271] due to both propionate and succinate) but also alters the flavour profile and runs the risk of vinegary flavour (excessive acetate) and 'late blowing' of aged cheese (excessive CO_2 production).[191]

In many cheese types, yeasts, fungi, and aerobic bacteria (adventitious or added as secondary flora) growing on the surface of the cheese are major contributors to cheese maturation. They can bring about significant deacidification *via* aerobically metabolizing lactate.[272–274]

2.3.3.5 Flavour Production from Milk Fat

Lipolytic enzymes (carboxylic-ester hydrolases, EC class 3.1.1) are typically described as lipases (most active on long-chain triglycerides at water–lipid interfaces) or esterases (most active on short-chain triglycerides and on the products of lipases, in aqueous environments), though this distinction is not always clearly made. The lipoprotein lipase (EC 3.1.1.34) native to milk is active in raw milk cheeses but has little activity after heat treatments such as milk pasteurization or cheesemaking involving a high-temperature cook step.[275–277] Some cheese types are made with a coagulant preparation that has lipolytic activity (see Section 2.2.3.1) or, in non-traditional cheesemaking, have added lipase.

Starter and non-starter LAB have low lipolytic activity but may be significant contributors of these activities in maturing cheeses which do not have other sources.[276,278,279] Their enzymes are most active on monoglycerides, generating free fatty acids and glycerol, and also generate esters *via* acyltransferase activity.[279] In some cases this has been attributed to intracellular esterases, and so it has been suggested that autolysis facilitates access to their substrates. Aliphatic carboxylic acids ('fatty acids' but not derived from milk fat) can also be derived from amino acids (Section 2.3.3.6) or from pyruvate.[280]

The degradation of milk lipids contributes to cheese flavour directly through liberation of free fatty acids and indirectly *via* subsequent fatty acid metabolism which generates compounds such as thioesters, ethyl esters, lactones, methyl ketones, and secondary alcohols.[278] These compounds tend to be prominent in cheeses with secondary microflora that includes fungi (moulds and yeasts, *e.g. Penicillium roqueforti*, *Penicillium camemberti*,

Geotrichum candidum, *Yarrowia lipolytica*, *Kluyveromyces lactis*), smear bacteria (*e.g.* *Brevibacterium linens*), or propionibacteria, but are low in cheeses made solely with LAB.[278,281]

2.3.3.6 *Peptidolysis, Amino Acid Metabolism and Flavour*

As with coagulant proteinases, starter lactocepins vary in their cleavage specificities, and so vary in their propensity to generate bitter-tasting peptides from caseins. Starter species and strains also vary in their complement of intracellular peptidases (aminopeptidases, endopeptidases, dipeptidases, tripeptidases, proline peptidases, *etc.*).[252,282] Mechanistically, these include cysteine peptidases, serine peptidases and metallopeptidases. The peptidases of *Lc. lactis*, for example, span at least ten groups (MEROPS families or EC designations). Starter peptidases also vary in their level of expression and the rate at which autolysis might release them into the cheese matrix, and so they differ in their ability to 'de-bitter' cheese by degrading bitter peptides or, more generally, to generate free amino acids.[261] With the exception of bitter peptides, peptides and free amino acids are not generally considered to contribute more than background flavour to cheese.[221] Free amino acids have mostly sweet/sour/bitter flavours, or (particularly glutamate) contribute an umami note to cheese.[221,271]

In hard and semi-hard cheeses which are matured for months or years, bacterial survival (starter and non-starter) depends largely on the catabolism of amino acids. Many of the resulting intermediates and terminal metabolites are significant cheese flavour compounds.[222,223,280] Just some of the pathways are discussed below.

In addition to the processes previously described, LAB commonly have aminotransferases (pyridoxal-5′-phosphate-dependent transaminases, EC class 2.6.1) that catalyse the transfer of an amino group from an amino acid (generating the corresponding α-keto acid) to a recipient α-keto acid (generating the corresponding amino acid). The dominant amino acceptor in cheese seems to be α-ketoglutarate. The newly-generated keto acids from this process (or from simple deamination) are themselves flavour compounds, but can be further metabolized to corresponding alcohols, aldehydes, esters, carboxylic (fatty) acids, *etc.*[222,223,280] These and related pathways provide opportunities for amino acid interconversion, generation of ATP and options for management of acetyl-CoA levels and of the NADH/NAD^+ ratio. Deamination of alanine or aspartate generates pyruvate or oxaloacetate, respectively, which can also feed directly into the pathways described in Section 2.3.3.4. Methionine, aromatic amino acids and branched-chain amino acids are the most significant source amino acids for flavour development. Sulphur compounds derived from methionine (such as methanethiol, methional, and dimethyl sulphide) are particularly important in Cheddar and other matured hard and semi-hard cheeses.

Some starter strains can produce α-ketoglutarate (notably *S. thermophilus*, but also some strains of *Lc. lactis*, *Lb. helveticus*, *Lb. delbrueckii*, *etc.*), but in cheese it is believed that the glutamate dehydrogenase activity of non-starter lactobacilli is a major contributor[283,284] (*e.g. Lactiplantibacillus* (*Lactobacillus*) *plantarum*, *Lacticas. paracasei*, *Lacticas. rhamnosus*). Metabolites may diffuse within the cheese, such that different cells, strains and species may contribute both individually and collectively to overall flavour development and complexity.[224,225,285] This presents opportunities for deliberately combining starter and adjunct cultures, an area of culture science in which advances continue to be made.

References

1. M. Salque, P. I. Bogucki, J. Pyzel, I. Sobkowiak-Tabaka, R. Grygiel, M. Szmyt and R. P. Evershed, *Nature*, 2013, **493**, 522.
2. P. L. H. McSweeney, G. Ottogalli and P. F. Fox, in *Cheese: Chemistry, Physics and Microbiology*, ed. P. L. H. McSweeney, P. F. Fox, P. D. Cotter and D. W. Everett, Elsevier Academic Press, London, 4th edn, 2017, vol. 1, General Aspects, p. 781.
3. T. Bintsis and P. Papademas, in *Global Cheesemaking Technology: Cheese Quality and Characteristics*, ed. P. Papademas and T. Bintsis, Wiley, Chichester, 1st edn, 2018, p. 120.
4. R. A. Wilbey, J. E. Scott and R. K. Robinson, *Cheesemaking Practice*, Springer, New York, 1998.
5. M. Gobbetti, E. Neviani and P. Fox, *The Cheeses of Italy: Science and Technology*, Springer, Cham, 2018.
6. Codex Alimentarius Commission, Codex General Standard for Cheese – CODEX STAN 283-1978, World Health Organization/Food and Agriculture Organization of the United Nations, Rome, 2013.
7. O. T. Oftedal, in *Advanced Dairy Chemistry*, ed. P. L. H. McSweeney and P. F. Fox, Springer Science + Business Media, New York, 4th edn, 2013, vol. 1A: Proteins: Basic Aspects, p. 1.
8. A. K. H. MacGibbon, in *Advanced Dairy Chemistry*, ed. P. L. H. McSweeney, P. F. Fox and J. A. O'Mahony, Springer Nature Switzerland, Cham, 3rd edn, 2020, vol. 2, p. 1.
9. T. Huppertz, T. Uniacke-Lowe and A. L. Kelly, in *Advanced Dairy Chemistry*, ed. P. L. H. McSweeney, P. F. Fox and J. A. O'Mahony, Springer Nature Switzerland, Cham, 3rd edn, 2020, vol. 2, p. 133.
10. H. M. Farrell, R. Jimenez-Flores, G. T. Bleck, E. M. Brown, J. E. Butler, L. K. Creamer, C. L. Hicks, C. M. Hollar, K. F. Ng-Kwai-Hang and H. E. Swaisgood, *J. Dairy Sci.*, 2004, **87**, 1641.
11. D. A. Goulding, P. F. Fox, J. A. O'Mahony, in *Milk Proteins: From Expression to Food*, ed. M. Boland and H. Singh, Academic Press, London, 3rd edn, 2020, p. 21.
12. C. Holt, *Int. Dairy J.*, 2016, **60**, 2.
13. J. Manguy and D. C. Shields, *R. Soc. Open Sci.*, 2019, **6**, 190939.
14. M. Madende and G. Osthoff, *J. Dairy Res.*, 2019, **86**, 323.

15. P. F. Fox, T. Uniacke-Lowe, P. L. H. McSweeney and J. A. O'Mahony, in *Dairy Chemistry and Biochemistry*, Springer, Cham, 2nd edn, 2015, p. 145.
16. P. Martin, C. Cebo and G. Miranda, in *Advanced Dairy Chemistry*, ed. P. L. H. McSweeney and P. F. Fox, Springer, New York, 4th edn, 2013, vol. 1A: Proteins: Basic Aspects, p. 387.
17. E. Bijl, J. W. Holland and M. Boland, in *Milk Proteins: From Expression to Food*, ed. M. Boland and H. Singh, Academic Press, London, 3rd edn, 2020, p. 173.
18. N. A. Poulsen, H. B. Jensen and L. B. Larsen, *J. Dairy Sci.*, 2016, **99**, 3325.
19. T. Huppertz, I. Gazi, H. Luyten, H. Nieuwenhuijse, A. Alting and E. Schokker, *Int. Dairy J.*, 2017, **74**, 1.
20. R. K. Dewan, A. Chudgar, R. Mead, V. A. Bloomfield and C. V. Morr, *Biochim. Biophys. Acta*, 1974, **342**, 313.
21. C. G. de Kruif, T. Huppertz, V. S. Urban and A. V. Petukhov, *Adv. Colloid Interface Sci.*, 2012, **171–172**, 36.
22. R. Trejo, T. Dokland, J. Jurat-Fuentes and F. Harte, *J. Dairy Sci.*, 2011, **94**, 5770.
23. E. Bijl, T. Huppertz, H. van Velenberg and C. Holt, *Eur. Biophys. J.*, 2019, **48**, 45.
24. D. S. Horne, in *Milk Proteins: From Expression to Food*, ed. M. Boland and H. Singh, Elsevier Academic Press, 3rd edn, 2020, p. 213.
25. D. G. Dalgleish and M. Corredig, *Annu. Rev. Food Sci. Technol.*, 2012, **3**, 449.
26. H. M. Farrell, E. M. Brown and E. L. Malin, in *Advances in Dairy Chemistry*, ed. P. L. H. McSweeney and P. F. Fox, Springer, New York, 4th edn, 2013, vol. 1A: Proteins: Basic Aspects, p. 161.
27. D. J. McMahon and B. S. Oommen, in *Advances in Dairy Chemistry*, ed. P. L. H. McSweeney and P. F. Fox, Springer, New York, 4th edn, 2013, vol. 1A: Proteins: Basic Aspects, p. 185.
28. J. A. Lucey and D. S. Horne, *Int. Dairy J.*, 2018, **85**, 56.
29. L. K. Creamer, J. E. Plowman, M. J. Liddell, M. H. Smith and J. P. Hill, *J. Dairy Sci.*, 1998, **81**, 3004.
30. T. F. Kumosinski, E. M. Brown and H. M. Farrell, *J. Dairy Sci.*, 1991, **74**, 2879.
31. P. F. Fox, T. P. Guinee, T. M. Cogan and P. L. H. McSweeney, in *Fundamentals of Cheese Science*, Springer, New York, 2nd edn, 2017, p. 543.
32. K. Hartmann, F. Berthier and C. Licitra, in *Global Cheesemaking Technology: Cheese Quality and Characteristics*, ed. P. Papademas and T. Bintsis, Wiley, Chichester, 1st edn, 2018, p. 436.
33. Food Safety and Standards Authority of India, Food Safety and Standards (Food Products Standards and Food Additives) Regulations, 2011 (Version XXI, 31.01.2022), 2022.
34. J. A. Lucey, in *Cheese: Chemistry, Physics and Microbiology*, ed. P. L. H. McSweeney, P. F. Fox, P. D. Cotter and D. W. Everett, Academic Press, London, 4th edn, 2017, p. 179.
35. S. G. Anema, *J. Food Process. Preserv.*, 2008, **32**, 1016.

36. D. J. McMahon, H. Du, W. R. McManus and K. M. Larsen, *J. Dairy Sci.*, 2009, **92**, 5854.
37. D. G. Dalgleish, M. Alexander and M. Corredig, in *Food Colloids: Interactions, Microstructure and Processing*, ed. E. Dickinson, The Royal Society of Chemistry, Cambridge, 2005, p. 16.
38. H. Sinaga, N. Bansal and B. Bhandari, *Int. J. Food Prop.*, 2017, **20**, 179.
39. H. C. Deeth and M. J. Lewis, in *High Temperature Processing of Milk and Milk Products*, Wiley, Chichester, 1st edn, 2017, p. 177.
40. K. Raynal-Ljutovac, Y. W. Park, F. Gaucheron and S. Bouhallab, *Small Rumin. Res.*, 2007, **68**, 207.
41. S. G. Anema, *Int. Dairy J.*, 2021, **122**, 105136.
42. T. Nicolai and C. Chassenieux, *Food Hydrocolloids*, 2021, **118**, 106755.
43. T. Huppertz, in *Advanced Dairy Chemistry Volume 1B: Proteins: Applied Aspects*, ed. P. L. H. McSweeney and J. A. O'Mahony, Springer, New York, 4th edn, 2016, p. 179.
44. S. G. Anema, in *Milk Proteins*, ed. M. Boland and H. Singh, Academic Press, London, 3rd edn, 2020, p. 325.
45. L. Donato and F. Guyomarc'h, *Dairy Sci. Technol.*, 2009, **89**, 3.
46. A. J. Vasbinder, A. C. Alting and K. G. de Kruif, *Colloids Surf., B*, 2001, **21**, 245.
47. N. Y. Farkye, in *Cheese: Chemistry, Physics and Microbiology*, ed. P. L. H. McSweeney, P. F. Fox, P. D. Cotter and D. W. Everett, Academic Press, London, 4th edn, 2017, p. 1111.
48. S. U. Khan and M. A. Pal, *J. Food Sci. Technol.*, 2011, **48**, 645.
49. G. Moatsou, E. Moschopoulou, E. Zoidou, A. Kamvysi, D. Liaskou, V. Tsigkou and L. Sakkas, *Dairy*, 2021, **2**, 410.
50. M. B. Pesic, M. B. Barac, S. P. Stanojevic and M. M. Vrvic, *Int. Dairy J.*, 2014, **39**, 178.
51. K. Raynal and F. Remeuf, *Int. Dairy J.*, 1998, **8**, 695.
52. P. Papademas, T. Bintsis, E. Alichanidis and Y. Ardõ, in *Global Cheese-making Technology: Cheese Quality and Characteristics*, ed. P. Papademas and T. Bintsis, Wiley, Chichester, 1st edn, 2018, p. 446.
53. M. E. Pintado, A. C. Macedo and F. X. Malcata, *Food Sci. Technol. Int.*, 2001, **7**, 105.
54. A. Andrén, in *Agents of Change: Enzymes in Milk and Dairy Products*, ed. A. L. Kelly and L. B. Larsen, Springer, New York, 2021, p. 349.
55. M. Harboe, M. L. Broe and K. B. Qvist, in *Technology of Cheesemaking*, ed. B. A. Law and A. Y. Tamime, Blackwell, Oxford, 2nd edn, 2010, p. 98.
56. D. Jaros and H. Rohm, in *Cheese: Chemistry, Physics and Microbiology*, ed. P. L. H. McSweeney, P. F. Fox, P. D. Cotter and D. W. Everett, Academic Press, London, 4th edn, 2017, vol. 1, General Aspects, p. 53.
57. T. Uniacke-Lowe and P. F. Fox, in *Cheese: Chemistry, Physics and Microbiology*, ed. P. L. H. McSweeney, P. F. Fox, P. D. Cotter and D. W. Everett, Academic Press, London, 4th edn, 2017, vol. 1, General Aspects, p. 69.

58. D. S. Horne and J. A. Lucey, in *Cheese: Chemistry, Physics and Microbiology*, ed. P. L. H. McSweeney, P. F. Fox, P. D. Cotter and D. W. Everett, Academic Press, London, 4th edn, 2017, vol. 1, General Aspects, p. 115.
59. T. Janhøj and K. B. Qvist, in *Technology of Cheesemaking*, ed. B. A. Law and A. Y. Tamime, Wiley-Blackwell, Chichester, 2nd edn, 2010, p. 130.
60. P. F. Fox and P. L. H. McSweeney, in *Cheese: Chemistry, Physics and Microbiology*, ed. P. L. H. McSweeney, P. F. Fox, P. D. Cotter and D. W. Everett, Elsevier Academic Press, London, 4th edn, 2017, vol. 1, General Aspects, p. 5.
61. A. Chang, L. Jeske, S. Ulbrich, J. Hofmann, J. Koblitz, I. Schomburg, M. Neumann-Schaal, D. Jahn and D. Schomburg, *Nucleic Acids Res.*, 2021, **49**, D498.
62. F. Claverie-Martín and M. C. Vega-Hernández, in *Industrial Enzymes: Structure, Function and Applications*, ed. J. Polaina and A. P. MacCabe, Springer, Dordrecht, 2007, p. 207.
63. A. Wlodawer, A. Gustchina and M. N. G. James, in *Handbook of Proteolytic Enzymes*, ed. N. D. Rawlings and G. Salvesen, Academic Press, London, 3rd edn, 2013, vol. 1, p. 19.
64. S. Yegin and P. Dekker, *Dairy Sci. Technol.*, 2013, **93**, 565.
65. N. D. Rawlings and A. J. Barrett, in *Handbook of Proteolytic Enzymes*, ed. N. D. Rawlings and G. Salvesen, Academic Press, London, 2013, vol. 1, p. 3.
66. International Organization for Standardization and International Dairy Federation, ISO 11815:2007 [IDF 157:2007]: Milk – Determination of total milk-clotting activity of bovine rennets, ISO, Geneva, 2007.
67. International Organization for Standardization and International Dairy Federation, ISO 15174:2012 [IDF 176:2012] Milk and milk products – Microbial coagulants – Determination of total milk-clotting activity, ISO, Geneva 2012.
68. P. F. Fox, T. P. Guinee, T. M. Cogan and P. L. H. McSweeney, in *Fundamentals of Cheese Science*, Springer, New York, 2nd edn, 2017, p. 315.
69. M. Corredig and E. Salvatore, in *Advanced Dairy Chemistry*, ed. P. L. H. McSweeney and P. F. Fox, Springer Science + Business Media, New York, 4th edn, 2016, vol. 1B: Proteins: Applied Aspects, p. 287.
70. X. T. Liu, H. Zhang, F. Wang, J. Luo, H. Y. Guo and F. Z. Ren, *J. Dairy Sci.*, 2014, **97**, 3292.
71. C. L. C. Esteves, J. A. Lucey, T. Wang and E. M. V. Pires, *J. Dairy Sci.*, 2003, **86**, 2558.
72. Chr. Hansen, CHY-MAX Plus Product Information, version 2 PI-GLOB-EN 07-22-2013, 2013.
73. J. A. O'Mahony, J. A. Lucey and P. L. H. McSweeney, *J. Dairy Sci.*, 2005, **88**, 3101.
74. P. F. Fox, T. P. Guinee, T. M. Cogan and P. L. H. McSweeney, *Fundamentals of Cheese Science*, Springer, New York, 2nd edn, 2017.
75. M. W. Børsting, K. B. Qvist and Y. Ardö, *Int. Dairy J.*, 2014, **38**, 133.

76. P. B. Szecsi and M. Harboe, in *Handbook of Proteolytic Enzymes*, ed. N. D. Rawlings and G. Salvesen, Academic Press, London, 3rd edn, 2013, vol. 1, p. 37.
77. M. C. Broome, X. Xu and J. J. Mayes, *Aust. J. Dairy Technol.*, 2006, **61**, 85.
78. P. Garnot and D. Molle, *J. Food Sci.*, 1987, **52**, 75.
79. M. G. Hayes, J. C. Oliveira, P. L. H. McSweeney and A. L. Kelly, *J. Dairy Res.*, 2002, **69**, 269.
80. L. M. Costabel, C. V. Bergamini, L. Pozza, F. Cuffia, M. C. Candioti and E. Hynes, *J. Dairy Res.*, 2015, **82**, 375.
81. T. P. Guinee and M. G. Wilkinson, *J. Soc. Dairy Technol.*, 1992, **45**, 94.
82. N. Bansal, P. F. Fox and P. L. H. McSweeney, *J. Agric. Food Chem.*, 2007, **55**, 9219.
83. A. Dalby, *Cheese: A Global History*, Reaktion Books, London, 2009.
84. V. Cheke, *The Story of Cheese-Making in Britain*, Routledge and Kegan Paul, London, 1959.
85. C. L. Flint, *Milch Cows and Dairy Farming*, Phillips, Sampson and Co., Boston, 1859.
86. G. K. Y. Limsowtin, I. B. Powell, E. Parente, in *Dairy Starter Cultures*, ed. T. M. Cogan and J.-P. Accolas, VCH, New York, 1996, p. 101.
87. J. Auclair and J.-P. Accolas, *Antonie van Leeuwenhoek*, 1983, **49**, 313.
88. E. Moschopoulou, *Small Rumin. Res.*, 2011, **101**, 188.
89. M. Bustamante, F. Chávarri, A. Santisteban, G. Ceballos, I. Hernández, M. J. Miguélez, I. Aranburu, L. J. R. Barrón, M. Virto and M. De Renobales, *J. Dairy Res.*, 2000, **67**, 393.
90. N. D. Rawlings, A. J. Barrett, P. D. Thomas, X. Huang, A. Bateman and R. D. Finn, *Nucleic Acids Res.*, 2018, **46**, D624.
91. J. Tang, in *Handbook of Proteolytic Enzymes*, ed. N. D. Rawlings and G. Salvesen, Academic Press, London, 3rd edn, 2013, vol. 1, p. 28.
92. S. Lilla, S. Caira, P. Ferranti and F. Addeo, *Rapid Commun. Mass Spectrom.*, 2001, **15**, 1101.
93. V. B. Pedersen, K. A. Christensen and B. Foltmann, *Eur. J. Biochem.*, 1979, **94**, 573.
94. M. Rampilli, R. Larsen and M. Harboe, *Int. Dairy J.*, 2005, **15**, 1130.
95. S. Visser, C. J. Slangen and R. J. Van Rooijen, *Biochem. J.*, 1987, **244**, 553.
96. D. S. Palmer, A. U. Christensen, J. Sørensen, L. Celik, K. B. Qvist and B. Schiøtt, *Biochemistry*, 2010, **49**, 2563.
97. E. Bijl, H. van Valenberg, S. Sikkes, S. Jumelet, G. Sala, K. Olieman, T. van Hooijdonk and T. Huppertz, *Int. Dairy J.*, 2014, **39**, 215.
98. C. Carles and B. R. Dumas, *FEBS Lett.*, 1985, **185**, 282.
99. P. Martin, M.-N. Raymond, E. Bricas and B. Ribadeau Dumas, *Biochim. Biophys. Acta*, 1980, **612**, 410.
100. R. Tofalo, M. Schirone, G. Fasoli, G. Perpetuini, F. Patrignani, A. C. Manetta, R. Lanciotti, A. Corsetti, G. Martino and G. Suzzi, *Food Chem.*, 2015, **175**, 121.
101. D. B. Emmons and M. Binns, *J. Dairy Sci.*, 1990, **73**, 2028.
102. D. M. Barbano and R. R. Rasmussen, *J. Dairy Sci.*, 1992, **75**, 1.

103. J. H. Nelson, *J. Dairy Sci.*, 1975, **58**, 1739.
104. D. Trono, in *Advances in Enzyme Technology: Biomass, Biofuels, Biochemicals*, ed. R. S. Singh, R. R. Singhania, A. Pandey and C. Larroche, Elsevier, Amsterdam, 1st edn, 2019, p. 349.
105. M. E. Johnson, *J. Dairy Sci.*, 2017, **100**, 9952.
106. J. L. Jensen, J. Jacobsen, M. L. Moss, F. Rasmussen, K. B. Qvist, S. Larsen and J. M. van den Brink, *J. Dairy Sci.*, 2015, **98**, 2853.
107. S. R. Kappeler, H. J. M. van den Brink, H. Rahbek-Nielsen, Z. Farah, Z. Puhan, E. B. Hansen and E. Johansen, *Biochem. Biophys. Res. Commun.*, 2006, **342**, 647.
108. J. L. Jensen, A. Mølgaard, J.-C. N. Poulsen, M. K. Harboe, J. B. Simonsen, A. M. Lorentzen, K. Hjernø, J. M. van den Brink, K. B. Qvist and S. Larsen, *Acta Crystallogr., Sect. D: Biol. Crystallogr.*, 2013, **D69**, 901.
109. K. K. Møller, F. P. Rattray and Y. Ardö, *J. Agric. Food Chem.*, 2012, **60**, 11421.
110. J. A. Beltrán-Espinoza, B. Domínguez-Lujan, N. Gutiérrez-Méndez, D. R. Chávez-Garay, C. Nájera-Domínguez and M. Y. Leal-Ramos, *Int. J. Dairy Technol.*, 2021, **74**, 297.
111. N. Bansal, M. A. Drake, P. Piraino, M. L. Broe, M. Harboe, P. F. Fox and P. L. H. McSweeney, *Int. Dairy J.*, 2009, **19**, 510.
112. P. J. T. Dekker, R. M. De Jong, M. D. Tabeling and C. M. Muijlwijk, *WIPO Pat.*, WO/2013/164481, 2013.
113. J. M. van den Brink, M. Lund and J. Jacobsen, *WIPO Pat.*, WO/2013/174840, 2013.
114. P. Dekker, in *Industrial Enzyme Applications*, ed. A. Vogel and O. May, Wiley-VCH, Weinheim, 1st edn, 2019, p. 143.
115. L. Feijoo-Siota, L. Blasco, J. L. Rodríguez-Rama, J. Barros-Velázquez, T. de Miguel, A. Sánchez-Pérez and T. G. Villa, *Recent Adv. DNA Gene Sequences*, 2014, **8**, 44.
116. J. B. Cooper, in *Handbook of Proteolytic Enzymes*, ed. N. D. Rawlings and G. Salvesen, Academic Press, London, 3rd edn, 2013, vol. 1, p. 147.
117. H. B. Drøhse and B. Foltmann, *Biochim. Biophys. Acta*, 1989, **995**, 221.
118. S. Awad, Q.-Q. Lüthi-Peng and Z. Puhan, *J. Agric. Food Chem.*, 1999, **47**, 3632.
119. T. Beppu and M. Nishiyama, in *Handbook of Proteolytic Enzymes*, ed. N. D. Rawlings and G. Salvesen, Academic Press, London, 3rd edn, 2013, vol. 1, p. 154.
120. B. A. Law, in *Enzymes in Food Technology*, ed. R. J. Whitehurst and M. van Oort, Blackwell, Chichester, 2nd edn, 2010, p. 88.
121. M. Jacob, D. Jaros and H. Rohm, *Int. J. Dairy Technol.*, 2021, **63**, 370.
122. M. Jacob, D. Jaros and H. Rohm, *Int. J. Dairy Technol.*, 2010, **64**, 14.
123. F. Alavi and S. Momen, *Int. Dairy J.*, 2020, **107**, 104709.
124. A. Ben Amira, S. Besbes, H. Attia and C. Blecker, *Int. J. Food Prop.*, 2017, **20**, S76.

125. M. A. Mazorra-Manzano, J. M. Moreno-Hernández and J. C. Ramírez-Suarez, in *Biotechnological Applications of Plant Proteolytic Enzymes*, ed. M. G. Guevara and G. R. Daleo, Springer Nature, Cham, 2018, p. 21.
126. M. A. Shah, S. A. Mir and M. A. Paray, *Dairy Sci. Technol.*, 2014, **94**, 5.
127. C. M. Almeida and I. Simões, *Appl. Microbiol. Biotechnol.*, 2018, **102**, 4675.
128. S. V. Cavalli, D. Lufrano, M. L. Colombo and N. Priolo, *Phytochemistry*, 2013, **92**, 16.
129. E. Ordiales, A. Martín, M. J. Benito, S. Ruiz-Moyano, G. Gallardo and M. De Guia Córdoba, *Int. J. Dairy Technol.*, 2016, **69**, 272.
130. A. C. Sarmento, H. Lopes, C. S. Oliveira, R. Vitorino, B. Samyn, K. Seargeant, G. Debyser, J. Van Beeumen, P. Domingues, F. Amando, E. Pires, M. R. M. Domingues and M. T. Barros, *Plants*, 2009, **230**, 429.
131. I. Q. Macedo, C. J. Faro and E. M. Pires, *J. Agric. Food Chem.*, 1993, **41**, 1537.
132. J. Kervinen and A. Wlodawer, in *Handbook of Proteolytic Enzymes*, ed. N. D. Rawlings and G. Salvesen, Academic Press, London, 3rd edn, 2013, vol. 1, p. 118.
133. M. Murlidhar, R. Anusha and O. S. Bindhu, in *Dairy Engineering: Advanced Technologies and their Applications*, ed. M. Meghwal, M. R. Goyal and R. S. Chavan, Apple Academic, Boca Raton, 1st edn, 2017, p. 3.
134. P. Barracosa, N. Rosa, M. Barros and E. Pires, *Chem. Biodiversity*, 2018, **15**, e1800110.
135. M. J. Lewis, *Int. J. Dairy Technol.*, 2011, **64**, 1.
136. L. Ong, K. Soodam, S. E. Kentish, I. B. Powell and S. L. Gras, *Int. Dairy J.*, 2015, **46**, 53.
137. S. Sandra, M. Ho, M. Alexander and M. Corredig, *J. Dairy Sci.*, 2012, **95**, 75.
138. N. A. Bringe and J. E. Kinsella, *J. Dairy Res.*, 1986, **53**, 371.
139. G. Balakrishnan, J. V. C. Silva, T. Nicolai, C. Chassenieux, C. Bovay, J. Buczkowski and C. Schmitt, *Colloids Surf., B*, 2018, **163**, 218.
140. L. Ramasubramanian, B. D'Arcy and H. C. Deeth, *Int. J. Dairy Technol.*, 2021, **65**, 183.
141. Q. Wang and Y. Ma, *Int. Dairy J.*, 2020, **110**, 104713.
142. N. A. Bringe and J. E. Kinsella, *Food Hydrocolloids*, 1993, **7**, 113.
143. E. Romeih and G. Walker, *Trends Food Sci. Technol.*, 2017, **62**, 133.
144. F. Masotti, S. Cattaneo, M. Stuknyte and I. De Noni, *Trends Food Sci. Technol.*, 2017, **64**, 102.
145. J. Long and J. Benson, *Cheese and Cheese-making, Butter and Milk with Special Reference to Continental Fancy Cheeses*, Chapman and Hall, London, 1896.
146. J. Lister, *Transactions of the Pathological Society of London*, 1878, vol. 29, p. 425.
147. International Organization for Standardization and International Dairy Federation, ISO 27205:2010 (IDF 149:2010) Fermented milk products – Bacterial starter cultures – Standard of identity, ISO, Geneva, 2010.

148. E. Parente, T. M. Cogan and I. B. Powell, in *Cheese: Chemistry, Physics & Microbiology*, ed. P. F. Fox, P. L. H. McSweeney, P. Cotter and D. W. Everett, Academic Press, London, 4th edn, 2017, p. 201.
149. F. Irlinger, S. Helinck and J. L. Jany, in *Cheese: Chemistry, Physics and Microbiology*, ed. P. L. H. McSweeney, P. F. Fox, P. D. Cotter and D. W. Everett, Academic Press, London, 4th edn, 2017, p. 273.
150. W. Bockelmann, in *Technology of Cheesemaking*, ed. B. A. Law and A. Y. Tamime, Wiley-Blackwell, Chichester, 2nd edn, 2010, p. 193.
151. M. Gobbetti, R. Di Cagno, M. Calasso, E. Neviani, P. F. Fox and M. De Angelis, *Trends Food Sci. Technol.*, 2018, **78**, 244.
152. J. V. Chambers, in *Dairy Microbiology Handbook*, ed. R. K. Robinson, Wiley, New York, 3rd edn, 2002, p. 39.
153. G. Moatsou and E. Moschopoulou, in *Dairy Microbiology and Biochemistry – Recent Developments*, ed. B. H. Özer and G. Akdemir-Evrendilek, CRC Press, Boca Raton, 2015, p. 1.
154. A. M. Olajide and G. Lapointe, in *Encyclopedia of Dairy Sciences*, ed. P. L. H. McSweeney and J. P. McNamara, Elsevier, Academic Press, Cambridge MA, 3rd edn, 2022, p. 319.
155. E. Parente, A. Ricciardi and T. Zotta, *Int. Dairy J.*, 2020, **107**, 104714.
156. B. Tilocca, N. Costanzo, V. M. Morittu, A. A. Spina, A. Soggiu, D. Britti, P. Roncada and C. Piras, *J. Proteomics*, 2020, **210**, 103534.
157. N. C. Prévost-Bouré, B. Karimi, S. Sadet-Bourgeteau, C. Djemiel, M. Brie, J. Dumont, M. Campedelli, V. Nowak, P. Guyot, C. Letourneur, V. Manneville, F. Gillet and Y. Bouton, *Sci. Rep.*, 2021, **11**, 18144.
158. G. Licitra, M. Caccamo and S. Lortal, in *Raw Milk: Balance Between Hazards and Benefits*, ed. L. A. Nero and A. F. De Carvalho, Academic Press, London, 2019, p. 175.
159. S. Lortal, A. Di Blasi, M.-N. Madec, C. Pediliggieri, L. Tuminello, G. Tanguy, J. Fauquant, Y. Lecuona, P. Campo, S. Carpino and G. Licitra, *Int. J. Food Microbiol.*, 2009, **132**, 1.
160. R. Didienne, C. Defargues, C. Callon, T. Meylheuc, S. Hulin and M.-C. Montel, *Int. J. Food Microbiol.*, 2012, **156**, 91.
161. N. A. Bokulich and D. A. Mills, *Appl. Environ. Microbiol.*, 2013, **79**, 5214.
162. A. von Wright and L. Axelsson, in *Lactic Acid Bacteria: Microbiological and Functional Aspects*, ed. G. Vinderola, A. Ouwehand, S. Salminen and A. von Wright, CRC Press, Boca Raton, 5th edn, 2019, p. 1.
163. W. H. Holzapfel and B. J. B. Wood, in *Lactic Acid Bacteria: Biodiversity and Taxonomy*, Wiley, Chichester, UK, 2014, p. 1.
164. M. B. Pedersen, P. Gaudu, D. Lechardeur, M.-A. Petit and A. Gruss, *Annu. Rev. Food Sci. Technol.*, 2012, **3**, 37.
165. T. Zotta, E. Parente and A. Ricciardi, *J. Appl. Microbiol.*, 2017, **122**, 857.
166. M. Wels, R. Siezen, S. van Hijum, W. J. Kelly and H. Bachmann, *Front. Microbiol.*, 2019, **10**, 4.
167. M. Kleerebezem, H. Bachmann, E. van Pelt-KleinJan, S. Douwenga, E. J. Smid, B. Teusink and O. van Mastrigt, *FEMS Microbiol. Rev.*, 2020, **44**, 804.

168. H. El Kafsi, J. Binesse, V. Loux, J. Buratti, S. Boudebbouze, R. Dervyn, S. Kennedy, N. Galleron, B. Quinquis, J.-M. Batto, B. Moumen, E. Maguin and M. van de Guchte, *BMC Genomics*, 2014, **15**, 407.
169. P. Kelleher, J. Mahony, F. Bottacini, G. A. Lugli, M. Ventura and D. van Sinderen, *Front. Microbiol.*, 2019, **10**, 707.
170. EFSA Panel on Biological Hazards (BIOHAZ), *EFSA J.*, 2021, **19**, 6377.
171. F. Bourdichon, I. Alper, R. Bibiloni, A. Dubois, S. Laulund, M. Miks, L. Morelli, V. Zuliani and S. Yao, *Bull. Int. Dairy Fed.*, 2018, **495**, 5.
172. H. R. Whitehead, *Bacteriol. Rev.*, 1953, **17**, 109.
173. M. Briggiler Marcó, S. Moineau and A. Quiberoni, *Bacteriophage*, 2012, **2**, 149.
174. G. Giraffa, M. Zago and D. Carminati, in *Microbiology in Dairy Processing: Challenges and Opportunities*, ed. P. Poltronieri, Wiley, Hoboken, NJ, 2018, p. 233.
175. J. Lyne, in *Encyclopedia of Dairy Sciences*, ed. P. L. H. McSweeney and J. P. McNamara, Elsevier, Academic Press, Cambridge MA, 3rd edn, 2022, vol. 3, p. 80.
176. J. Murphy, J. Mahony, G. F. Fitzgerald and D. van Sinderen, in *Cheese: Chemistry, Physics and Microbiology*, ed. P. L. H. McSweeney, P. F. Fox, P. D. Cotter and D. W. Everett, Academic Press, London, 4th edn, 2017, p. 249.
177. I. B. Powell, M. C. Broome and G. K. Y. Limsowtin, in *Encyclopedia of Dairy Sciences*, ed. P. L. H. McSweeney and J. P. McNamara, Elsevier, Academic Press, Cambridge MA, 3rd edn, 2022, vol. 3, p. 358.
178. M. Gobbetti and R. Di Cagno, in *Cheese: Chemistry, Physics & Microbiology*, ed. P. F. Fox, P. L. H. McSweeney, P. Cotter and D. W. Everett, Academic Press, London, 4th edn, 2017, p. 809.
179. T. Zotta, A. Ricciardi, N. Condelli and E. Parente, *Crit. Rev. Food Sci. Nutr.*, 2022, **62**, 3898.
180. E. Parente, A. Guidone, A. Matera, F. de Filippis, G. Mauriello and A. Ricciardi, *Int. J. Food Microbiol.*, 2016, **217**, 59.
181. M. Gatti, B. Bottari, C. Lazzi, E. Neviani and G. Mucchetti, *J. Dairy Sci.*, 2014, **97**, 573.
182. E. Parente, *Aust. J. Dairy Technol.*, 2006, **61**, 108.
183. G. Giraffa, *Foods*, 2021, **10**, 2632.
184. F. de Filippis, A. La Storia, G. Stellato, M. Gatti and D. Ercolini, *PLoS One*, 2014, **9**, e89680.
185. J. Zheng, S. Wittouck, E. Salvetti, C. M. A. P. Franz, H. M. B. Harris, P. Mattarelli, P. W. O'Toole, B. Pot, P. Vandamme, J. Walter, K. Watanabe, S. Wuyts, G. E. Felis, M. G. Gänzle and S. Lebeer, *Int. J. Syst. Evol. Microbiol.*, 2020, **70**, 2782.
186. F. B. Linfield, *Bulletin No. 73: Experiments in Butter-making and Cheese-making*, The Agricultural College of Utah, Logan, 1901.
187. M. H. Meyer, *A Treatise on Commercial Starters in Butter and Cheese Making*, Chr. Hansen's Laboratory, Little Falls, NY, 1909.
188. S. Knudsen, *J. Dairy Res.*, 1931, **2**, 137.

189. J. J. Leisner, *Food Policy*, 2005, **30**, 419.
190. K. Hartmann, E. Eugster-Meier, M.-T. Fröhlich-Wyder, E. Jakob, D. Wechsler, Y. Ardö, E.-M. Düsterhöft, W. Engels, T. Huppertz, E. R. Hynes, M. C. Perotti and C. V. Bergamini, in *Global Cheesemaking Technology: Cheese Quality and Characteristics*, ed. P. Papademas and T. Bintsis, Wiley, Chichester, 1st edn, 2018, p. 336.
191. M.-T. Fröhlich-Wyder, W. Bisig, D. Guggisberg, E. Jakob, M. Turgay and D. Wechsler, in *Cheese: Chemistry, Physics & Microbiology*, ed. P. F. Fox, P. L. H. McSweeney, P. Cotter and D. W. Everett, Academic Press, London, 4th edn, 2017, p. 889.
192. D. Isolini, A. Roetschi and E. Eugster, *Rohmischkulturen sind sehr vielfältig*, Alimenta, Publi-lactis, Bern, 2010, vol. 8, p. 24.
193. W. Bisig, M.-T. Fröhlich-Wyder, E. Jakob and D. Wechsler, *Aust. J. Dairy Technol.*, 2010, **65**, 206.
194. Préparation d'une culture d'exploitation, https://www.liebefeld-kulturen.ch/wp-content/uploads/2021/06/Preparation_culture_d_exploitation_V2.pdf (accessed November 2021).
195. Kulturen/Cultures, https://www.liebefeld-kulturen.ch/fr/cultures/ (accessed November 2021).
196. I. Powell, in *Improving the Flavour of Cheese*, ed. B. C. Weimer, Woodhead, Cambridge, 2007, p. 300.
197. W. E. Sandine, in *Dairy Starter Cultures*, ed. T. M. Cogan and J.-P. Accolas, VCH, New York, 1996, p. 191.
198. E.-M. Düsterhöft, W. Engels and T. Huppertz, in *Cheese: Chemistry, Physics and Microbiology*, ed. P. L. H. McSweeney, P. F. Fox, P. D. Cotter and D. W. Everett, Academic Press, London, 4th edn, 2017, p. 865.
199. E.-M. Düsterhöft, W. Engels and T. Huppertz, in *Global Cheesemaking Technology: Cheese Quality and Characteristics*, ed. P. Papademas and T. Bintsis, Wiley, Chichester, 1st edn, 2018, p. 326.
200. E. J. Smid, O. Erkus, M. Spus, J. C. M. Wolkers-Rooijackers, S. Alexeeva and M. Kleerebezem, *Microb. Cell Fact.*, 2014, **13**(Suppl. 1), S2.
201. C. A. Frantzen, H. P. Kleppen and H. Holo, *Appl. Environ. Microbiol.*, 2018, **84**, e02199-17.
202. C. A. Frantzen, W. Kot, T. B. Pedersen, Y. M. Ardö, J. R. Broadbent, H. Neve, L. H. Hansen, F. Dal Bello, H. M. Østlie, H. P. Kleppen, F. K. Vogensen and H. Holo, *Front. Microbiol.*, 2017, **8**, 132.
203. T. A. Lodics and L. R. Steenson, *J. Dairy Sci.*, 1990, **73**, 2685.
204. F. Bissonnette, S. Labrie, H. Deveau, M. Lamoureux and S. Moineau, *J. Dairy Sci.*, 2000, **83**, 620.
205. E. W. Nielsen, *Int. Dairy J.*, 1998, **8**, 1003.
206. D. K. O'Toole, *Int. J. Dairy Technol.*, 2004, **57**, 53.
207. G. K. Y. Limsowtin, P. G. Bruinenberg and I. B. Powell, *J. Microbiol. Biotechnol.*, 1997, **7**, 1.
208. J. Czulak, D. J. Bant, S. C. Blyth and J. B. Crace, *Dairy Ind. Int.*, 1979, **44**(2), 17.
209. H. A. Heap, *Aust. J. Dairy Technol.*, 1998, **53**, 74.

210. P. Timmons, M. Hurley, F. Drinan, C. Daly and T. M. Cogan, *Int. J. Dairy Technol.*, 1988, **41**, 49.
211. L. Ong, R. C. Lawrence, J. Gilles, L. K. Creamer, V. L. Crow, H. A. Heap, C. G. Honoré, K. A. Johnston, P. K. Samal, I. B. Powell and S. L. Gras, in *Cheese: Chemistry, Physics and Microbiology*, ed. P. L. H. McSweeney, P. F. Fox, P. D. Cotter and D. W. Everett, Academic Press, London, 4th edn, 2017, p. 829.
212. P. M. F. Derkx, T. Janzen, K. I. Sørensen, J. E. Christensen, B. Stuer-Lauridsen and E. Johansen, *Microb. Cell Fact.*, 2014, **13**(Suppl 1), S5.
213. E. Johansen, *Annu. Rev. Food Sci. Technol.*, 2018, **9**, 1.
214. D. A. Romero, D. Magill, A. Millen, P. Horvath and C. Fremaux, *FEMS Microbiol. Rev.*, 2020, **44**, 909.
215. D. J. McMahon and C. J. Oberg, in *Cheese: Chemistry, Physics and Microbiology*, ed. P. L. H. McSweeney, P. F. Fox, P. D. Cotter and D. W. Everett, Academic Press, London, 4th edn, 2017, p. 1041.
216. W. C. Frazier, L. A. Burkey, K. J. Matheson and P. D. Watson, *J. Dairy Sci.*, 1933, **16**, 387.
217. J. M. Sherman, *J. Bacteriol.*, 1921, **6**, 379.
218. D. Stokes, R. P. Ross, G. F. Fitzgerald and A. Coffey, *Lait*, 2001, **81**, 327.
219. B. Percival and F. Percival, *Reinventing the Wheel: Milk, Microbes and the Fight for Real Cheese*, Bloomsbury, London, 2017.
220. R. W. Hutkins, in *Applied Dairy Microbiology*, ed. E. H. Marth and J. L. Steele, Marcel Dekker, New York, 2nd edn, 2001, p. 207.
221. P. L. H. McSweeney and M. J. Sousa, *Lait*, 2000, **80**, 293.
222. T. K. Singh, M. A. Drake and K. R. Cadwallader, *Compr. Rev. Food Sci. Food Saf.*, 2003, **2**, 166.
223. G. Smit, B. A. Smit and W. Engels, *FEMS Microbiol. Rev.*, 2005, **29**, 591.
224. E. J. Smid and C. Lacroix, *Curr. Opin. Biotechnol*, 2013, **24**, 148.
225. J. Blaya, Z. Barzideh and G. Lapointe, *J. Dairy Sci.*, 2018, **101**, 3611.
226. B. Mayo, J. Rodríguez, L. Vázquez and A. B. Flórez, *Foods*, 2021, **10**, 602.
227. K. Papadimitriou, A. Alegría, P. A. Bron, M. De Angelis, M. Gobbetti, M. Kleerebezem, J. A. Lemos, D. M. Linares, P. Ross, C. Stanton, F. Turroni, D. van Sinderen, P. Varmanen, M. Ventura, M. Zúñiga, E. Tsakalidou and J. Kok, *Microbiol. Mol. Biol. Rev.*, 2016, **80**, 844.
228. M. van de Guchte, P. Serror, C. Chervaux, T. Smokvina, S. D. Ehrlich and E. Maguin, *Antonie van Leeuwenhoek*, 2002, **82**, 187.
229. B. Poolman, *FEMS Microbiol. Rev.*, 1993, **12**, 125.
230. M. Cocaign-Bousquet, S. Even, N. D. Lindley and P. Loubière, *Appl. Microbiol. Biotechnol.*, 2002, **60**, 24.
231. R. W. Hutkins and N. L. Nannen, *J. Dairy Sci.*, 1993, **76**, 2354.
232. J. R. Broadbent and J. L. Steele, in *Handbook of Proteolytic Enzymes*, ed. N. D. Rawlings and G. Salvesen, Academic Press, London, 3rd edn, 2013, vol. 1, p. 3188.
233. D. Ji, J. Ma, M. Xu and D. Agyei, *Compr. Rev. Food Sci. Food Saf.*, 2021, **20**, 369.

234. K. Savijoki, H. Ingmer and P. Varmanen, *Appl. Microbiol. Biotechnol.*, 2006, **71**, 394.
235. E. B. Hansen and P. Marcatili, *Front. Bioeng. Biotechnol.*, 2020, **8**, 613986.
236. M. W. Børsting, K. B. Qvist, E. Brockmann, J. Vindeløv, T. L. Pedersen, F. K. Vogensen and Y. Ardö, *J. Dairy Sci.*, 2015, **98**, 68.
237. V. Juillard, S. Furlan, C. Foucaud and J. Richard, *J. Dairy Sci.*, 1996, **79**, 964.
238. E. J. Smid and M. Kleerebezem, *Annu. Rev. Food Sci. Technol.*, 2014, **5**, 313.
239. P. Bellengier, J. Richard and C. Foucaud, *J. Dairy Sci.*, 1997, **80**, 1527.
240. M. Spus, M. Li, S. Alexeeva, J. C. M. Wolkers-Rooijackers, M. H. Zwietering, T. Abee and E. J. Smid, *J. Dairy Sci.*, 2015, **98**, 5173.
241. E. F. O'Shea, P. D. Cotter, R. P. Ross and C. Hill, *Curr. Opin. Biotechnol*, 2013, **24**, 130.
242. S. Siedler, R. Balti and A. R. Neves, *Curr. Opin. Biotechnol*, 2019, **56**, 138.
243. E. B. Hansen, *Curr. Opin. Food Sci.*, 2018, **19**, 98.
244. S. Tachon, J. B. Brandsma and M. Yvon, *Appl. Environ. Microbiol.*, 2010, **76**, 1131.
245. V. Caldeo and P. L. H. McSweeney, *Int. Dairy J.*, 2012, **25**, 16.
246. S. Abraham, R. Cachon, B. Colas, G. Feron and J. De Coninck, *Int. Dairy J.*, 2007, **17**, 954.
247. M. Brasca, S. Morandi, R. Lodi and A. Tamburini, *J. Appl. Microbiol.*, 2007, **103**, 1516.
248. C. M. A. P. Franz and W. H. Holzapfel, in *Stress Responses of Lactic Acid Bacteria*, ed. E. Tsakalidou and K. Papadimitriou, Springer, New York, 2011, p. 3.
249. A. D. W. Nugroho, M. Kleerebezem and H. Bachmann, *Curr. Opin. Food Sci.*, 2021, **39**, 22.
250. B. C. Weimer, in *Stress Responses of Lactic Acid Bacteria*, ed. E. Tsakalidou and K. Papadimitriou, Springer, New York, 2011, p. 129.
251. J. Zhang, C. Wu, F. Xue, G. Du and J. Chen, in *Stress Responses of Lactic Acid Bacteria*, ed. E. Tsakalidou and K. Papadimitriou, Springer, New York, 2011, p. 199.
252. J. E. Christensen, E. G. Dudley, J. A. Pederson and J. L. Steele, *Antonie van Leeuwenhoek*, 1999, **76**, 217.
253. P. Lamichhane, P. Sharma, A. L. Kelly, J. Risbo, F. P. Rattray and J. J. Sheehan, *Food Chem.*, 2021, **336**, 127625.
254. C. Wenzel, S. Irmler, W. Bisig, D. Guggisberg, A. Roetschi, R. Portmann, D. Wechsler and M.-T. Fröhlich-Wyder, *Int. Dairy J.*, 2018, **85**, 191.
255. F. Barbieri, C. Montanari, F. Gardini and G. Tabanelli, *Foods*, 2019, **8**, 17.
256. F. A. Zuljan, P. Mortera, S. H. Alarcón, V. C. Blancato, M. Espariz and C. Magni, *Int. Dairy J.*, 2016, **62**, 53.
257. EFSA Panel on Biological Hazards (BIOHAZ), *EFSA J.*, 2011, **9**, 2393.
258. M. Johnson, *Dairy Pipeline (Center for Dairy Research)*, 2017, **29**(3), 6.

259. M.-T. Fröhlich-Wyder, D. Guggisberg, R. Badertscher, D. Wechsler, A. Wittwer and S. Irmler, *Int. Dairy J.*, 2013, **33**, 120.
260. M. G. Wilkinson and G. Lapointe, *J. Dairy Sci.*, 2020, **103**, 10963.
261. V. L. Crow, T. Coolbear, P. K. Gopal, F. G. Martley, L. L. McKay and H. Riepe, *Int. Dairy J.*, 1995, **5**, 855.
262. F. A. Zuljan, G. D. Repizo, S. H. Alarcon and C. Magni, *Int. J. Food Microbiol.*, 2014, **188**, 99.
263. J. Hugenholtz, *FEMS Microbiol. Rev.*, 1993, **12**, 165.
264. F. Lopez de Felipe and P. Gaudu, *Appl. Microbiol. Biotechnol.*, 2009, **82**, 1115.
265. M. Mataragas, *Food Microbiol.*, 2020, **87**, 103392.
266. N. Bassit, C.-Y. Boquien, D. Picque and G. Corrieu, *Appl. Environ. Microbiol.*, 1993, **59**, 1893.
267. O. van Mastrigt, E. E. Mager, C. Jamin, T. Abee and E. J. Smid, *Microb. Biotechnol.*, 2018, **11**, 369.
268. J. Hugenholtz, *Aust. J. Dairy Technol.*, 1995, **50**, 25.
269. Y. Jo, D. M. Benoist, A. Ameerally and M. A. Drake, *J. Dairy Sci.*, 2018, **101**, 1967.
270. M. Turgay, H.-P. Bachmann, S. Irmler, U. von Ah, M.-T. Fröhlich-Wyder, H. Falentin, S.-M. Deutsch, G. Jan and A. Thierry, in *Encyclopedia of Dairy Sciences*, ed. P. L. H. McSweeney and J. P. McNamara, Elsevier, Academic Press, Cambridge MA, 3rd edn, 2022, vol. 4, p. 34.
271. S. L. Drake, M. E. Curunchia Whetstine, M. A. Drake, P. Courtney, K. Fligner, J. Jenkins and C. Pruitt, *J. Food Sci.*, 2007, **72**, S360.
272. C. Monnet, S. Landaud, P. Bonnarme and D. Swennen, *FEMS Microbiol. Lett.*, 2014, **263**, 1.
273. T. M. Cogan, S. Goerges, R. Gelsomino, S. Larpin, M. Hohenegger, N. Bora, E. Jamet, M. C. Rea, J. Mounier, M. Vancanneyt, M. Guéguen, N. Desmasures, J. Swings, M. Goodfellow, A. C. Ward, H. Sebastiani, F. Irlinger, J.-F. Chamba, R. Beduhn and S. Scherer, *Microbiol. Spectrum*, 2014, **2**, CM-0010-2012.
274. E. Dugat-Bony, C. Straub, A. Teissandier, D. Onésime, V. Loux, C. Monnet, F. Irlinger, S. Landaud, N. Leclercq-Periat, P. Bento, S. Fraud, J.-F. Gibrat, J. Aubert, F. Fer, E. Guédon, N. Pons, S. Kennedy, M. Beckerich, D. Swennen and P. Bonnarme, *PLoS One*, 2015, **10**, e0124360.
275. C. Lopez, M.-B. Maillard, V. Briard-Bion, B. Camier and J. A. Hannon, *J. Agric. Food Chem.*, 2006, **54**, 5855.
276. D. K. Hickey, K. N. Kilcawley, T. P. Beresford and M. G. Wilkinson, *J. Dairy Sci.*, 2007, **90**, 47.
277. F. M. Driessen, DAgSci thesis, Landbouwhogeschool te Wageningen, 1983.
278. A. Thierry, Y. F. Collins, M. C. A. Mudski, P. L. H. McSweeney, M. G. Wilkinson and H. E. Spinnler, in *Cheese: Chemistry, Physics and Microbiology*, ed. P. L. H. McSweeney, P. F. Fox, P. D. Cotter and D. W. Everett, Academic Press, London, 4th edn, 2017, p. 423.

279. R. Holland, S.-Q. Liu, V. L. Crow, M.-L. Delabre, M. Lubbers, M. Bennett and G. Norris, *Int. Dairy J.*, 2005, **15**, 711.
280. B. Ganesan and B. Weimer, in *Cheese: Chemistry, Physics and Microbiology*, ed. P. L. H. McSweeney, P. F. Fox, P. D. Cotter and D. W. Everett, Academic Press, London, 4th edn, 2017, p. 483.
281. Y. F. Collins, P. L. H. McSweeney and M. G. Wilkinson, *Int. Dairy J.*, 2003, **13**, 841.
282. M. Liu, J. R. Bayjanov, B. Renckens, A. Nauta and R. J. Siezen, *BMC Genomics*, 2010, **11**, 36.
283. C. Tanous, A. Kieronczyk, S. Helinck, E. Chambellon and M. Yvon, *Antonie van Leeuwenhoek*, 2002, **82**, 271.
284. M. De Angelis, M. Calasso, R. Di Cagno, S. Siragusa, F. Minervini and M. Gobbetti, *J. Appl. Microbiol.*, 2010, **109**, 1763.
285. E. Parente, T. Zotta and A. Ricciardi, *Int. J. Food Microbiol.*, 2022, **368**, 109618.
286. T. M. Cogan and C. Hill, in *Cheese: Chemistry, Physics and Microbiology*, ed. P. F. Fox, Chapman and Hall, London, 2nd edn, 1993, vol. 1, p. 193.
287. K. H. Schleifer, J. Kraus, C. Dvorak, R. Kilpper-Bälz, M. D. Collins and W. Fischer, *Syst. Appl. Microbiol.*, 1985, **6**, 183.
288. T. T. Li, W. L. Tian and C. T. Gu, *Int. J. Syst. Evol. Microbiol.*, 2021, **71**, 004727.
289. V. Alexandraki, M. Kazou, J. Blom, B. Pot, K. Papadimitriou and E. Tsakalidou, *Front. Microbiol.*, 2019, **10**, 2916.
290. M. I. Torino, M. P. Taranto and G. Font de Valdez, *Biotechnol. Lett.*, 2001, **23**, 1799.
291. M. W. Hickey, A. J. Hillier and G. R. Jago, *Appl. Environ. Microbiol.*, 1986, **51**, 825.
292. J. Björkroth, L. M. T. Dicks, A. Endo and W. H. Holzapfel, in *Lactic Acid Bacteria: Biodiversity and Taxonomy*, Wiley, Chichester, 1st edn, 2014, p. 391.

CHAPTER 3

Nonstarter Lactic Acid Bacteria in Cheese

LING GUO* AND BIQI LIU

Key Laboratory of Dairy Science, Ministry of Education, Food College, Northeast Agricultural University, Harbin, Heilongjiang Province, China
*Email: guoling@neau.edu.cn

3.1 Introduction

Cheese is a food based on fermented milk and is an important part of the world's dairy industry. About 35% of raw milk is processed into cheese every year in the world.[1] According to the data from the Food and Agriculture Organization of the United Nations (FAO), the global cheese market share is expected to grow to US$124 billion by 2022.[2] Contrary to most dairy products with biological, biochemical, and physicochemical stability, cheese is unstable due to its microbial and biochemical reactions, and the manufacturing and maturation process requires a lot of microbial interventions. When these interventions are balanced, they can produce pleasant sensory properties, but if they are not controlled, they can produce peculiar defects.[3] The microbial ecology of cheese is provided by the rich and complex interaction between SLAB (starter lactic acid bacteria) and NSLAB (non-starter lactic acid bacteria);[4,5] NSLAB play a pivotal role in the maturation of cheese and will further affect the flavour and texture of the cheese as well as the final sensory characteristics. Therefore, it is very important to explore the impact of NSLAB on cheese quality. This article summarizes the classification, source, screening, and functional characteristics of non-starter lactic

Food Chemistry, Function and Analysis No. 40
Handbook of Cheese Chemistry
Edited by Michael H. Tunick

Published by the Royal Society of Chemistry, www.rsc.org

acid bacteria in cheese, their advantages as an auxiliary starter and their influence on cheese quality, and their application prospects.

3.2 Main Microbial Flora in Cheese

The special flavour and nutritional characteristics of cheese are the result of the combined action of the milk source, climate environment, processing technology, and microorganisms, among which microorganisms play an important role in cheese quality.[6] These microorganisms enter milk from a variety of sources. Once they enter the milk system, they can play a variety of roles, such as promoting milk fermentation, regulating human health, causing cheese spoilage and causing food-borne diseases.[7]

Cheese microorganisms are mainly divided into two categories: SLAB and secondary microorganisms. The main function of SLAB is to produce acid, which is a key point in the cheese production process. They can convert sugars and other substances into organic acids. The content of organic acids reflects the type of fermentation and the deviation of the expected ripening process that may cause defects.[8,9] At the same time, it can also secrete rennet to coagulate the protein and help the initial maturation process of cheese.[10] At present, the secondary microorganisms that have been isolated mainly include NSLAB, propionibacterium, yeast, and mold, among which NSLAB are the main secondary microorganisms.[11] Although the NSLAB do not produce acid or rarely produce acid, they have outstanding contributions to the change of cheese flavour and texture during the maturation process.[12,13] The final quality of fermented dairy products depends on the interaction between SLAB and NSLAB in the fermentation process, and because of the different raw materials, production environment and processing technology, the types and quantities of microorganisms are quite different. Even if the same fermented dairy product is produced, the types and distribution of microorganisms in different parts of the same operation link will be significantly different, thus forming a complex and dynamic microbial system.

3.3 NSLAB

3.3.1 Classification of NSLAB

The current general definition of NSLAB is "lactic acid bacteria that are not part of the starter in cheese, that is, they do not produce acid in the cheese production process", mainly Lactobacillus, including *Lactobacillus casei*, *Lactobacillus paracasei*, *Lactobacillus plantarum*, *Lactobacillus curvatus*, *Lactobacillus rhamnosus*, *etc.*[14] NSLAB are usually composed of various mesophilic lactobacilli, including *Pediococcus*, *Enterococcus*, and *Leuconostoc.*[15] Among them, mesophilic lactobacilli are the most common NSLAB flora and play an important role in the cheese maturation process because they can ferment citrate and may participate in proteolysis and other enzymatic processes that occur during cheese ripening.[16] In most cases, these microbial strains are necessary to create the unique taste, flavour, aroma, and texture of cheese, and depending on the kind of strains, rather than the number of strains, different kinds of dominant bacteria will directly have different effects on the quality and flavour of cheese. The types of non-starter lactobacilli appearing in common cheese are shown in Table 3.1.

Table 3.1 The main types of NSLAB appearing in common cheese.

Classification	Name	Country	Types of non-starter lactic acid bacteria	Ref.
Fresh	Feta	Greece	*Lb. brevis*, *Lb. paracasei*, *Lb. plantarum*	17
	Galotyri	Greece	*Lb. plantarum*, *Lb. rhamnosus*	18
	Mozzarella	Italy	*Lb. hilgardii*, *Lb. plantarum*	19
	Mozzarella	Brazil	*Lb. fermentum*, *Lb. paracasei*	20
	Queso Fresco	America	*Ln. mesenteroides*	21
	Ricotta	Italy	*Lb. brevis*, *Lb. casei*, *Lb. paracasei*	22
White mold	Brie	France	*Lb. acidophilus*, *Lb. paracasei*, *Lb. plantarum*	23
	Camembert	France	*Lb. fermentum*, *Ln. mesenteroides*, *Lb. paracasei*	24
Blue	Cabrales	Spain	*Lb. brevis*, *Lb. casei*, *Lb. curvatus*	25
	Gamonedo	Spain	*Lb. casei*, *Lb. plantarum*, *Ln. mesenteroides*	26
	Gorgonzola	Italy	*Lb. brevis*, *Lb. plantarum*, *Ln. mesenteroides*	27
	Stilton	England	*Lb. brevis*, *Lb. plantarum*	28
	Valdeœn	Spain	*Lb. paracasei*, *Lb. plantarum*	29
Semi-hard	Abbaye	France	*Lb. brevis*, *Lb. paracasei*	30
	Batzos	Greece	*Lb. brevis*, *Lb. curvatus*, *Lb. paracasei*, *Lb. plantarum*	31
	Caciocavallo	Italy	*Lb. paracasei*, *Lb. plantarum*, *Lb. fermentum*	32
	Caciotta	Italy	*Lb. paracasei*	33
	Castelmagno	Italy	*Lb. casei*, *Lb. paracasei*, *Lb. plantarum*	34
	Danbo	Denmark	*Lb. curvatus*, *Lb. paracasei*, *Lb. plantarum*	35
	Edam	Holland	*Lb. casei*, *Lb. curvatus*, *Lb. paracasei*, *Lb. plantarum*	36
	Fossa	Italy	*Lb. casei*, *Lb. curvatus*, *Lb. paracasei*, *Lb. rhamnosus*	37
	Gouda	Holland	*Lb. brevis*, *Lb. paracasei*, *Lb. plantarum*, *Lb. rhamnosus*	38
	Gouda	Norway	*Lb. casei*, *Lb. curvatus*, *Lb. paracasei*, *Lb. plantarum*	39
	Maasdam	Holland	*Lb. casei*, *Lb. paracasei*, *Lb. plantarum*	40
	Raclette	Switzerland	*Lb. brevis*, *Lb. fermentum*, *Lb. parabuchneri*	41
	Scamorza	Italy	*Lb. fermentum*, *Lb. paracasei*	42

Table 3.1 (*Continued*)

Classification	Name	Country	Types of non-starter lactic acid bacteria	Ref.
Hard	Cheddar	New Zealand	*Lb. paracasei*, *Lb. rhamnosus*	43
	Cheddar	Australia	*Lb. casei*, *Lb. plantarum*	44
	Cheddar	America	*Lb. fermentum*, *Lb. paracasei*	45
	Cheddar	England	*Lb. brevis*, *Lb. curvatus*, *Lb. paracasei*, *Lb. plantarum*	46
	Fossa	Italy	*Lb. casei*, *Lb. paracasei*, *Lb. plantarum*, *Lb. rhamnosus*	47
	Grana Padano	Italy	*Lb. casei*, *Lb. paracasei*, *Lb. rhamnosus*	48
	Grana Trentino	Italy	*Lb. paracasei*, *Lb. rhamnosus*	49
	Graviera Naxou	Greece	*Lb. brevis*	50
	Graviera Kritis	Greece	*Lb. brevis*, *Lb. paracasei*	51
	Gruyere	Switzerland	*Lb. brevis*, *Lb. fermentum*	52
	Parmigiano Reggiano	Italy	*Lb. casei*, *Lb. fermentum*, *Lb. paracasei*, *Lb. rhamnosus*	53
	Pecorino Romano	Italy	*Lb. curvatus*, *Lb. fermentum*, *Lb. plantarum*	54
	Romano	Italy	*Lb. casei*, *Lb. plantarum*, *Lb. rhamnosus*	55

3.3.2 Source of NSLAB

The main source of NSLAB is the contamination of milk, that is, the NSLAB in raw milk that can survive the pasteurization process in cheese making and the environmental microorganisms found in the production process,[56] for example, from equipment, intermediate environment, or in raw materials. The level of these NSLAB will be very low during the initial preparation process, but gradually dominates as the fermentation progresses.[57]

A unique and stable flora is not directly related to the factory environment for cheese production but is inseparable from the difference in the flora of raw milk. Casey *et al.*[58] conducted genotype identification of NSLAB in Swiss Gruyère cheese and found that almost all the gene types of the non-starter lactic acid bacteria isolated from the cheese were derived from raw milk. Sgarbi *et al.*[59] used the cell lysate of other lactic acid bacteria as the only source of nutrition to study the growth ability of non-fermenting lactic acid bacteria in dairy products and found that the non-fermenting lactic acid bacteria were mainly derived from raw milk and the environment. Nikoloudaki *et al.*[60] studied how a variety of culture conditions are related to the composition of the microbial community in raw milk and found that different raw milks contain multiple NSLAB. All these indicate that raw milk is the main source of non-fermenting lactic acid bacteria, and other environmental microorganisms are the secondary source of non-fermenting lactic acid bacteria.

3.3.3 Advantages of NSLAB as Adjunct Cultures

The International Codex Alimentarius Commission (CAC), the European Union, the United States, Canada and other international organizations, countries, and regions have improved the content of relevant laws, regulations, and standards for cheese, such as CAC's "General Standard for Cheese" (CODEX STAN 283-1978), the EU's "Special Regulations on Food Hygiene" [(EC) No. 853/2004], "Maximum Limits of Contaminants in Foods" [(EC) No. 1881/2006] and "Food Microbiological Standards" [(EC) No. 2073/2005], US Code of Federal Regulations, Part 109 and Part 133, *etc.*, as well as the wide application of pasteurization technology. The raw milk can almost reach a sterile state, which is beneficial to the safety of consumers eating dairy products, but it also leads to the disappearance and decrease of some NSLAB that are beneficial to product quality. Therefore, to accelerate the maturation of cheese and strengthen their functions and quality characteristics, cheese producers and researchers began to study and use auxiliary starters. Most auxiliary starters are derived from SLAB and NSLAB in existing cheeses. SLAB-based adjuvants are developed from starters that produce good flavours, which can effectively enhance the flavour, while starters based on NSLAB are easier to use and have varying degrees of impact on the flavour and texture of cheese.[61] Lactobacilli are the most common adjunct cultures in cheese production. They have the functions of inhibiting mold growth, extending shelf life, improving ACE inhibitory activity and antioxidant activity, reducing the concentration of biogenic amines, and producing a variety of flavour substances in cheese.[62–70] The effect of NSLAB in cheese is shown in Table 3.2.

3.4 Influence of NSLAB on Cheese Maturation

Maturation is a key process in cheese production and consists of a series of biochemical and microbiological reactions.[71] The starter microorganisms and non-starter microorganisms are dynamic throughout cheese processing. The starter microorganisms are at a high level in the early stage of cheese maturation. They convert the easily absorbed nutrients such as sugars into organic acids, which can grow and produce enzymes for various microorganisms, and exhibit metabolic activities to create a suitable environment; in the late stage of cheese maturation, the number of non-starter microorganisms reaches a high level. They secrete a variety of proteases to degrade casein and peptides, and produce a large number of amino acids, which are precursors to the metabolic reaction of volatile flavour substances.[72] Therefore, the quality of cheese depends on the complex changes that occur during the maturation process.

3.4.1 Influence of NSLAB on Cheese Flavour

The research on cheese flavour is mainly related to the market's acceptance of cheese and consumers' perception of flavour. Moisture, fat, amino acids, minerals, salt, microflora, and other trace components make cheese a complex food. The component balance theory points out that the mixture of the right chemicals at appropriate levels will give cheese aroma and texture,

Table 3.2 The influence of NSLAB in cheese.

Classification	Adjunct cultures	Effect	Ref.
Caciotta cheese	*Lb. paracasei*	Promotes the production of multiple flavours	62
Cheddar cheese	*Lb. plantarum*	Has a beneficial effect on cheese texture and flavour	63
Cheddar cheese	*Lb. paracasei*, *Lb. rhamnosus*	Colonize in the intestines and develop the potential of probiotics	56
Cremoso cheese	*Lb. paracasei*, *Lb. plantarum*	Have a probiotic effect and help in standardization of cheese	64
Dutch-type cheese	*Lb. casei*, *Lb. plantarum*	Reduce the concentration of biogenic amines (BAs) in cheese	65
Fresh cheese	*Lb. casei*, *Lb. plantarum*	Control the growth of mold in cheese, with antifungal activity	66
Manchego cheese	*Lb. paracasei*, *Lb. plantarum*	Improve cheese flavour and maintain its sensory characteristics	67
Tosèla cheese	*Lb. paracasei*	Inhibits the growth of coliforms and staphylococci	68
White-brined cheese	*Lb. casei*, *Lb. plantarum*	Improve ACE inhibitory activity and antioxidant activity	69
Sheep cheese	*Lb. casei*, *Lb. plantarum*	Increase the content of biologically active fatty acids in cheese	70

and the aroma of cheese is the result of the combined action of various volatile components.[73] These volatile substances mainly come from the metabolites in the process of cheese maturation. At the same time, the microorganisms that play a major role in the flavour of cheese during the maturation process are NSLAB instead of SLAB.[74]

Therefore, changes in cheese flavour are closely related to changes in NSLAB during cheese maturation, and the occurrence of these pathways is directly related to the metabolic activities of NSLAB in cheese. Several biochemical events occur during cheese maturation. These events will change the physical and chemical properties and flavour of cheese. They mainly include three main reactions: metabolism of residual lactose, lactate, and citrate, lipolysis, and proteolysis.[75] Analyzing these three main reactions is of great significance for understanding the influence of NSLAB on the flavour of cheese.

3.4.1.1 Metabolism of Residual Lactose, Lactate and Citrate

Cheese is a fermented dairy product. The lactose in the raw milk can be metabolized into lactic acid and CO_2 by a specific lactic acid bacterial culture, and lactic acid can be further metabolized to produce a variety of flavour compounds, such as diacetyl, acetic acid, and propionic acid. When the cheese is maturing, the pH value decreases due to acid production. The energy required for the growth of SLAB is insufficient, the vitality is inhibited, and the survival time becomes shorter so that 1–2% of lactose will not be metabolized. SLAB can still grow normally under low water activity or low pH conditions. Therefore, when the activity of starter strains is limited, the NSLAB will become the main metabolic productivity substance, and the remaining 1–2% lactose can be metabolized into DL-lactate by NSLAB and then racemized into L-lactate, thus finally forming acetate (Figure 3.1). Because the types and quantities of NSLAB are different, the level of acetate formed is also different.

Citrate is an important precursor of some flavour compounds, such as diacetyl, acetoin (3-hydroxy-2-butanone), and 2,3-butanediol. In some cheeses, citrate is rapidly metabolized in the presence of fermentable carbohydrates to produce acetic acid, diacetyl, and CO_2. However, this citrate metabolism is the main cause of floating curd in cheese, which leads to extremely poor cheese texture. However, about 90% citrate in milk is soluble, most of which is lost in whey, and the remaining citrate can be fully metabolized by NSLAB to produce corresponding flavour substances.

3.4.1.2 *Effect of NSLAB on Lipolysis*

Fat decomposition is usually considered as the accumulation of free fatty acids (FFA) during cheese ripening. Fat in cheese undergoes fat decomposition or oxidative degradation under the action of lipases and esterases, in which the hydrolysis of triacylglycerides is the most important biochemical transformation during cheese ripening, generating large amounts of FFA. The FFA are further decomposed to generate volatile substances, which has an important influence on the development of the flavour and aroma of cheese during ripening.[76] The lipase in cheese comes from lipase, chymosin paste, starter, adjunct starter, NSLAB, and exogenous lipase in raw milk. In short, the fat decomposition and oxidative degradation of cheese mainly rely on its lipase, and the NSLAB is an important source of lipase, which makes an important contribution to its catabolism and thus plays a vital role in cheese maturation.

3.4.1.3 *Effect of NSLAB on Proteolysis*

Protein hydrolysis is the most complicated and important event in the cheese ripening process, which mainly involves different kinds of casein hydrolysis, and is usually related to raw milk, SLAB, NSLAB, and related proteolytic enzymes. Among them, protease and peptidase produced by NSLAB can decompose casein to produce free amino acids. The increase of these free amino acids will strengthen the flavour of cheese and gradually dominate, making it play an active role in forming unique flavours of different types of cheese.

Amino acids are the precursor of volatile aromatic compounds, and their increasing levels will directly affect the flavour of cheese, making some flavour components play a leading role in cheese, and playing a positive role in forming unique flavours of different characteristic cheeses.

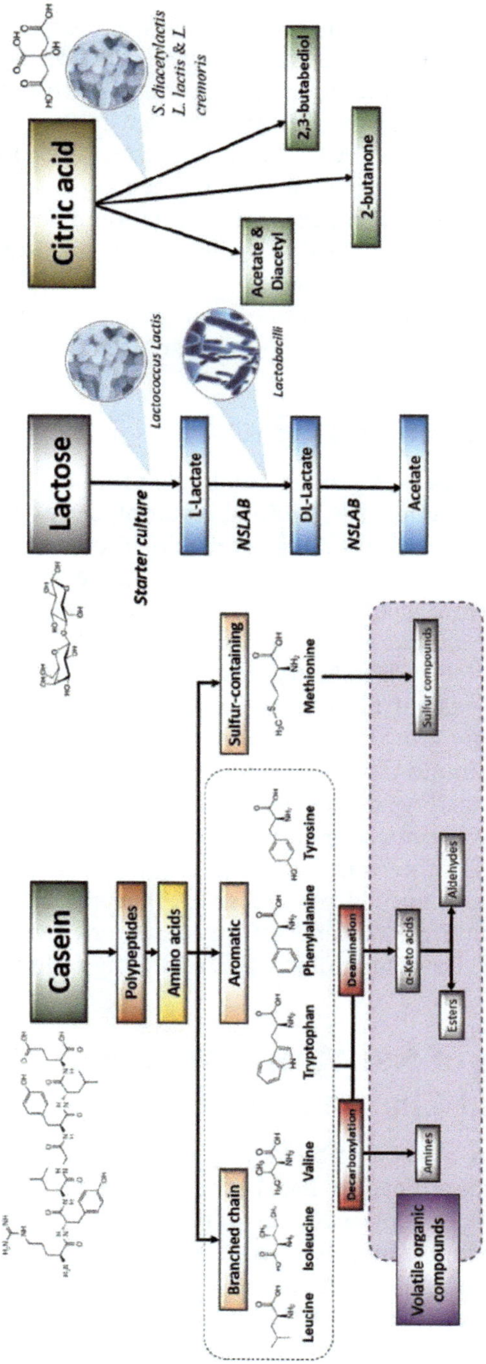

Figure 3.1 Metabolism and proteolysis of residual lactose, lactate, and citrate during the formation of cheese flavour. Reproduced from ref. 76 with permission from Elsevier, Copyright 2019.

The decomposition and metabolism of free amino acids by NSLAB includes deamination, decarboxylation, and desulphurization, which leads to the formation of amines, aldehydes, alcohols, indoles, carboxylic acids, phenols, and sulphur-containing components. In cheese with or without auxiliary materials, the range of volatile components from decomposition and metabolism may be similar, but their relative concentrations increase significantly with the addition of NSLAB.[77] For a long time, the total amount and composition of the amino acid mixture has been used as an indicator of cheese maturation. Among them, the transamination reaction, which is the main degradation pathway, will lead to the production of α-keto acids, which are further degraded to produce volatile aromatic compounds through different pathways.[78] The volatile aromatic compounds produced by these pathways are critical to the formation of cheese flavour (Figure 3.1).

3.4.2 Effect of NSLAB on Texture

NSLAB can cause defects such as gas and calcium lactate crystals on the cheese surface. These crystals are the final result of lactic acid racemization in which some NSLAB converts L(+)-lactic acid into the less soluble D(−) isomer. In a study of the effects of NSLAB and aging temperature on the crystallization of calcium lactate in cheddar cheese, Chou *et al.*[79] found that adding *Lactobacillus curvatus* to raw milk to produce cheese would promote the appearance of calcium lactate crystals on the surface of cheese. At the same time, aging cheese at high temperatures to accelerate ripening may promote the secondary growth of NSLAB and the formation of calcium lactate. Agarwal *et al.*[80] found that not all NSLAB could racemize enough L(+)-lactic acid into D(−)-lactic acid to cause lactate in cheese, but some NSLAB strains with positive racemase could cause lactate in cheese. Racemase-positive lactobacillus is the cause of lactate formation in cheese. Although calcium lactate crystals formed on cheese are harmless, consumers often mistake them for mold growth and refuse to accept cheese, which brings economic burden to cheese manufacturers and dairy farmers.[81]

3.5 Screening of NSLAB and Mining of Their Functional Characteristics

With the increase in molecular biology research in the 21st century, the accuracy of strain identification has become higher, and the traditional culture method has been replaced by modern molecular biotechnology. It is of great significance to use molecular methods to select appropriate NSLAB to improve cheese flavour, texture, and other functional characteristics. Uymaz *et al.*[82] analyzed the 16S rRNA gene sequence of NSLAB in Ezine cheese and found that Lactobacillus accounted for 36.65%, *Enterococcus* accounted for 25.65%, *Lactococcus* accounted for 23.56%, and *Leuconostoc* accounted for 7.33%. Gantzias *et al.*[83] used MALDI-TOF MS and pheS sequence analysis to detect the NSLAB strains of four traditional cheeses produced on Naxos and found that *Lb. brevis* and *Lb. plantarum* were dominant species. At present, the development of high-throughput or next-generation sequencing (HTS or NGS) technology makes it possible to explore microbiota and genome information.[2] Margalho *et al.*[84] used high-throughput sequencing to screen starter and auxiliary cultures that can be used as fermented dairy products in Brazilian handmade cheese, and found that *L. plantarum*, *L. paracasei,* and *L. brevis* are the most promising lactic acid bacteria isolated from cheese.

In recent years, "foodomics" has gradually become a powerful tool for people to study the composition and functional characteristics of food. It mainly combines metabolomics, genomics, protein-omics, and other different omics technologies, and uses analytical technologies to explore and promote the fields of food, nutrition, and health.[85] Studies have shown that analyzing the data obtained from metabolomics, metatranscriptomics, and metagenomics in cheese can be used to better understand the contribution of NSLAB to the maturation process. These studies will also increase the direct impact of maturation temperature on the expression of proteolysis- and lipolysis-related genes and further reveal its microbial community dynamics and potential functional characteristics.[10]

3.6 Outlook

NSLAB is a large and complex microbial group, which plays a significant role in the process of cheese maturation. The analysis of its mechanism of action on cheese maturation and the influence of different strains on cheese has guiding significance for cheese production and processing. With the increase of research and the development of molecular biology, the research on non-fermenter lactic acid bacteria has shifted from the limited research on finished cheese to the application of molecular biology in bacterial species identification. In addition, studying the community structure and diversity of NSLAB is of great significance for the protection of microbial resources, the development and utilization of lactic acid bacteria resources, and the inheritance and optimization of traditional cheese technology.

References

1. C. D. Hickey, M. A. E. Auty, M. G. Wilkinson and J. J. Sheehan, *Int. Dairy J.*, 2017, **69**, 9.
2. J. Choi, S. In Lee, B. Rackerby, R. Frojen, L. Goddik, S. D. Ha and S. H. Park, *Appl. Microbiol. Biotechnol.*, 2020, **104**, 6249.
3. M. Gobbetti, R. Di Cagno, M. Calasso, E. Neviani, P. F. Fox and M. De Angelis, *Trends Food Sci. Technol.*, 2018, **78**, 244.
4. L. Alessia, G. Bertani, B. Bottari, V. Bernini, C. Lazzi, M. Gatti and E. Neviani, *Curr. Opin. Food Sci.*, 2021, **38**, 131.

5. A. Valentina, F. Ilario, D. Francesca, F. Mauro, R. Kalliopi, E. Danilo and C. Luca, *Appl. Environ. Microbiol.*, 2016, **82**, 3988.
6. Z. Zijian, N. Chao, C. Long, Z. Yujuan, Y. Ge, W. Chao, C. Naishi, Z. Zhaoye and L. Shengyu, *Food Res. Int.*, 2021, **148**, 110600.
7. R. H. Mallappa, C. Balasubramaniam, B. H. Nataraj, C. Ramesh, S. Kadyan, D. Pradhan, S. K. Muniyappa and S. Grover, *Int. Dairy J.*, 2021, **114**, 104941.
8. T. Czárán, F. P. Rattray, O. D. Cleide and B. Christensen, *Int. Dairy J.*, 2018, **80**, 35.
9. L. Ong and N. P. Shah, *LWT–Food Sci. Technol.*, 2009, **42**, 1260.
10. J. Blaya, Z. Barzideh and G. LaPointe, *J. Dairy Sci.*, 2018, **101**, 3611.
11. L. Zhu, C. Zeng, S. Yang, Z. Hou, Y. Wang, X. Hu, K. Senoo and W. Wei, *MicrobiologyOpen*, 2020, **9**, e1066.
12. E. Stefanovic, K. N. Kilcawley, C. Roces, M. C. Rea, M. O'Sullivan, J. J. Sheehan and O. McAuliffe, *Front. Microbiol.*, 2018, **9**, 1506.
13. A. Levante, F. De Filippis, A. La Storia, M. Gatti, E. Neviani, D. Ercolini and C. Lazzi, *Int. J. Food Microbiol.*, 2017, **257**, 276.
14. É. Desfossés-Foucault, G. LaPointe and D. Roy, *Int. J. Food Microbiol.*, 2014, **178**, 76.
15. R. Saiki, T. Hagi, T. Narita, M. Kobayashi, K. Sasaki, Y. Asahina, A. Tajima and M. Nomura, *Food Sci. Technol. Res.*, 2018, **24**, 299.
16. G. Licitra and S. Carpino, *Microbiol. Spectrum*, 2014, **2**, DOI: 10.1128/microbiolspec.CM-0007-2012.
17. D. Bozoudi, C. Kotzamanidis, M. Hatzikamari, N. Tzanetakis, G. Menexes and E. Litopoulou-Tzanetaki, *Int. J. Food Microbiol.*, 2015, **200**, 87.
18. J. Samelis, A. I. Doulgeraki, V. Bikouli, D. Pappas and A. Kakouri, *Fermentation*, 2021, **7**, 67.
19. M. Marilena, G. D. de Wittenau, E. Saccà, F. Cattonaro, A. Spadotto, N. Innocente, S. Radovic, E. Piasentier and F. Marroni, *Food Microbiol.*, 2019, **79**, 123.
20. L. F. Silva, T. Casella, E. S. Gomes, M. C. L. Nogueira, D. D. Linder and A. L. B. Penna, *J. Food Sci.*, 2015, **80**, M411.
21. T. S. Lusk, A. R. Ottesen, J. R. White, M. W. Allard, E. W. Brown and J. A. Kase, *BMC Microbiol.*, 2012, **12**, 1.
22. M. L. Scatassa, I. Mancuso, S. Sciortino, G. Macaluso, M. Palmeri, L. Arcuri, M. Todaro and C. Cardamone, *Ital. J. Food Saf.*, 2018, **7**, 6911.
23. R. Unno, M. Matsutani, T. Suzuki, K. Kodama, H. Matsushita, K. Yamasato, Y. Koizumi and M. Ishikawa, *Int. Dairy J.*, 2020, **109**, 104757.
24. O. Firmesse, E. Alvaro, A. Mogenet, J. L. Bresson, R. Lemée, P. Le Ruyet, C. Bonhomme, D. Lambert, C. Andrieux, J. Doré and G. Corthier, *Int. J. Food Microbiol.*, 2008, **125**, 176.
25. V. Ladero, M. Fernández, I. Cuesta and M. A. Alvarez, *Food Microbiol.*, 2010, **27**, 933.
26. B. Redruello, A. Szwengiel, V. Ladero, B. del Rio and M. A. Alvarez, *LWT*, 2020, **130**, 109603.
27. C. Fontana, F. Cappa, A. Rebecchi and P. S. Cocconcelli, *Int. J. Food Microbiol.*, 2010, **138**, 205.

28. D. Mugampoza, K. Gkatzionis, R. S. T. Linforth and C. E. Dodd, *Food Chem.*, 2019, **287**, 222.
29. I. Diezhandino, D. Fernández, L. González, P. L. H. McSweeney and J. M. Fresno, *Food Chem.*, 2015, **168**, 134.
30. E. Dugat-Bony, L. Garnier, J. Denonfoux, S. Ferreira, A. S. Sarthou, P. Bonnarme and F. Irlinger, *Int. J. Food Microbiol.*, 2016, **238**, 265.
31. L. Psoni, N. Tzanetakis and E. Litopoulou-Tzanetaki, *Food Microbiol.*, 2003, **20**, 575.
32. P. Piraino, T. Zotta, A. Ricciardi and E. Parente, *Int. Dairy J.*, 2005, **15**, 1138.
33. P. G. Burns, F. Patrignani, G. Tabanelli, G. C. Vinderola, L. Siroli, J. A. Reinheimer, F. Gardini and R. Lanciotti, *J. Funct. Foods*, 2015, **13**, 126.
34. P. Dolci, V. Alessandria, K. Rantsiou, M. Bertilono and L. U. Cocolin, *Int. J. Food Microbiol.*, 2010, **143**, 71.
35. M. Antonsson, G. Molin and Y. Ardö, *Int. J. Food Microbiol.*, 2003, **85**, 159.
36. L. Buňková, F. Buňka, G. Mantlová, A. Čablová, I. Sedláček, P. Švec, V. Pachlová and S. Kráčmar, *Food Microbiol.*, 2010, **27**, 880.
37. E. Barbieri, G. F. Schiavano, M. De Santi, L. Vallorani, L. Casadei, M. Guescini, A. M. Gioacchini, L. Rinaldi, V. Stocchi and G. Brandi, *Int. Dairy J.*, 2011, **23**, 62.
38. K. Van Hoorde, T. Verstraete, P. Vandamme and G. Huys, *Food Microbiol.*, 2008, **25**, 929.
39. D. Porcellato, H. M. Østlie, K. H. Liland, K. Rudi, T. Isaksson and S. B. Skeie, *J. Dairy Sci.*, 2012, **95**, 4804.
40. P. Lamichhane, A. Pietrzyk, C. Feehily, P. D. Cotter, D. T. Mannion, K. N. Kilcawley, A. L. Kelly and J. J. Sheehan, *J. Dairy Sci.*, 2018, **101**, 5738.
41. D. Wechsler, S. Irmler, H. Berthoud, R. Portmann, R. Badertscher, W. Bisig, K. Schafroth and M. T. Fröhlich-Wyder, *Int. Dairy J.*, 2021, **113**, 104883.
42. F. Baruzzi, A. Matarante, M. Morea and P. S. Cocconcelli, *J. Dairy Sci.*, 2002, **85**, 1390.
43. V. Crow, B. Curry and M. Hayes, *Int. Dairy J.*, 2001, **11**, 275.
44. P. S. Chandry, S. C. Moore and A. J. Hillier, *Aust. J. Dairy Technol.*, 2002, **57**, 106.
45. P. A. Swearingen, D. J. O'Sullivan and J. J. Warthesen, *J. Dairy Sci.*, 2001, **84**, 50.
46. A. G. Williams, S. C. Choi and J. M. Banks, *Food Res. Int.*, 2002, **35**, 483.
47. E. Barbieri, G. F. Schiavano, M. De Santi, L. Vallorani, L. Casadei, M. Guescini, A. M. Gioacchini, L. Rinaldi, V. Stocchi and G. Brandi, *Int. Dairy J.*, 2011, **23**, 62.
48. M. Gatti, B. Bottari, C. Lazzi, E. Neviani and G. Mucchetti, *J. Dairy Sci.*, 2014, **97**, 573.
49. L. Monfredini, L. Settanni, E. Poznanski, A. Cavazza and E. Franciosi, *Syst. Appl. Microbiol.*, 2012, **35**, 54.

50. D. Bozoudi, S. Pavlidou, C. Kotzamanidis, P. Georgakopoulos, S. Torriani, E. Kondyli, S. Claps, S. Belibasaki and E. Litopoulou-Tzanetaki, *Small Rumin. Res.*, 2016, **136**, 161.
51. T. Panagiota, D. Tsafrakidou, S. Bozoudi, C. Pavlidou, M. Kotzamanidis, A. Hatzikamari, E. Zdragas and E. Litopoulou-Tzanetaki, *LWT–Food Sci. Technol.*, 2016, **68**, 681.
52. C. Wenzel, S. Irmler, W. Bisig, D. Guggisberg, A. Roetschi, R. Portmann, D. Wechsler and M. T. Fröhlich-Wyder, *Int. Dairy J.*, 2018, **85**, 191.
53. L. Solieri, A. Bianchi, G. Mottolese, F. Lemmetti and P. Giudici, *Food Microbiol.*, 2014, **38**, 240.
54. R. Di Cagno, C. C. Lòpez, R. Tofalo, G. Gallo, M. De Angelis, A. Paparella, W. P. Hammes and M. Gobbetti, *Meat Sci.*, 2008, **79**, 224.
55. G. Blaiotta, R. Marrone, M. Aponte, M. F. Peruzy, G. Smaldone, L. Vollano and N. Murru, *Int. Dairy J.*, 2021, **120**, 105077.
56. N. Leeuwendaal, C. Stanton, P. W. O'Toole and T. P. Beresford, *J. Funct. Foods*, 2021, **83**, 104425.
57. I. Shankar and A. Usha, *Trends Food Sci. Technol.*, 2021, **118**, 617.
58. M. G. Casey, J. P. Häni, J. Gruskovnjak, W. Schaeren and D. Wechsler, *Lait*, 2006, **86**, 407.
59. E. Sgarbi, B. Bottari, M. Gatti and E. Neviani, *Int. J. Dairy Technol.*, 2014, **67**, 342.
60. O. Nikoloudaki, W. J. Lemos Jr., L. Borruso, S. Campanaro, M. De Angelis, R. F. Vogel, R. Di Cagno and M. Gobbetti, *Environ. Microbiol.*, 2021, **23**, 1702.
61. B. Law, *Aust. J. Dairy Technol.*, 2010, **65**, 459.
62. E. Bancalari, C. Montanari, A. Levante, M. Alinovi, E. Neviani, F. Gardini and M. Gatti, *Food Res. Int.*, 2020, **135**, 109284.
63. J. Wang, X. Fang, T. Wu, W. Min and Z. Yang, *LWT*, 2018, **97**, 419.
64. P. Burns, F. Cuffia, M. Milesi, G. Vinderola, C. Meinardi, N. Sabbag and E. Hynes, *Food Microbiol.*, 2012, **30**, 45.
65. R. Adámek, V. Pachlová, R. N. Salek, I. Němečková, F. Buňka and L. Buňková, *LWT*, 2021, **152**, 112397.
66. H. Sedaghat, M. H. Eskandari, M. Moosavi-Nasab and S. S. Shekarforoush, *Int. Dairy J.*, 2016, **56**, 8.
67. P. Nieto-Arribas, J. M. Poveda, S. Seseña, L. Palop and L. Cabezas, *Food Control*, 2009, **20**, 1092.
68. L. Settanni, E. Franciosi, A. Cavazza, P. S. Cocconcelli and E. Pozanski, *Food Microbiol.*, 2011, **28**, 883.
69. A. Kocak, T. Sanli, E. A. Anli and A. A. Hayaloglu, *LWT*, 2020, **123**, 109127.
70. E. Renes, P. Gómez-Cortés, M. A. de la Fuente, D. M. Linares, M. E. Tornadijo and J. M. Fresno, *Food Res. Int.*, 2019, **116**, 819.
71. B. Nalepa and L. H. Markiewicz, *LWT*, 2017, **84**, 168.
72. M. G. Wilkinson and G. LaPointe, *J. Dairy Sci.*, 2020, **103**, 10963.
73. M. A. Murtaza, S. Ur-Rehman, F. M. Anjum, N. Huma and I. Hafiz, *Crit. Rev. Food Sci. Nutr.*, 2014, **54**, 1309.
74. L. Settanni and G. Moschetti, *Food Microbiol.*, 2010, **27**, 691.

75. X. Zheng, X. Shi and B. Wang, *Front. Microbiol.*, 2021, **12**, 703284.
76. A. R. Khattab, H. A. Guirguis, S. M. Tawfik and M. A. Farag, *Trends Food Sci. Technol.*, 2019, **88**, 343.
77. M. Gobbetti, M. De Angelis, R. DiCagno, L. Mancini and P. F. Fox, *Trends Food Sci. Technol.*, 2015, **45**, 167.
78. J. Steele, J. Broadbent and J. Kok, *Curr. Opin. Biotechnol.*, 2013, **24**, 135.
79. Y.-E. Chou, C. G. Edwards, L. O. Luedecke, M. P. Bates and S. Clark, *J. Dairy Sci.*, 2003, **86**, 2516.
80. S. Agarwal, K. Sharma, B. G. Swanson, G.Ü. Yüksel and S. Clark, *J. Dairy Sci.*, 2006, **89**, 1452.
81. E. B. Somers, M. E. Johnson and A. C. L. Wong, *J. Dairy Sci.*, 2001, **84**, 1926.
82. B. Uymaz, N. Akçelik and Z. Yüksel, *Food Sci. Anim. Resour.*, 2019, **39**, 804.
83. C. Gantzias, I. K. Lappa, M. Aerts, M. Georgalaki, E. Manolopoulou, K. Papadimitriou, E. De Brandt, E. Tsakalidou and P. Vandamme, *Int. J. Food Microbiol.*, 2020, **323**, 108586.
84. L. P. Margalho, B. A. Kamimura, R. P. Brexó, V. O. Alvarenga, A. S. Cebeci, P. W. Janssen, A. Dijkstra, M. J. Starrenburg, R. S. Sheombarsing, A. G. Cruz and W. Alkema, *Food Microbiol.*, 2021, **100**, 103872.
85. B. Palkir, K. Kemahlioglu and U. Yucel, *Trends Food Sci. Technol.*, 2021, **108**, 49.

CHAPTER 4

Cheese Manufacture

BOZHAO LI, ALAN L. KELLY* AND PAUL L. H. McSWEENEY

School of Food and Nutritional Sciences, University College Cork,
College Road, Cork T12 K8AF, Ireland
*Email: a.kelly@ucc.ie

4.1 Introduction

Cheese is a common and convenient dairy product in markets all over the world; it can be served as a dessert, snack, food ingredient or sandwich filler. All types of cheese are manufactured by the coagulation of milk.

In the bovine milk system, casein, which accounts for 80% of milk protein, exists in the form of casein micelles. Casein micelles in milk consist of four types of casein: κ-casein, α_{s1}-casein, α_{s2}-casein, and β-casein, typically found with a ratio of 1.5 : 4 : 1 : 3.5.[1] Casein micelles are distributed evenly in the milk system and are prevented from aggregating through steric stabilisation and electrostatic repulsion. For most types of cheese, casein micelles undergo rennet coagulation or acid coagulation which eliminates the electrostatic repulsion and removes steric stabilisation, leading to destabilisation of the micelles and the formation of a gel structure.

The essential process of cheese manufacture is transforming milk to curd, which involves about a 10-fold concentration of casein and fat, while the whey proteins, soluble salts, and lactose are separated in the cheese whey. Normally, milk for cheese manufacture is standardised and pasteurised. Cheese manufacture involves steps including gelation, acidification, cutting, heating, salting, pressing, and ripening, and occasionally other processes depending on the cheese type. The cheese family and cheese-related

Food Chemistry, Function and Analysis No. 40
Handbook of Cheese Chemistry
Edited by Michael H. Tunick

Published by the Royal Society of Chemistry, www.rsc.org

products are complex and can be classified according to their coagulation method (see Figure 4.1).

Rennet, an enzyme preparation consisting of one or more proteinases, initiates coagulation through cleavage of the κ-casein on the surface of casein micelles, producing *para*-κ-casein and glycomacropeptide (GMP)[2] in the first phase of rennet coagulation. In the second phase, without the negative charges on κ-casein attached to casein micelles, at a suitable temperature (>~15 °C), casein micelles aggregate with each other to form a gel network.[3] Many types of rennet, including those from bovine, camel,[4] plant,[5] and microbial[6] sources, can be used for rennet coagulation. This enzymatic coagulation can be used for the manufacture of many types of rennet-coagulated cheeses, *e.g.*, Cheddar, Gouda, and Mozzarella.

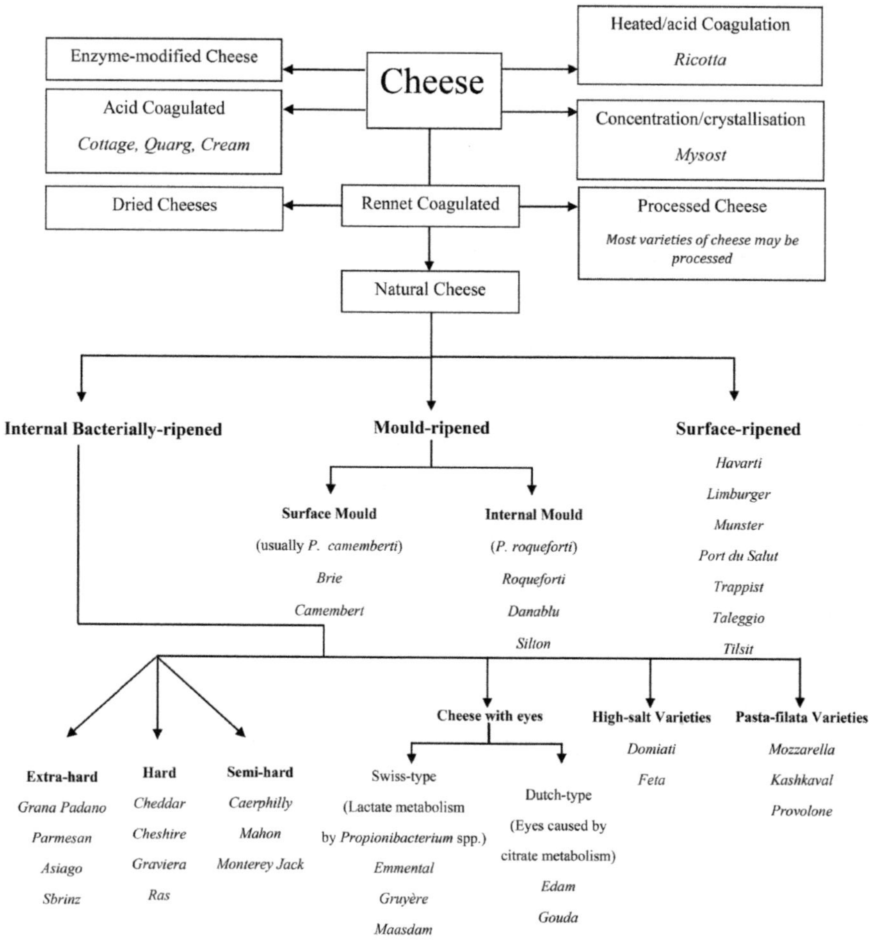

Figure 4.1 Cheese family scheme based on the coagulation method and ripening conditions. Reproduced from ref. 42 with permission from Elsevier, Copyright 2004.

Different methods are used to describe the rennet coagulation properties of milk such as rennet coagulation time (RCT)[7] and rheometry.[8] The coagulation properties of milk are influenced by temperature,[9,10] pH,[11] Ca^{2+},[12,13] preheating treatment,[14] enzymatic concentration, levels of non-micellar casein,[15,16] the presence of whey proteins,[17] and fat globules.[18] Syneresis of the coagulum is a related phenomenon following rennet coagulation and is essential for cheese manufacture.[19]

The principle of acid coagulation is that when the pH decreases to 4.6 – the isoelectric point of the caseins – negative charges on the surface of casein micelles are neutralised, which initiates the aggregation of casein micelles. Acidification usually occurs through fermentation of lactose to lactic acid by LAB, but on occasion it can be achieved by the introduction of mineral acids, the addition of an acidogen such as glucono-delta-lactone (GDL), or by using carbon dioxide.[20] Acid-coagulated cheeses such as Cottage, Quarg, and cream cheeses are manufactured by fermentation of skim milk, milk, and cream, respectively. Preheating of milk,[21] temperature,[22] and the buffering capacity of milk affect the acid coagulation properties of milk.[23] Acid-coagulated cheeses can be consumed fresh, while most rennet-coagulated cheeses undergo a ripening period from 4 weeks to 2 years to develop the desired texture, flavour, and aroma of the cheese.[24] A small amount of rennet may be added during the manufacture of acid-coagulated cheese such as Quarg to make the gel firmer.[25]

As a relatively recent development in milk pretreatment for cheese-making, membrane filtration technology can be used to fractionate milk contents according to their molecular sizes. Membrane filtration technologies include microfiltration (MF), ultrafiltration (UF), nanofiltration (NF), and reverse osmosis (RO). These four types of membrane filtration are classified according to the membrane pore size and the pressure that is applied.[26] Membrane filtration has been used for the standardisation of cheese milk and the concentration of milk for cheese manufacture.[27] Through membrane filtration, the buffering capacity, rennet coagulation properties, and rheological properties of milk can be affected; different consequences of cheese manufacture from membrane-filtered milk are reported.[28–30] Membrane filtration can be applied in cheese manufacture for standardising cheese milk,[31] concentrating cheese milk,[32] and using membrane filtration retentates for cheese manufacture.[33,34]

This chapter will discuss the principles and protocols of rennet and acid-coagulated cheese manufacture and the applications of membrane filtration for cheese manufacture.

4.2 The Casein System

In the bovine milk system, the casein micelles are polydisperse and include the four types of casein, *i.e.*, κ-casein, α_{s1}-casein, α_{s2}-casein, and β-casein, occluded water, and calcium phosphate nanoclusters. Casein micelles provide and transfer the energy, calcium, phosphate, and protein that are required during the growth of mammals.[35,36] The dual-binding model hypothesises that caseins within the micelle are bound through both hydrophobic and electrostatic interactions.[37] The hydrophobic interactions are from the phosphate groups on the casein molecules. In all models of the casein micelle, including the recent and widely accepted Holt/Dalgleish model, κ-casein is located on the surface of casein micelles; in the interior of casein micelles, calcium phosphate nanoclusters are attached to the caseins. β-Casein is bound through hydrophobic interactions and may dissociate from casein micelles at low temperatures as water channels exist inside the micelle.[38] κ-Casein, as a protruding molecule from the casein micelle, reaches out to the neighbouring environment.[39] Through the effect of the negative charges on κ-casein and the "hairy layer" of this protein on the surface of casein micelles, casein micelles are prevented from aggregation with each other through electrostatic repulsion and steric stabilisation.

4.3 Pretreatment of Cheese Milk

Cheese manufacture is a process that transforms the original material, milk, into the final product, cheese, by concentrating the fat and casein in milk about 10-fold. Bovine milk is the most common starting material for cheese manufacture. Low heat skim milk powder, micellar casein concentrate, whey-protein-depleted cheese milk, and other dairy ingredients are used as raw materials for cheese manufacture on occasion.[33,34,40]

The composition of the milk is frequently standardised to a specific protein : fat ratio (PFR) or fat : casein ratio to achieve the desired target cheese composition. Milk can be separated into skim milk (0.04–0.07% fat) and cream by separation, and, after determining the PFR of milk, either cream or skim milk should be used to adjust the PFR of milk to the desired value. Standardisation can be facilitated using rapid compositional analysis through spectroscopic techniques such as Fourier-transform infrared (FTIR) methods; for example, the MilkoScan system can be used to measure the protein, fat, and lactose contents of milk simultaneously in a short time based on the infrared light-scattering profile of milk samples and is commonly used in dairy laboratories and the dairy industry today. The PFR has a significant effect on the composition, component recoveries and yield.[41] The typical casein-to-fat ratio used for Cheddar cheese manufacture is 0.7 : 1.0. Standardisation of milk for Cheddar cheese manufacture to a PFR of 0.7 gives higher levels of moisture, protein, Ca, and P and lower moisture-in-non-fat-substances, fat-in-dry-matter, and salt-in-moisture compared to cheeses made from milk with a higher PFR. In some cases, skim milk or milk membrane filtration permeate may be used to increase the protein content to reach the target PFR value.

Raw milk may carry pathogens like *Escherichia coli*, *Listeria monocytogenes*, *Mycobacterium tuberculosis* or *Staphylococcus aureus*, which could cause disease to humans, and spoilage organisms like *Pseudomonas fluorescens* that can produce heat-stable proteinases and lipases that affect milk quality.

Therefore, pasteurisation, an essential process in milk and dairy product manufacture, is usually practised. Commonly, two kinds of pasteurisation methods are used: continuous pasteurisation and batch pasteurisation. Continuous pasteurisation involves heating the milk at 72 °C for 15 seconds, in what is called the high temperature short time (HTST) process, using a plate heat exchanger, while batch pasteurisation involves heating milk to 63–65 °C for 30 minutes and is typically only found today in small-scale (*e.g.*, artisanal, farmhouse) cheese plants. Although pasteurisation inactivates vegetative cells, bacterial endospores with higher thermal resistance will not be killed and may germinate and be active again after pasteurisation. Also, not all spoilage organisms can be destroyed during pasteurisation; therefore, it is necessary to keep the pasteurised milk at 4 °C until the beginning of cheese manufacture if it is not used immediately.

4.4 Rennets and Coagulants

Rennet-coagulated cheeses constitute perhaps ~75% of the world's entire cheese manufacture, and most rennet-coagulated cheese varieties are ripened.[42] The manufacture of rennet-induced cheeses involves various steps including standardisation, pasteurisation, acidification, gelling, cutting, heating, pressing, ripening, and other processes depending on the type of cheese to be manufactured. The protocols for some common rennet-coagulated cheeses are summarised in Figure 4.2a–d.[43]

Rennet is the common name for preparations from the abomasa of ruminant mammals used to coagulate milk that contain enzymes including chymosin, pepsin, and, on occasion, active lipase. It is the main enzyme source that is used to make cheese by coagulating casein micelles.[44] The cleavage site of rennet on κ-casein normally is the Phe_{105}–Met_{106} peptide bond; this hydrolysis produces *para*-κ-casein and glycomacropeptide (GMP) containing *N*-acetyl neuraminic acid and other carbohydrate moieties,[2] which initiates the coagulation of milk.[45] On reducing the negative charges on κ-casein and removing the physical barrier of steric stabilisation, casein micelles coagulate and associate to form a gel network.[15]

There are many sources of coagulant, including animal, vegetable, and microbial rennets. Animal rennet originates from the fourth stomach (abomasum) of young ruminants. Depending on the age of the animal, the chymosin content in rennet as a percentage of coagulating enzyme activity varies from 50 to 95%; the remainder is mainly pepsin.[46] Other than the most widely used bovine chymosin, camel chymosin and lamb chymosin are also commercially available. Fermentation-produced chymosin (FPC), which is bovine or camel chymosin secreted by genetically modified microorganisms, is also used for cheese manufacture. Since the development of genetic engineering, animal enzymes like chymosin can be produced by *Aspergillus niger* or *Kluyveromyces lactis* to obtain better and more consistent quality rather than extracting it from the abomasum of the animal.[47] Camel chymosin shares 85% of its amino acid sequence with calf chymosin and has

70% higher rennet coagulation activity than calf chymosin and 25% of proteolytic activity of calf chymosin.[38,45] Different to calf chymosin, camel chymosin has two additional positively charged residues that attract the negative charge in κ-casein, which therefore makes camel chymosin bind more readily to casein micelles; this higher electrostatic interaction of camel chymosin makes the milk coagulation activity of camel chymosin higher

Figure 4.2 Methods for manufacture of Cheddar (a), Gouda (b), Emmental (c) and Parmigiano-Reggiano (d). Reproduced from ref. 43 with permission from Springer Nature, Copyright 2015.

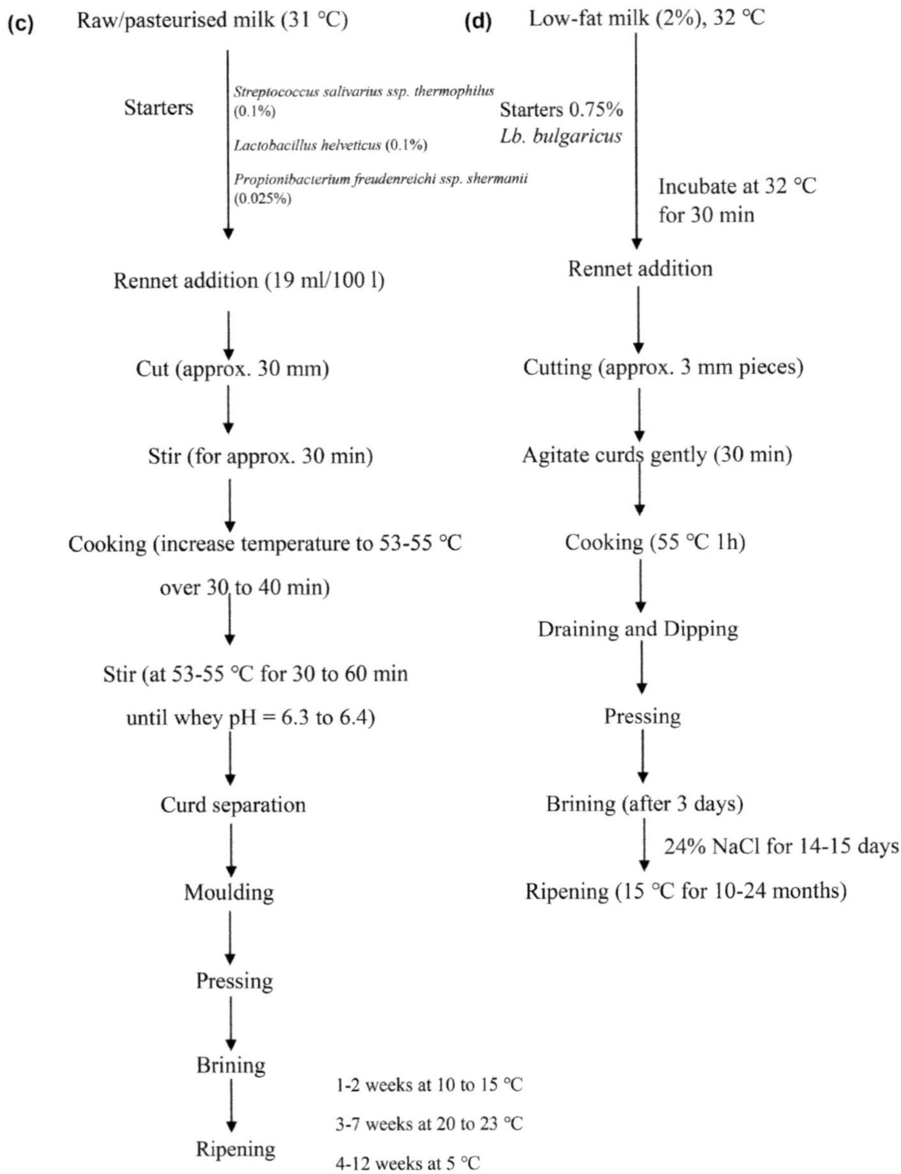

Figure 4.2 Continued.

than that of bovine chymosin.[47,48] The proteolytic specificity of camel chymosin, which has been studied in the context of Cheddar and Mozzarella cheese manufacture,[4,8] shows the same cleavage site on κ-CN, *i.e.*, the Phe_{105}–Met_{106} bond.[49]

Enzymes naturally produced by microorganisms (*e.g.*, *Rhizomucor miehei*) are also sometimes used for cheese manufacture.[6] Coagulants from plants such as *Cynara cardunculus*, *Ficus carica*, *Arctium minus* and *Solanum dubium* are very occasionally used for traditional cheese manufacture.[5] The proteolytic activity of plant rennet is usually much higher than that of rennet extracted from animals, and the preference of plant rennets could be for other bonds on κ-casein rather than Phe_{105}–Met_{106}, leading to a reduction in cheese yield or flavour problems.[50]

4.4.1 Rennet Coagulation

The enzymatic coagulation of milk for the manufacture of many varieties (Cheddar, Gouda, Mozzarella, *etc.*) is also called rennet coagulation. Rennet coagulation is a biochemical reaction that includes two stages. In the first stage, due to the action of rennet on the κ-casein bond Phe_{105}–Met_{106}, the κ-casein is cleaved into *para*-κ-casein and glycomacropeptide (GMP), which is the peptide that is cleaved off from the casein micelle. GMP is a promising ingredient in the food area due to its biological activities beneficial to human health and ease of recovery from whey, as GMP mainly partitions into the whey during cheese manufacture.[51] The casein with GMP cleaved off is called *para*-casein.

para-Casein aggregates in the second stage of rennet coagulation. During this second stage, when around 85% of the total κ-casein is hydrolysed, the stability of the casein micelle is reduced. This is due to the reduced negative charge on the casein micelle (the zeta potential of the casein micelle is reduced from −20 mV to −10 mV) and removal of the steric stabilising layer, both of which promote aggregation of casein micelles through hydrophobic interactions. The calcium ions in serum also help the aggregation of casein micelles by crosslinking casein micelles. The second stage of rennet coagulation can only happen in the presence of calcium ions when the temperature is greater than 20 °C, while the first stage takes place at temperatures >0 °C.[3] Through aggregation, casein micelles form a gel network (Figure 4.3).

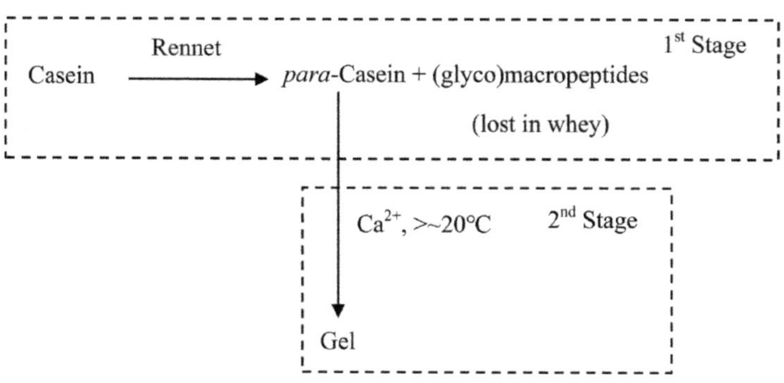

Figure 4.3 The first and second stages of enzymatic coagulation of milk. Reproduced from ref. 43 with permission from Springer Nature, Copyright 2015.

4.4.2 Factors That Influence Milk Coagulation Properties

Several common factors influence milk coagulation properties: temperature, pH, enzymatic concentration, preheating, and concentrations of Ca^{2+} and casein. As the pH decreases (*e.g.*, from 6.8 to 6.0), the first stage of rennet coagulation will be promoted and the RCT decreases; this is because the optimum pH of rennet for cleavage of the Phe–Met bond is around 5 and the second stage is also facilitated somewhat as electrostatic repulsion between micelles is reduced at low pH.[11,52] The optimum temperature for chymosin is around 40 °C and, therefore, the RCT is the lowest when the coagulation temperature is close to 40 °C;[9] Espinoza-Molina *et al.* reported that the optimum temperature of bovine chymosin is between 35 and 40 °C.[10] However, the most common coagulation temperature in cheese manufacture is about 30 °C as this favours mesophilic starter bacteria. The RCT is inversely correlated with enzyme concentration. Preheating (<70 °C) of milk slightly decreases RCT because heating causes precipitation of colloidal calcium phosphate (CCP), which influences the stability of casein micelles and decreases the pH. Preheating at over 70 °C causes denaturation of the whey protein β-lactoglobulin (β-lg), which interacts with κ-casein through disulphide bonds, therefore inhibiting rennet coagulation.[14] The concentration of Ca^{2+} (<10 mM) has a positive effect, mainly on the second stage of rennet coagulation, and influences gel firmness.[12,13]

Other particles, including non-micellar casein, whey proteins, and fat globules, can also influence rennet-induced coagulation and should also be considered during cheese manufacture. The addition of non-micellar casein, *e.g.*, sodium caseinate, inhibits the rennet coagulation of casein micelles by attaching to rennet-hydrolysed casein micelles, which delays their aggregation through steric repulsion forces, while the rennet coagulation properties can be restored through adding Ca^{2+}, although some inhibition by sodium caseinate still occurs.[15,16]

Whey protein denatured by a severe heating process suppresses rennet coagulation of casein micelles, but native whey proteins also inhibit casein micelle aggregation; the increase of the whey protein-to-casein ratio in milk thus inhibits rennet-induced coagulation of milk as whey proteins create a physical barrier that prevents *para*-casein aggregation.[17]

The fat content of milk has a positive effect on rennet-induced coagulation of casein micelles. If the casein content is fixed, increasing the fat content from 0.1 to 10% (w/w) causes a decrease in rennet coagulation time and in the time required to cut (20 Pa). However, the decreasing effect of increasing fat content is less than that of increasing protein content.[18]

4.4.3 Measurement of Milk Coagulation Properties

The international milk chymosin unit (IMCU) is a defined measure of chymosin or rennet activity. One IMCU per mL is the activity of chymosin required to coagulate 10 mL of bovine milk, at pH 6.5, within 100 s at 30 °C.[53]

The concentration of commercial chymosin is about 200–600 IMCU per mL which is quite concentrated for rennet coagulation. The common dosage of bovine chymosin for Cheddar cheese manufacture is around 60 IMCU per L milk.

Different methods may be used to describe or quantify milk coagulation properties (MCP). The Berridge method is used to measure the rennet coagulation time, which is the time for milk to start to coagulate after the addition of rennet. IDF Standard 157:2007[7] is the updated standard method for measuring RCT. In the Berridge method, a milk sample is placed in a rotating bottle, the temperature of which is controlled through a water bath. When rennet is added to the sample, a timer is set and the sample in the rotating bottle should be carefully observed. The RCT is determined when white flecks can be observed visually. The Berridge method is subjective as it relies on the experience and judgement of the operator to identify the time when flecks appear; different operators may observe different RCT from the same sample.

An optical method that measures the infrared light backscatter at 880 nm of milk during rennet coagulation has been developed, which is objective and as reliable as the Berridge method.[54] The use of Fourier-transform mid-infrared (FT-MIR) spectroscopy for the prediction of the MCP of ovine, buffalo, and bovine milk has been studied. Although the prediction accuracy of the FT-MIR spectroscopy method is low, it can be used as a tool to approximately distinguish the coagulation stages of milk (early-stage [<10 min], mid-stage [10–20 min], and late-stage [20–30 min]) to increase the efficiency of cheese manufacture.[55,56]

Curd firmness after milk coagulation is also an important aspect of MCP. A rheometer uses tiny strain rheological tests like dynamic low amplitude oscillatory shear rheology which will not damage a sample.[57–60] Rheometers can precisely and continuously measure the storage modulus (G', the energy stored during each oscillatory shear cycle), loss modulus (G'', the energy transformed to heat during each oscillatory shear cycle), and loss tangent ($\tan\delta$, the ratio of the loss modulus to storage modulus) of a sample by applying sinusoidal oscillatory shear to the sample.[61] As rennet coagulation proceeds, the storage modulus, which represents gel firmness, increases. Researchers have used the gel firmness to indicate when to cut the curd;[4,8] this method can also be used to indicate the gelation time, which is often defined as the time when G' is >1 Pa.[8] The following parameters related to cheese manufacture are used to describe MCP as well: maximum curd-firming rate, A_{40} and $\tan\delta$ 40 (the G' and $\tan\delta$ values after 40 min of rennet addition, respectively), K_{35} and K_{70} (the time required for the sample to reach 35 and 70 Pa, respectively), and the cutting window (calculated from K_{35} and K_{70}).[62]

There are several in-vat methods to determine the curd strength during the manufacture, *e.g.*, knife test, hot wire sensor, and near infra-red (NIR) reflectance. The knife test uses a sharp, flat, and clean tool to cut the curd at an angle of 45° to around 5 cm depth and then lift the gap with the knife vertically. If there is a clean break of the curd with whey drainage, the curd is ready to be cut; otherwise, the coagulation requires more time to complete.

The hot wire sensor uses a wire to pass a current through the milk sample which generates heat that dissipates easily by convection currents; when the milk coagulates, the heat dissipates less easily, increasing the temperature and conductivity of the wire. NIR reflectance sensors determine the cutting time of the gel by measuring light reflectance through milk coagulation.[63]

The microstructure of gels, which consists of a coarse molecular network of casein formed by chains, strips, and brunches of aggregated casein micelles, can be observed through confocal laser scanning microscopy (CLSM) or electron microscopy (EM).[64] CLSM is an optical technology that uses spatial pinholes to block the defocused light during imaging, thereby improving the optical resolution and contrast of photomicrographs. This technique requires minimal preparation for the sample and users can dye protein and fat with different coloured probes to distinguish them. Unlike CLSM, samples for EM should be carefully dehydrated, fixed, embedded, sectioned, and stained before observation. Several studies have used EM to observe the microstructure of acid and rennet gels.[65–69]

4.4.4 Acidification

The pH plays an important role in all types of cheese manufacture. The initial pH of cheese milk is around 6.6. For rennet cheese manufacture, the pH decreases to about 5 within 6 to 24 h. The rate of pH decrease depends on the buffering capacity of milk;[23] buffering capacity is the ability of a system to resist changes in pH upon the addition of an acid or an alkali.[70] In milk, the buffering capacity is contributed by salts (like inorganic phosphate, citrate), organic acids and milk proteins (caseins and whey proteins) and is affected by physical or chemical treatments during manufacture, such as heat treatment, high-pressure treatment, salt addition, and membrane filtration.

The enzyme activity and proteolysis are related to pH. The chymosin activity increases when the pH of milk decreases from 6.9 to 5.0.[10] Milk pH also affects the strength of rennet-coagulated gels. When the pH becomes lower (6.0), the coagulation time becomes shorter and the rate of gel formation increases, regardless of whether chymosin or plant coagulants are used.[71] A lower pH gives better syneresis which is related to the moisture content of cheese and whey drainage.[72] Microbial growth is also affected by pH as the environmental pH is related to the energy yield and redox reactions of microorganisms, therefore affecting microbial respiration.[73] The change of pH influences the calcium equilibrium in the system; when the pH is below 5.6, CCP is dissolved by hydrogen ions to calcium ions and hydrogen phosphate, and Ca^{2+} is then released from the casein micelle to the milk serum phase.[74] At the whey drainage stage, the released Ca^{2+} is lost in whey.

4.4.5 Syneresis

Syneresis is a phenomenon that happens after the coagulation of milk. Local stresses that occur in the gel structure cause local rearrangements of the protein matrix, leading to the expulsion of whey, resulting in syneresis.[19] Syneresis is related to MCP and is traditionally measured through physical separation or tracer methods. Physical separation methods measure the weight of whey and gel to indicate the extent of syneresis, which requires cautious handling to minimize the expulsion of whey during measurement.[19] Tracer methods submerge the curd into dyed ultrafiltration milk permeate and measure syneresis through the dilution of tracers determined by absorbance.[75] Low-field nuclear magnetic resonance (LF-NMR) is a spectroscopic technique that quantifies and monitors the state or location of water in the food system, with the advantage that it is not destructive nor invasive to sample and requires no or very little sample preparation. LF-NMR could detect the release of protons from water and, therefore, it is also used to continuously measure endogenous syneresis (syneresis that happens due to the pressure from gel network formation) or mechanical factors.[19]

4.4.6 Other Steps

After cutting, the curds tend to sink to the bottom of the cheese vat and fuse into a large curd when the whey is drained. "Cheddaring" is the unique process for Cheddar cheese that traditionally involves cutting the drained curd into large chunks and stacking the chunks one over another to press the curd by its weight, which also keeps the curd pieces warm, thus promoting uniform acidification. During this process, physicochemical changes happen to the curd. Additional lactose is transformed to lactic acid, causing a pH decrease from 6.15 to 5.2. The decrease in pH leads to the solubilisation of CCP, increasing the composition of Ca^{2+} and decreasing the casein-bound Ca, which, therefore, increases *para*-casein hydration. The pressure of the weight and the *para*-casein hydration result in a limited flow of the *para*-casein network.[76]

Besides cheddaring, various steps like cooking, stirring, whey drainage, and pressing can be applied to the curd to promote dehydration and acidification during the cheese manufacture, depending on the type of cheese. In the production of Dutch cheese varieties such as Edam and Gouda, the curd is cooked by washing with warm water to reduce the lactose content and control the pH of cheese.

Salting in cheese contributes to cheese flavour and functionality and initiates preservative action. NaCl increases saltiness, which is a favoured flavour by the customers. With the effect of salt, the osmotic pressure of food is increased, and the water activity is depressed; therefore, the growth of the bacteria is terminated or inhibited. Besides, salting affects the flavour of cheese directly and through influencing the metabolic and enzymatic activities that are related to cheese pH, flavour compounds, and degradation of casein and fats. Different amounts of salt (from 0.4% for Emmental to 5% for blue cheeses to greater than 10% for Feta) is added to cheese usually through brine or as dry salt mixed with the milled curds.[77,78]

Most cheeses are pressed in moulds of characteristic size and shape and may be vacuum-packed for ripening. Some hard cheeses are waxed to protect the cheese from airborne bacteria and moulds and to prevent moisture evaporation during ripening.

4.5 Acid-coagulated Cheeses

Acid-coagulated cheeses are manufactured through the acidification of milk, skim milk or cream. The principle of the manufacture of acid-coagulated cheeses is different to that of rennet-coagulated cheeses. Unlike induction of coagulation by rennet at about pH 6.4–6.6, acid coagulation occurs when the pH is close to the isoelectric point of casein, which is pH 4.6.[79]

Acid coagulation, following a decrease in milk pH, is critical to the manufacture of certain cheeses, yoghurt, and other cultured dairy products. Due to the different compositions and different amino acid sequences, each protein has zero net charge at a unique pH, called the isoelectric point. The isoelectric point of the caseins is at pH 4.6. As mentioned in the rennet coagulation section, the casein micelle consists of four different casein proteins and calcium phosphate nanoclusters held together through both hydrophobic and electrostatic interactions.[38,80]

The principle of acid coagulation is to decrease the pH of the milk to 4.6, which initiates a series of interactions that coagulate the casein micelles. When the milk is acidified to a pH value ~5.1 at temperatures >20 °C, physicochemical interactions occur in the casein micelles; most of the CCP in the casein micelles is dissolved, Ca^{2+} is released to the milk serum, and the charge on each casein is changed, while the ionic strength of milk increases. On acidification of milk to pH ~ 4.6, the negative charges of the casein micelles are neutralised and the repulsion reactions between micelles are reduced. These changes initiate the aggregation of casein micelles. The coagulation starts from the formation of chains and chains are linked together to form the 3D network structure.[81] Acid casein gels can be formed from milk or sodium caseinate, and the acid casein gels usually have similar properties, which indicates that CCP does not play an important role in acid coagulation since it is lost from the micelle before the milk reaches pH 4.6.[82]

Acidification can be achieved in various ways: (i) the addition of mineral acids, (ii) the addition of LAB starters, (iii) the addition of GDL, (iv) injection of carbon dioxide, or a combination of these methods.[20] The addition of mineral acids such as hydrochloric or sulfuric acids to milk introduces hydrogen ions into the environment and immediately decreases the pH of the milk, which is called fast acidification. Fast acidification typically reduces the pH of the milk to the range of 4.2–4.6 and the caseins precipitate to produce a coarse acid casein curd. After cooking and washing, whey proteins, minerals, and lactose will be removed from acid curd. The acid casein curd could be further dried and milled to acid casein powder, which is used to produce caseinates.[83,84]

Unlike fast acidification, the use of LAB or GDL gradually generates acid to decrease pH over a much longer time. Mesophilic LAB such as *Lactococcus lactis* subsp. *lactis* and *Lactococcus lactis* subsp. *cremoris* are widely used for acidification and flavour development in acid cheese manufacture. At the optimum temperature, LAB transform lactose into lactic acid to generate energy. As the total number of LAB increases, more lactic acid is secreted and the pH decreases slowly to the isoelectric point of casein, causing coagulation of casein and therefore forming an acid gel. For acid coagulation initiated by LAB, lactic acid is continuously produced to reach a low pH, which eventually actually inhibits the growth of LAB. The acidification by LAB becomes more efficient as the temperature is closer to the optimum temperature for the growth of the starter. The optimum growth temperature range for LAB is between 20 and 35 °C.[22]

GDL is a pH-neutral acidifier that generates gluconic acid gradually when GDL is hydrolysed by water. Since the acid production by GDL is more consistent and controllable compared to that of LAB, it is often used to study acid coagulation of milk. For milk acidified with GDL, the higher the acidifying temperature, the faster the acidification, as the hydrolysis of GDL is promoted.

The differences between GDL and LAB are in the pH decrease rate and final pH. At the same temperature, GDL initiates a faster drop in pH, while LAB requires a period to grow and multiply before decreasing pH. The final pH can be controlled by the level of GDL.

To describe the acid coagulation of milk, the pH of the milk sample could be measured as the acidifying time progresses. Acid coagulation properties can be monitored by dynamic small-amplitude oscillatory rheology. Parameters like storage modulus, loss modulus and loss tangent are recorded to reflect the hardness change and acid coagulation time. During acidification, the measurement of pH and rheology may be performed simultaneously to obtain plots of pH *vs.* G' over the course of time.[85]

Preheating of milk, the temperature of acidification and the buffering capacity of milk are the key factors that affect acid coagulation of milk. Unlike rennet coagulation, preheating of milk promotes acid coagulation, as whey protein will be denatured by heat treatment and associated with κ-casein. When the pH decreases, heated milk is easier to coagulate as the electrostatic repulsion is reduced by denatured whey protein through hydrophobic interactions and disulphide bonds. The gel formed from unheated milk could undergo particle rearrangement during coagulation which makes the casein network hard to provide gel firmness and results in a low G' gel.[21] Gels coagulated from heated milk were found to be firmer and more stable due to the crosslinks from denatured whey proteins.[86]

The manufacture of acid cheeses involves pretreatment of milk, slow quiescent acidification, coagulation, whey drainage, and curd treatments (Figure 4.4). In the manufacture of some cheeses like Quarg, Cottage, and Fromage Frais, the use of low levels of rennet is common and thus rennet coagulation and acid coagulation of milk are combined to form a firmer coagulum and reduce the loss of casein during whey separation.[25,79] In the combined coagulation of milk, the dosage of rennet is substantially less compared with cheese made using rennet coagulation. From a protein structure perspective, acidification dissolves CCP and alters the inner and

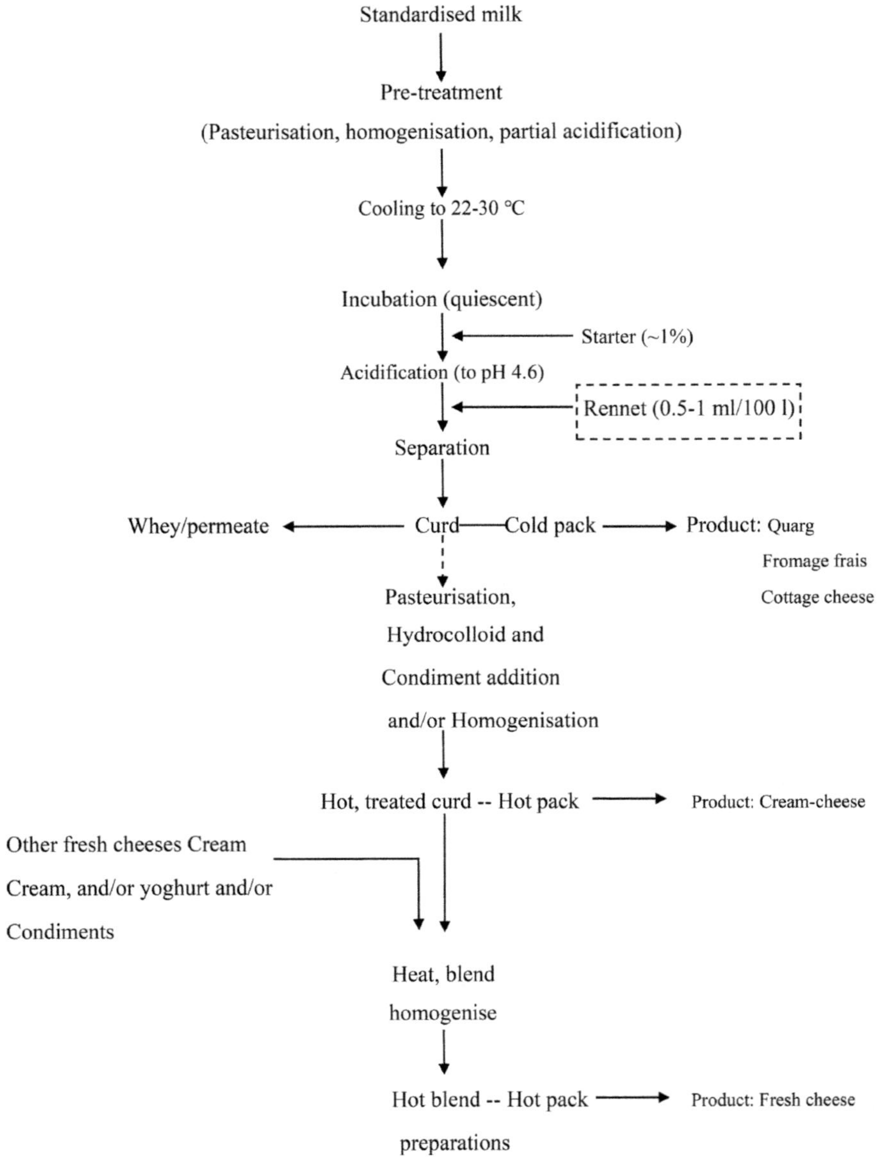

Figure 4.4 Methods for acid-coagulated cheese manufacture. Reproduced from ref. 79 with permission from Elsevier, Copyright 1996.

outer casein micelle structures, and acidification promotes rennet coagulation as the rennet activity increases with decreasing pH (from pH 6.7 to ~5).[36] Rennet coagulation facilitates the acid coagulation of milk through hydrolysing κ-casein and reducing the surface negative charge of casein micelles. The acidification of partially renneted casein micelles may result in a higher gelation pH and a higher final G' value.[87]

4.6 Cheese Manufacture with Membrane Filtration

Membrane separation technology has been used in different aspects of dairy processing. With the application of membrane filtration, cheese plant efficiency and cheese yield can be increased, and many plants worldwide use membrane filtration to manufacture various cheeses.[88]

4.6.1 Membrane Filtration

Membrane filtration processes including microfiltration (MF), ultrafiltration (UF), nanofiltration (NF), and reverse osmosis (RO) can be used to separate skim milk into different streams according to molecular size.[32] Different membrane filtration processes are characterised by different pore sizes of the membrane and different pressures are applied to the membranes to fractionate streams (Table 4.1).

MF, with a pore size of 0.08–0.20 μm, allows serum protein, lactose, minerals, and other small molecules to pass through the membrane with the pressure range of 1–3 bar.[26,34,89] The retentate of MF can be used to standardise cheese milk[31] or to purify, enrich, and recover micellar casein.[90] The main content of MF permeate is serum protein and is considered as native whey that is free from CMP, bacterial activity, fat, suspended cheese, and casein micelles compared to cheese whey.[91] MF can also be used to remove bacteria from cheese milk.[92]

UF is a process that uses higher pressures (3–10 bar) to force a permeate through a semi-permeable membrane.[26] The pore size of UF is within the range of 0.1 to 2 μm, which allows lactose, minerals, non-protein nitrogen and water in milk to pass through the membrane.[26,89] The concentration factor is the ratio between the initial feed and the final retentate. The low concentration factor UF (1.5× maximum) process or addition of UF retentate can concentrate the cheese milk, standardise the casein-to-fat ratio to

Table 4.1 The pore size, origin of selectivity, pressure gradient, and elemental operation of the main types of membrane filtration processes. Reproduced from ref. 26 with permission from Elsevier, Copyright 2008.

Process	Pore size	Origin of selectivity	Pressure gradient	Elemental operation
Microfiltration, MF	0.1–10 μm	Sieving effect	1–3 bar	Clarification, debacterisation, separation
Ultrafiltration, UF	2 nm–0.1 μm	Sieving effect	3–10 bar	Clarification, purification, concentration
Nanofiltration, NF	<2 nm	Sieving and specific interactions with the membrane	10–40 bar	Purification, water softening, separation, concentration

defined levels, and increase yield per vat. Higher concentration factor UF has been used for Camembert cheese and cast-Feta cheese manufacture.[93]

NF with a pore size less than 2 nm using a pressure of 10–40 bar can selectively remove small minerals.[26] RO uses a dense membrane and high applied pressure, greater than that used for osmotic pressure, which allows RO to remove water from milk.[26,89]

4.6.2 Effect of Membrane Filtration on Milk

The properties of UF-, RO- or MF-processed milk are different to those of standardised cheese milk, besides the concentration of fat and protein. The key properties affected by membrane filtration of significance to cheese-making include the buffering capacity, rennet coagulation properties, and rheological characteristics of the cheese milk.[94]

The buffering capacity of milk is increased after the milk is ultrafiltered or microfiltered as the casein and colloidal salts like Ca, Mg, and P are concentrated. The consequence is changes in acidification kinetics by LAB, rennet coagulation properties, enzymatic activities, rheological properties of the curd, the final pH of cheese,[95] growth of spoilage bacteria,[96] and water activity of the ripening cheese.[94] Therefore, more lactic acid from LAB is required to achieve the final pH of cheese (~5.2 for hard cheese and 4.6 for soft and fresh cheeses) for concentrated milk,[28] which results in sour-tasting cheese.[97] The acid dissolves calcium phosphate salts and releases calcium ions into the aqueous phase during acidification in cheese manufacture, which influences casein micelle aggregation. The resulting cheese can have a sandy and brittle texture[29] and the spreadability and stretchability properties of the cheese are unsatisfactory.[30]

In terms of rennet coagulation properties, if the same quantity of rennet is added to the same volume of milk and UF retentate, the RCT is not influenced by the increased protein content, while the time from coagulation to cutting is reduced.[98–100] The rate of enzymatic coagulation of milk increases with the substrate (casein) concentration.[98] At pH 6.6, about 80–90% of the κ-CN of milk is hydrolysed at the rennet coagulation point, while only 50% of hydrolysed κ-CN is required for coagulation of milk concentrated four-fold by UF.[99] The increase of protein content results in a dramatic increase in the velocity of casein micelle aggregation[98] and is directly related to the final gel strength.[101–105] The unique rennet coagulation properties of UF milk are due to the reduction in the surface hydrophobicity of casein during the UF process.[106]

Milk is considered a Newtonian liquid as its shear stresses linearly increase with the shear rate.[107,108] As the concentration of protein increases and/or the temperature decreases, milk tends to behave in a more pseudoplastic manner.[109] The viscosity of UF-treated milk increases dramatically with protein content. When using high viscosity UF retentates for semi-hard cheese manufacture, it is important to remove any dissolved gases entrapped in the liquid and use special equipment designed to mix thoroughly the rennet, starter culture, and retentates.[27] Failure to remove dissolved gases through a vacuum may result in a spongy curd with poor appearance and texture.[94]

4.6.3 Application of Membrane Filtration in Cheese Manufacture

In the context of cheese manufacture, UF can be used for protein standardisation of milk, production of milk concentrates and production of liquid pre-cheeses. UF may be used for protein standardisation of milk in cheese manufacture and may be applied to most types of cheese. The benefits of this method are as follows: daily standardisation of cheese milk; higher rennet gel strength decreases casein loss in whey; increased cheese yield due to higher protein recovery; more cheese made per vat; and, except for UF equipment, no specialised cheese manufacture equipment is required.[94] Protein standardisation by low concentration factor UF is now an increasingly common practice in very large cheese factories.

Medium-concentration factor UF retentates have been used for several kinds of cheese ranging from soft to hard. Cheese made from two-fold to five-fold concentrated UF retentates requires a special machine to cut and handle the firm curd and diafiltration is needed with pure, salted, or acidified water.[26] Cheese vats are not required for manufacture and whey drainage is dramatically reduced compared to the standard cheese manufacture.[104,110] The concept was first used in the manufacture of Camembert cheese[110] and was developed further to be applied to the manufacture of fresh cheese like Quarg, as well as certain semi-hard cheeses.[111,112] The method is now mainly used for the manufacture of white brined ("cast Feta")-type cheese.[113]

RO can concentrate cheese milk, like thermal evaporation, although with less damage to whey proteins, which makes it more attractive for cheese manufacture. Cheddar cheese was manufactured using whole milk pre-concentrated by RO (20–25% solid) using the same traditional equipment and had an identical gross composition to that of cheese made from standardised milk,[114,115] while the dosages of the rennet and starter used were reduced to 60% and 50%, respectively.[114] With a reduction of milk volume of 20% by RO, it is possible to increase cheese yield by 2–3%.[115] The volume reduction of 15% is found to be most desirable.[116] With an 8% reduced volume of milk by RO, 5% additional cheese yield can be achieved in Cottage cheese manufacture.[117] This increase in yield is attributed to the entrapment of whey protein, lactose, and minerals by the casein network, while the incorporation of whey protein in cheese inhibits the proteolysis of cheese during ripening;[118] the increase in yield from whey protein can be offset by a reduction in the amount of whey products that can be made. In cheese manufactured using RO-treated milk, the residual lactose content is higher than that of normal cheese; this may cause additional lactic acid fermentation by the ripening bacteria after a period of ripening when adequate lactic acid has been metabolised. This phenomenon may impair the sensory qualities of cheese.[119]

The applications of microfiltration in cheese manufacture are relatively novel in the dairy industry. The use of MF in the dairy field is attracting increasing attention as the range of accessible pore size became wider, which enables MF to fractionate any milk protein particles.[27] MF can be used to remove bacteria from raw milk, enrich casein before the manufacture of cheese, and selectively fractionate globular milk fat.

Pasteurisation is generally used to control the microbiology of raw milk. With various temperature and time combinations, the desired microbiological effects can be achieved. However, heat treatment can induce physiochemical changes to the milk, therefore potentially negatively affecting the functionality and cheesemaking properties of milk.[120] With the use of MF for microbiological control, the temperature or time of heat treatment can be reduced. The degree of bacterial removal can achieve 99.9% using MF with a membrane of 0.8 or 1.4 μm pore size, which also retains fungal cells (yeasts and moulds).[92]

MF of skim milk using 0.1 μm pore size membranes allows enrichment of micellar casein.[121,122] The casein content of the retentate increases as the amount of microfiltrate increases. The applications of MF retentates are similar to those of UF retentates and give cheese manufacturers another pathway for the manufacture of cheese and cheese by-products.[122] The rennet coagulation properties of 20% casein-enriched cheese milk are enhanced,[122] resulting in a significant reduction of fat loss and curd loss in the whey, and the cheese yield is increased by 2–4%.[123] The stream that passes through 0.1 μm pore size membranes and undergoes diafiltration with water is a novel dairy product called micellar casein concentrate (MCC).[90] MCC has higher heat stability than skim milk because of the lower levels of whey protein.[124] MCC can be used for fortifying yoghurt and cheese milk, for the manufacture of processed cheese instead of cheese curd, for recovery of individual casein or glycomacropeptide,[122] or for the manufacture of Cheddar cheese.[33,34] Cheddar cheese manufactured from MCC has been found to have higher levels of proteolysis, plasmin and chymosin activity than Cheddar cheese made from standardised cheese milk, and shows higher springiness, cohesiveness, and meltability.[33]

The enrichment of casein through MF has been used for the manufacture of Cheddar and Mozzarella cheeses. With 0.2 μm membrane MF, a retentate with 17.9% casein concentration was obtained for Mozzarella cheese manufacture.[31] For Cheddar cheese manufacture, cheese milk was concentrated twofold to 4.2% casein, resulting in 87% casein as a percentage of true protein following MF.[125,126] MF influences protein recovery but does not affect the fat recovery of cheese. The use of MF for casein enrichment for low-fat Cheddar cheese was found to give softer cheese that lacked flavour compared to full-fat Cheddar cheese.[127] Cheese milk standardised using a MF retentate stream to a typical casein content (2.87%) had similar rennet coagulation properties, Cheddar cheese composition, yield, and texture to those of a control cheese.[34] Cheese milk supplemented with MCC to 1.5 times casein content showed a faster curd firming rate, and the Cheddar cheese made had lower cheese moisture, and higher pH, hardness, and cheese yield.[34]

In a cold environment, β-casein can partially dissociate from the casein micelle because of the weakening of hydrophobic interactions. When the temperature of the milk is decreased to 5 °C or less, β-casein will partition into the serum.[128] The solubilised β-casein can be fractionated from milk using MF with a 0.2 μm pore size[129] or 100 kDa cut-off UF membranes.[130] The retentate of such membrane filtration is MCC with a higher α_s/β-casein ratio, while the microfiltrate is the mixture of whey protein and β-casein.[129] This technology makes it possible to manufacture cheese from a base material with an adjusted α_s/β-casein ratio in the future.[129,131] There is a potential to make various new cheeses as the hydrolysis of each casein affects the generation of cheese flavour.[27]

4.7 Conclusion

Cheese manufacture involves many steps to achieve or improve the final cheese product and different varieties can be classified on the basis of rennet- and acid-based coagulation principles. The protocols of both types of cheese manufacture are summarised, and the critical point is identified as controlling the syneresis and acidification of cheese curd. A range of technologies may be applied to monitor or promote coagulation of milk, cheese manufacture, and cheese properties. In recent decades, new technological approaches have been applied to cheese manufacture, such as membrane filtration, which fractionates milk components according to their size, and which may be used to enhance cheese manufacture and increase cheese yield. Overall, the processes of cheese manufacture from milk are mature, while some novel techniques involving modified raw materials, coagulants, and procedures may change the yield and functionality of cheese products, leaving many areas of the future of cheese manufacture to be explored; the future for applications of these technologies for cheese manufacture holds much promise.

Acknowledgements

The authors gratefully acknowledge the fund of the Department of Agriculture, Food and the Marine, Ireland and the support from University College Cork.

References

1. D. G. Dalgleish and M. Corredig, *Annu. Rev. Food Sci. Technol.*, 2012, **3**, 449.
2. C. V. Morr, *J. Dairy Sci.*, 1975, **58**, 977.
3. M. Cheryan, P. J. Van Wyk, N. F. Olson and T. Richardson, *J. Dairy Sci.*, 1975, **58**, 477.
4. N. Bansal, M. A. Drake, P. Piraino, M. L. Broe, M. Harboe, P. F. Fox and P. L. H. McSweeney, *Int. Dairy J.*, 2009, **19**, 510.
5. R. K. Robinson and R. A. Wilbey, in *Cheesemaking Practice*, Springer, New York, 1998, p. 98.

6. S. Preetha and R. Boopathy, *World J. Microbiol. Biotechnol.*, 1997, **13**, 573.
7. IDF, *Determination of the total milk clotting activity of bovine rennets*, IDF standard 157, International Dairy Federation, Brussels, 2007.
8. A. C. Moynihan, S. Govindasamy-Lucey, J. J. Jaeggi, M. E. Johnson, J. A. Lucey and P. L. H. McSweeney, *J. Dairy Sci.*, 2014, **97**, 85.
9. Y. Ulusu, S. B. Sentuerk, H. Kudug and I. Goekce, *Prep. Biochem. Biotechnol.*, 2016, **46**, 596.
10. J. A. Espinoza-Molina, C. H. Acosta-Muñiz, D. R. Sepulveda, P. B. Zamudio-Flores and C. Rios-Velasco, *Mol. Biotechnol.*, 2016, **58**, 657.
11. A. I. Nájera, M. de Renobales and L. J. R. Barron, *Food Chem.*, 2003, **80**, 345.
12. E. Balcones, A. Olano and M. M. Calvo, *J. Agric. Food Chem.*, 1996, **44**, 1993.
13. R. S Patel and H. Reuter, *Lebensm.-Wiss. Technol.*, 1986, **19**, 288.
14. R. Karoui and M. Kamal, *J. Sci. Food Agric.*, 2017, **97**, 3948.
15. Z. Gaygadzhiev, V. Massel, M. Alexander and M. Corredig, *Food Hydrocolloids*, 2012, **26**, 405.
16. Y. Lin, A. L. Kelly, J. A. O'Mahony and T. P. Guinee, *J. Dairy Sci.*, 2017, **100**, 908.
17. C. J. Gamlath, T. S. H. Leong, M. Ashokkumar and G. J. O. Martin, *Food Chem.*, 2018, **244**, 36.
18. T. P. Guinee, C. B. Gorry, D. J. O'Callaghan, B. T. O'Kennedy, N. O'Brien and M. A. Fenelon, *Int. J. Dairy Technol.*, 1997, **50**, 99.
19. C. L. Hansen, Å. Rinnan, S. B. Engelsen, T. Janhøj, E. Micklander, U. Andersen and F. van den Berg, *J. Agric. Food Chem.*, 2010, **58**, 513.
20. J. A. Lucey, in *Advanced Dairy Chemistry*, Springer, New York, 2016, p. 309.
21. J. A. Lucey, C. T. Teo, P. A. Munro and H. Singh, *J. Dairy Res.*, 1997, **64**, 591.
22. T. Ahmed, K. Rashida and A. Najma, *Biotechnology*, 2006, **5**, 481.
23. J. A. Lucey, B. Hauth, C. Gorry and P. F. Fox, *Milchwissenschaft*, 1993, **48**, 268.
24. P. F. Fox, *J. Dairy Sci.*, 1989, **72**, 1379.
25. A. L. Kelly and H. J. O'Donnell, *Int. Dairy J.*, 1998, **8**, 295.
26. A. Ayral, A. Julbe, V. Rouessac, S. Roualdes and J. Durand, *Membr. Sci. Technol.*, 2008, **13**, 33.
27. V. V. Mistry and J. L. Maubois, in *Cheese*, Academic Press, London, 2017, p. 677.
28. V. V. Mistry and F. V. Kosikowski, *J. Dairy Sci.*, 1985, **68**, 1613.
29. G. Brulé, J. L. Maubois, J. Vandeweghe, J. Fauquant and H. Goudedranche, *Rev. Lait. Fr.*, 1975, **328**, 117.
30. M. L. Green and A. S. Grandison, in *Cheese: Chemistry, Physics and Microbiology*, ed. P. F. Fox, Springer, Boston, MA, 1993, vol. 1, p. 101.
31. R. Brandsma and S. Rizvi, *J. Dairy Sci.*, 1999, **82**, 2063.

32. S. Govindasamy-Lucey, J. J. Jaeggi, C. Martinelli, M. E. Johnson and J. A. Lucey, *J. Dairy Sci.*, 2011, **94**, 2719.
33. B. Li, D. S. Waldron, J. T. Tobin, S. Subhir, A. L. Kelly and P. L. H. McSweeney, *Int. Dairy J.*, 2020, **107**, 104711.
34. X. Xia, J. T. Tobin, P. Sharma, M. Fenelon, P. L. H. McSweeney and J. J. Sheehan, *Int. Dairy J.*, 2020, **110**, 104796.
35. D. J. McMahon and B. S. Oommen, in *Advanced Dairy Chemistry*, Springer, Boston, MA, 2012, p. 185.
36. Q. Li and Z. Zhao, *Food Chem.*, 2019, **291**, 231.
37. D. S. Horne, *Int. Dairy J.*, 1998, **8**, 171.
38. D. G. Dalgleish, *Soft Matter*, 2011, **7**, 2265.
39. D. J. McMahon and B. S. Oommen, *J. Dairy Sci.*, 2008, **91**, 1709.
40. T. R. Freeman, J. L. Bucy and Z. Hassan, *J. Dairy Sci.*, 1970, **53**, 727.
41. T. P. Guinee, E. O. Mulholland, J. Kelly and D. J. O. Callaghan, *J. Dairy Sci.*, 2007, **90**, 110.
42. P. L. H. McSweeney, G. Ottogalli and P. F. Fox, *Cheese: Chemistry, Physics and Microbiology*, ed. P. F. Fox, Springer, Boston, MA, 2004, vol. 2, p. 1.
43. P. F. Fox, T. Uniacke-Lowe, P. L. H. McSweeney and J. A. O'Mahony, in *Dairy Chemistry and Biochemistry*, Springer, New York, 2015, p. 499.
44. R. Singh, A. Singh and S. Sachan, in *Enzymes in Food Biotechnology*, Academic Press, Cambridge, MA, 2019, p. 827.
45. S. Sandra, M. Alexander and D. G. Dalgleish, *J. Colloid Interface Sci.*, 2007, **308**, 364.
46. M. Addis, G. Piredda and A. Pirisi, *Small Rumin. Res.*, 2008, **79**, 2.
47. S. R. Kappeler, H. J. M. van der Brink, H. Rahbek-Nielsen, Z. Farah, Z. Puhan, E. B. Hansen and E. Johansen, *Biochem. Biophys. Res. Commun.*, 2006, **342**, 647.
48. J. Langholm Jensen, A. Mølgaard, J. C. Navarro Poulsen, M. K. Harboe, J. B. Simonsen, A. M. Lorentzen, K. Hjernø, J. M. van den Brink, K. B. Qvist and S. Larsen, *Acta Crystallogr., Sect. D: Biol. Crystallogr.*, 2013, **69**, 901.
49. K. K. Møller, F. P. Rattray and Y. Ardö, *Food Chem.*, 2012, **60**, 11421.
50. A. R. Lo Piero, I. Puglisi and G. Petrone, *J. Agric. Food Chem.*, 2002, **50**, 2439.
51. J. Jauregui-Rincón, E. Salinas-Miralles, N. Chávez-Vela and M. Jiménez-Vargas, in *Whey – Biological Properties and Alternative Uses*, IntechOpen, 2018, DOI: 10.5772/intechopen.82144.
52. M. B. López, S. B. Lomholt and K. B. Qvist, *Int. Dairy J.*, 1998, **8**, 289.
53. P. L. H. McSweeney, P. F. Fox and N. F. Olson, *Int. Dairy J.*, 1995, **5**, 321.
54. N. Tabayehnejad, M. Castillo and F. A. Payne, *J. Food Eng.*, 2012, **108**, 549.
55. C. L. Manuelian, G. Visentin, C. Boselli, G. Giangolini, M. Cassandro and M. De Marchi, *J. Dairy Sci.*, 2017, **100**, 7083.
56. C. L. Manuelian, M. Penasa, G. Giangolini, C. Boselli, S. Currò and M. De Marchi, *J. Dairy Sci.*, 2019, **102**, 1927.

57. S. Gunasekaran and M. M. Ak, *Cheese Rheology and Texture*, CRC Press, Boca Raton, FL, 2003.
58. E. A. Foegeding and M. A. Drake, *J. Dairy Sci.*, 2007, **91**, 1611.
59. M. H. Tunick, *J. Agric. Food Chem.*, 2011, **59**, 1481.
60. W. J. Lee and M. R. Lee, *J. Agric. Life Sci.*, 2013, **47**, 197.
61. T. P. Guinee, E. P. Feeney, M. A. E. Auty and P. F. Fox, *J. Dairy Sci.*, 2002, **85**, 1655.
62. R. R. Panthi, A. L. Kelly, J. J. Sheehan, K. Bulbul, A. H. Vollmer and D. J. McMahon, *J. Dairy Sci.*, 2019, **102**, 177.
63. D. J. O'Callaghan, E. P. Mulholland, A. P. Duffy, C. P. O'Donnell and F. A. Payne, *Irish J. Agric. Food Res.*, 2001, **40**, 227.
64. M. Kalab, P. Allan-Wojtas and B. E. Phipps-Todd, *Food Microstruct.*, 1983, **2**, 51.
65. F. L. Davies, P. A. Shankar, B. E. Brooker and D. G. Hobbs, *J. Dairy Res.*, 1978, **45**, 53.
66. E. Parnell-Clunies, Y. Kakuda and A. K. Smith, *Milchwissenschaft*, 1987, **42**, 413.
67. J. Mottar, A. Bassier, M. Joniau and J. Baert, *J. Dairy Sci.*, 1989, **72**, 2247.
68. M. Mellema, J. W. M. Heesakkers, J. H. J. van Opheusden and T. van Vliet, *Langmuir*, 2000, **16**, 6847.
69. F. A. Benyahia-Krid and O. Aissaoui-Zitoun, *J. Vet. Sci. Technol.*, 2016, **7**, 375.
70. M. Allaby and C. Park, *A Dictionary of Environment and Conservation*, Oxford University Press, Oxford, UK, 2013, p. 51.
71. C. L. C. Esteves, J. A. Lucey, T. Wang and E. M. V. Pires, *J. Dairy Sci.*, 2003, **86**, 2558.
72. M. C. Patel, D. B. Lund and N. F. Olson, *J. Dairy Sci.*, 1972, **55**(7), 913.
73. Q. Jin and M. F. Kirk, *Front. Environ. Sci.*, 2018, **6**, 101.
74. J. A. Lucey and P. F. Fox, *J. Dairy Sci.*, 1993, **76**, 1714.
75. A. U. Grundelius, K. Lodaite, K. Östergren, M. Paulsson and P. Dejmek, *Int. Dairy J.*, 2000, **10**, 489.
76. T. P. Guinee, in *Advanced Dairy Chemistry*, Springer, New York, 2016, p. 347.
77. W. Bisig, *IDF Factsheet 001*, 2017, p. 1.
78. A. R. A. Hammam, R. Kapoor, P. Salunke and L. E. Metzger, *Foods*, 2022, **11**, 24.
79. P. F. Fox, T. P. O'Connor, P. L. H. McSweeney, T. P. Guinee and N. M. O'Brien, *Adv. Food Nutr. Res.*, 1996, **39**, 163.
80. D. D. Bogenrief and N. F. Olson, *Milchwissenschaft*, 1995, **50**, 678.
81. D. M. Mulvihill and M. B. Grufferty, in *Heat-Induced Changes in Milk*, ed. P. F. Fox, International Dairy Federation, Brussels, 2nd edn, 1995, p. 188.
82. J. A. Lucey, T. van Vliet, K. Grolle, T. Geurts and P. Walstra, *Int. Dairy J.*, 1997, **7**, 381.
83. D. M. Mulvihill and M. P. Ennis, in *Advanced Dairy Chemistry—1 Proteins*, ed. P. F. Fox and P. L. H. McSweeney, Springer, New York, 3rd edn, 2003, p. 1175.

84. A. Carr and M. Golding, in *Advanced Dairy Chemistry*, Springer, Boston, MA, 2016, p. 35.
85. G. H. Meletharayil, H. A. Patel, L. E. Metzger and T. Huppertz, *Int. Dairy J.*, 2016, **61**, 107.
86. J. A. Lucey, M. Tamehana, H. Singh and P. A. Munro, *J. Dairy Res.*, 2000, **67**, 415.
87. E. Gastaldi, N. Trial, C. Guillaume, E. Bourret, N. Gontard and J. L. Cuq, *J. Dairy Sci.*, 2003, **86**, 704.
88. A. Y. Tamime, R. K. Robinson and G. L. J. Kiers, in *Brined Cheeses*, ed. A. Y. Tamime, Blackwell Publishing, Oxford, UK, 2006, p. 77.
89. W. J. Koros, Y. H. Ma and T. Shimidzu, *Pure Appl. Chem.*, 1996, **68**(7), 1479.
90. P. Schuck, M. Piot, S. Mejean, Y. Le Graet, J. Fauquant and G. Brule, *Lait*, 1994, **74**, 375.
91. T. Bacher and P. Kønigsfeldt, *Scand. Dairy Info.*, 2000, **3**, 18.
92. L. V. Saboya and J. L. Maubois, *Lait*, 2000, **80**, 541.
93. F. V. Kosikowski, *J. Dairy Sci.*, 1974, **57**, 488.
94. V. V. Mistry, in *Membrane Processing: Dairy and Beverage Applications*, ed. A. Y. Tamime, John Wiley & Sons, Hoboken, NJ, 2012, p. 176.
95. L. V. Saboya, H. Goudédranche, J. L. Maubois, A. L. S. Lerayer and S. Lortal, *Lait*, 2001, **81**, 699.
96. K. E. Rash and F. V. Kosikowski, *J. Food Sci.*, 1982, **47**, 728.
97. J. L. Maubois, *Génie Rural*, 1979, **1**, 15.
98. P. Garnot, *Bull. Fédéra. Int. Lait.*, 1988, **225**, 11.
99. D. G. Dalgleish, *J. Dairy Res.*, 1980, **47**, 231.
100. P. Upreti, V. V. Mistry and M. R. Acharya, *Dairy Sci. Technol.*, 2011, **91**, 383.
101. C. Ferron-Baumy, J. L. Maubois, G. Garric and J. P. Quiblier, *Lait*, 1991, **71**, 423.
102. J. Korolczuk, J. L. Maubois and J. Fauquant, in *Proceedings of XXII International Dairy Congress*, D. Reidel, The Hague, 1986, p. 123.
103. F. V. Kosikowski, *Food Technol.*, 1986, **40**, 71.
104. J. L. Maubois and G. Mocquot, *Lait*, 1971, **51**, 495.
105. J. L. Maubois and G. Mocquot, *J. Dairy Sci.*, 1975, **58**, 1001.
106. Y. K. Erdem, *J. Food Eng.*, 2000, **44**, 63.
107. J. F. Vélez-Ruiz and G. V. Barbosa-Cánovas, *J. Food Eng.*, 1998, **35**, 177.
108. J. F. Vélez-Ruiz and G. V. Barbosa-Cánovas, *J. Texture Stud.*, 2000, **31**, 315.
109. J. Culioli, J. P. Bon, J. L. Maubois, H. Goudedranche and M. Piot, *Lait*, 1974, **54**, 481.
110. J. L. Maubois, presented in part at Proc. Symp. Cheese Biotechn. Intern. Food Dev., Cornell University, Ithaca, New York, 1987.
111. J. A. Lucey, in *Encyclopedia of Dairy Sciences*, ed. J. W. Fuquay, Academic Press, Cambridge, MA, 2nd edn, 2011, p. 698.
112. T. P. Guinee, P. D. Pudja, W. J. Reville, D. Harrington, E. O. Mulholland, M. Cotter and T. M. Cogan, *Int. Dairy J.*, 1995, **5**, 543.

113. R. Hansen, *North Eur. Dairy J.*, 1985, 153.
114. T. Agbevavi, D. Rouleau and R. Mayer, *J. Food Sci.*, 1983, **48**, 642.
115. D. M. Barbano and D. G. Bynum, *J. Dairy Sci.*, 1984, **67**, 2839.
116. D. M. Barbano, D. G. Bynum and G. F. Senyk, *J. Dairy Sci.*, 1983, **66**, 2447.
117. D. M. Barbano, presented in part at New Dairy Prod. *via* New Technol., Proc. IDF Seminar, Atlanta, GA, 1986.
118. J. Harper, M. Iyer, D. Knighton and J. Lelievre, *J. Dairy Sci.*, 1989, **72**(2), 333.
119. J. Richard, M. Desmazeaud, J. Lenoir, F. Remuef, N. Schneid, J. Hermier, O. Cerf, J. L. Maubois and J. Frankinet, in *Cheesemaking: from science to quality assurance*, ed. A. Eck and J. C. Gillis, Intercept, Paris, 2nd edn, 2000, p. 189.
120. H. Singh and A. Waungana, *Int. Dairy J.*, 2001, **11**, 543.
121. E. Hurt and D. M. Barbano, *J. Dairy Sci.*, 2010, **93**, 4928.
122. J. L. Maubois, J. Fauquant, M. H. Famelart and F. Caussin, *Proteins*, 2001, **96**, 46.
123. C. Daviau, PhD thesis, Université de Rennes, 2000.
124. I. R. T. Renhe and M. Corredig, *J. Dairy Sci.*, 2018, **101**, 8757.
125. M. Neocleous, D. M. Barbano and M. Rudan, *J. Dairy Sci.*, 2002, **85**, 2415.
126. M. Neocleous, D. M. Barbano and M. Rudan, *J. Dairy Sci.*, 2002, **85**, 2425.
127. I. Amelia, M. Drake, B. Nelson and D. M. Barbano, *J. Dairy Sci.*, 2013, **96**, 4870.
128. J. Zulewska, J. Kowalik and B. Dec, *J. Dairy Sci.*, 2018, **101**, 10831.
129. E. Terré, J. L. Maubois, G. Brulé and A. Pierre, *Fr. Pat.*, FR2592769A1, 1987.
130. J. M. Murphy and P. F. Fox, *Food Chem.*, 1991, **39**, 27.
131. J. M. Murphy and P. F. Fox, *Food Chem.*, 1991, **39**, 211.

CHAPTER 5

Chemistry of Cheese Ripening

MICHAEL H. TUNICK

Department of Food & Hospitality Management, College of Nursing & Health Professions, Drexel University, Philadelphia, PA 19104, USA
Email: mht39@drexel.edu

5.1 Introduction

Aging, curing, and maturing are synonyms for the process during which chemical, physical, and microbiological changes occur in cheese post-manufacture. Affinage, a French term that means "to refine", is applied to caring for cheese during storage, where an affineur observes the temperature, humidity, and other conditions that lead to the best possible product.[1] During storage, microorganisms and enzymes in cheese act upon various components to produce flavour and aroma compounds while altering the texture, and the affineur monitors these changes as well. A cheese is ripe when it has reached its prime, so this chapter will use ripening to describe the post-manufacture developments that lead to the finished product.

5.2 Metabolism of Carbohydrates

5.2.1 Lactose

The carbohydrates in cheese curd consist of citrate (the conjugate base of citric acid) and lactose. The lactose concentration in the milk of cows, buffalo, goats, and sheep is 42–50 $g L^{-1}$,[2] while the citrate concentration is around 1.9 $mg L^{-1}$.[3] Around 0.8–1.0% of curd consists of lactose, with the rest being lost in the whey during manufacture.[4] Carbohydrates provide a source of

Food Chemistry, Function and Analysis No. 40
Handbook of Cheese Chemistry
Edited by Michael H. Tunick

Published by the Royal Society of Chemistry, www.rsc.org

carbon and energy to lactic acid bacteria (LAB) in the starter cultures and propionic acid bacteria (PAB) in adjunct (secondary) cultures in Swiss-type cheeses, the most notable of which are Emmental (from Switzerland), Alpsberg and Felsberg (Germany), Grevé (Sweden), Jarlsberg (Norway), Maasdam (Netherlands), and Samsoe (Denmark).[5] LAB metabolize lactate (the conjugate base of lactic acid) to L-lactate, lowering the pH and affecting casein solubility. The reaction rate increases with temperature and decreases with the ratio of salt to moisture.[6] Mesophilic starter cultures, which grow best at 25–30 °C, typically contain strains of *Lactococcus lactis*, whereas thermophilic starters, which have an optimum range of 40–45 °C, usually include *Streptococcus thermophilus*. Metabolism by these bacteria nearly depletes the lactose in curd within a few days of manufacture[7] and adventitious nonstarter lactic acid bacteria (NSLAB; see Section 5.3.5) convert the remaining lactose to D-lactate by fermentation or by racemization of L-lactic acid.[8] Undesirable white crystals may form in cheese under some conditions, where NSLAB may generate an insoluble mixture of Ca D-lactate and Ca L-lactate.[9] LAB further metabolize lactate to several compounds important for flavour (Figure 5.1).

The holes in Swiss-type cheese varieties (known as eyes) result from the metabolism of lactate by PAB to propionate, acetate, CO_2, and water.[10] Dairy PAB include *Propionibacterium acidipropionici*, *Propionibacterium thoenii*, and *Propionibacterium jensenii*, but the characteristic eyes and nutty flavour come from certain strains of *Propionibacterium freudenreichii*.[5] As with LAB, PAB employs lactate as an energy source, converting it to pyruvate, but it also uses aspartate as an electron acceptor (Figure 5.2). Although most of the CO_2 created stays dissolved in the cheese mass or escapes to the atmosphere, some accumulates at weak spots in the matrix to form the eyes.[11] The propionate and acetate contribute to flavour. PAB grow during ripening, but the levels of growth and fermentation depend on NaCl concentration and other factors.[12]

Metabolism of lactate is also important in surface mould cheeses. *Geotrichum candidum* and *Penicillium camemberti* colonize the outside of varieties such as Camembert and metabolize lactic acid to CO_2 and O_2, causing a reduction in acid and allowing lactate to diffuse from the centre of the cheese. As the surface pH rises, calcium phosphate precipitates there, causing migration of more calcium phosphate from the interior. The inside of the cheese softens and the abundance of calcium in the exterior turns it white and somewhat firm.[4]

NSLAB is also responsible for conversion of lactate into acetate, ethanol, formate, and CO_2.[13] *Clostridium tyrobutyricum*, an undesirable NSLAB, anaerobically metabolizes lactate to butyrate and H_2, producing cracks in the cheese and off-flavours.[4]

5.2.2 Citrate

The concentration of citrate in cheese curd is around 2–5 $g\,kg^{-1}$.[13] Like lactose, most citrate is lost in the whey. Some LAB use citrate as a secondary energy source.[14] *Leuconostoc mesenteroides* is a citrate positive bacterium that

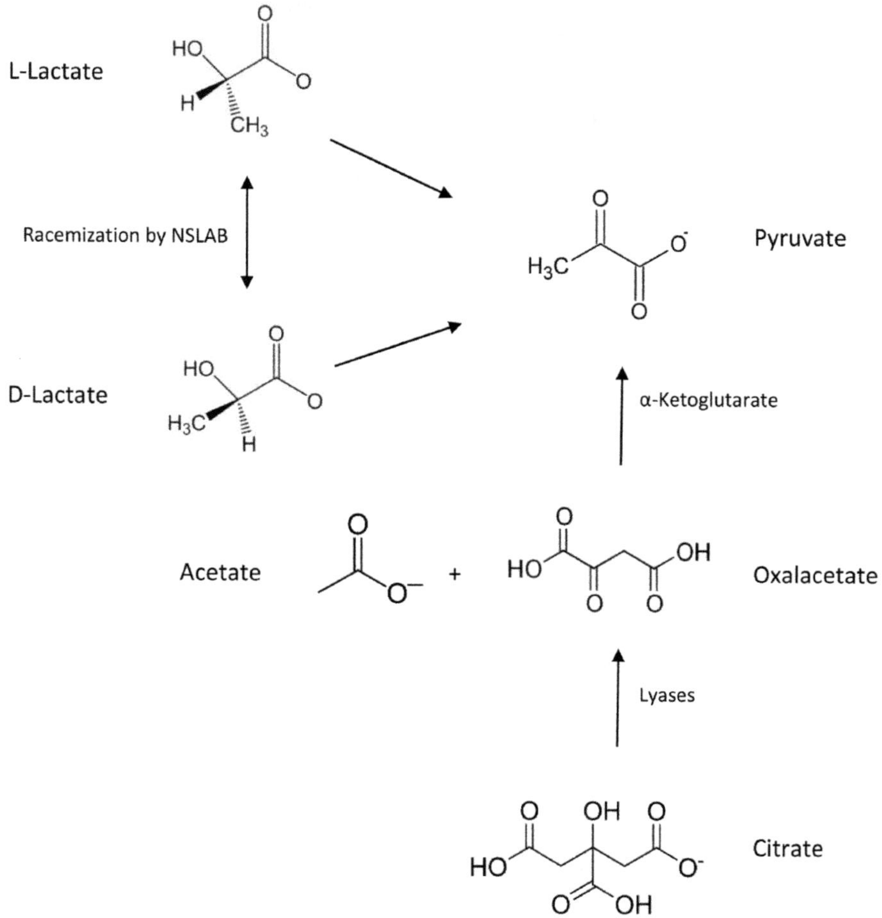

Figure 5.1 Metabolism of lactate and citrate to pyruvate in cheese.

produces the flavour compounds diacetyl and acetoin from citrate at acidic pH.[15] *Lactobacillus casei* and *Lactobacillus plantarum* can also metabolize citrate.[15] CO_2 is another product of citrate metabolism, which accounts for the small eyes in Dutch cheese varieties.[13] During ripening, *Lb. plantarum* and *Lactobacillus helveticus* metabolize the remaining citrate to succinic acid, a dicarboxylic acid with the formula $(CH_2)_2(CO_2H)_2$.[16] Figure 5.3 shows the steps involved in the metabolism of citrate.

5.3 Proteolysis

Protein breakdown is the most important and most complex series of biochemical events in cheese during ripening.[13] Proteolysis causes the release of amino acids and compounds responsible for aroma and flavour. The amino acids are responsible for some basic tastes, notably bitterness, and are themselves involved in further reactions that give rise to other flavours.[17]

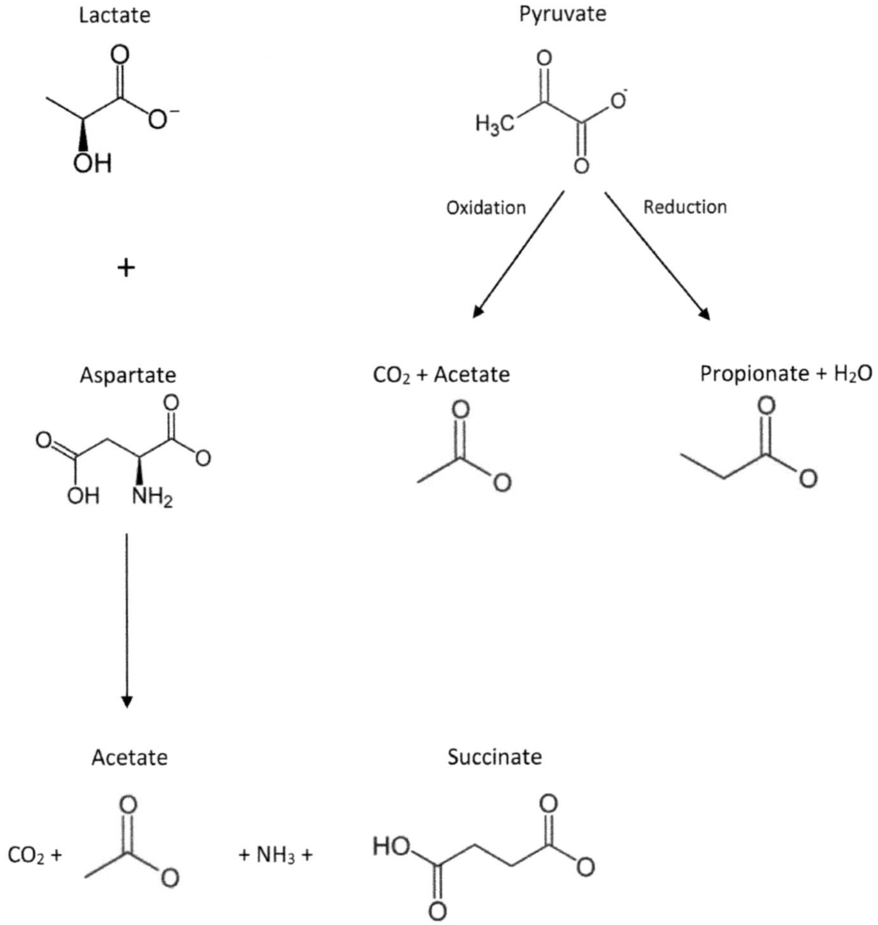

Figure 5.2 Simplified metabolic pathways of propionic acid bacteria in cheese.

Various proteinases and peptidases catalyse proteolysis in cheese. These enzymes originate in the milk, coagulant, starter culture, adjunct cultures, and NSLAB, as discussed in the subsections below. Cultures of bacteria and/or fungi added to the surface of smear-ripened cheese also contain these enzymes, and adding some proteinases and peptidases to the milk accelerates ripening. The coagulant and plasmin, the non-microbial proteinases in cheese, cause most of the proteolysis of intact caseins, especially α_{s1}- and β-caseins.[18]

Proteolysis also breaks the cross-links between casein molecules in the protein matrix, causing the structure of the cheese to weaken.[19] The firmness of cheese decreases as the amount of intact α_{s1}-casein declines because the breakdown products are mostly soluble in water.[20] At the same time, the breaking of protein molecules allows their carboxylic acid and amino acid groups to bind with water molecules, thereby decreasing the water activity of the curd and causing low moisture cheese to become harder with age.[13] Because of these competing effects, the cheese develops its characteristic texture.

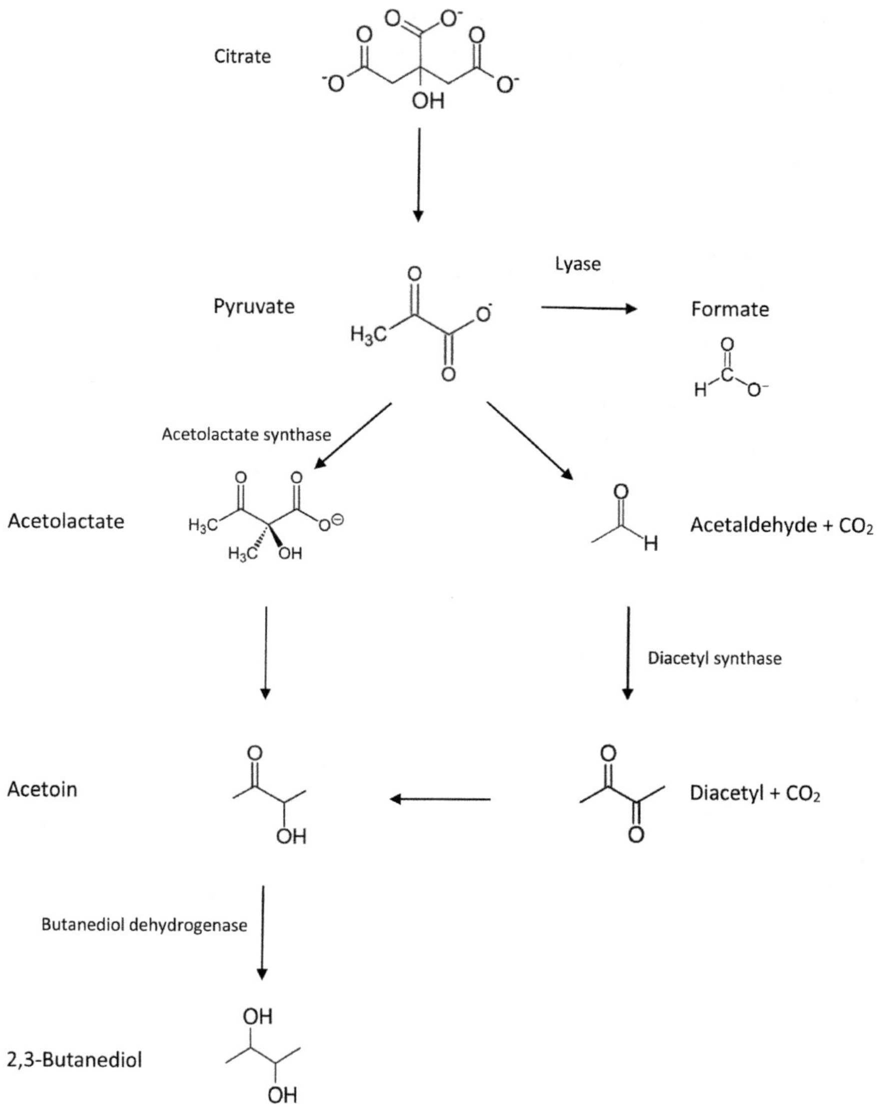

Figure 5.3 Metabolism of citrate in cheese.

5.3.1 Milk Enzymes

Milk contains some 60 enzymes including several proteinases, the most important of which is plasmin (EC 3.4.21.7). Plasmin functions in the body as a blood clot dissolver after being converted from its inactive form, plasminogen. This enzyme is heat resistant and in fact increases in activity in milk after pasteurisation, evidently due to inactivation of inhibitors. It is highly specific for peptide bonds containing Lys or Arg at the N-terminal portions of α_{s1}-, α_{s2}-, and β-casein. Plasmin (and plasminogen) associates with casein micelles in rennet-coagulated milk and contributes to proteolysis.[21] Curd washing, thermal treatment, and other factors affect the plasmin activity in cheese.[18]

5.3.2 Coagulant Enzymes

Rennet is the set of enzymes produced in the stomachs of ruminants, and the primary enzyme chymosin (EC 3.4.23.4) is often the coagulant in cheesemaking. Chymosin attacks α_{s1}-, β, and κ-casein at different bonds on the peptide chain. Most of the coagulant is lost in the whey but the curd holds some. High-cooked varieties retain almost no coagulant and high-moisture, low-pH cheese such as Camembert retains up to 50%.[22] Chymosin activity is primarily related to thermal treatment during manufacture, with as much as 30% of the chymosin remaining enzymatically active in the curd.[23] Decreasing cooking temperature increases chymosin activity. Other enzymes that may be present in the coagulant, such as pepsin, contribute to proteolysis.

5.3.3 Starter Culture Enzymes

In addition to carbohydrates, LAB need amino acids to survive, and they possess a number of proteinase and peptidase systems to liberate them. The three major components of the proteolytic system of LAB are proteinases, peptide transporters, and intracellular peptidases. The Prt group of proteinases are bound to the cell wall and begin the degradation of extracellular casein into oligopeptides. Opp, Dpp, and DtpT are peptide transporters that move the peptides into the cell. Intracellular aminopeptidases such as PepN and PepC break down the peptides into amino acids and smaller peptides.[24] Caseins are rich in proline, and at least six of the peptidases degrade peptides containing that amino acid.[25] Exopeptidases cleave the peptide bond adjacent to the amino terminus or carboxy terminus of the substrate and aminopeptidases cleave peptide bonds farther from the termini.[24] Peptidases are contained within the cell, which must be broken open before they act on peptides. LAB need peptidases for survival and growth, and all genomes encode for them.[25]

Lactobacillus bulgaricus and *Lb. helveticus* contain an array of proteolytic enzymes. As a result, *Lb. helveticus* serves as the main proteolytic bacterium in yogurt[26] and cheesemakers may use *Lb. helveticus* as an adjunct culture for degrading bitter peptides in cheese.[27]

Research has uncovered the targets of many of the enzymes; for example, PrtS in *Streptococcus thermophilus* uses α_{s1}, α_{s2}-, and β-casein as substrates, and PrtL in a strain of *Lactobacillus delbrueckii* subsp. *lactis* attacks α_{s1}-, β-, and κ-casein.[21] LAB have the ability to regulate the expression of their genes in response to their environmental conditions and to react to changes in the availability of nitrogen by regulating the activity of their proteolytic systems to ensure the proper nitrogen balance in their cells.[21] The most proteolytically active cocci species used in starter cultures are *Lactococcus salivarius* subsp. *thermophilus* and *Lc. lactis* subsp. *cremoris*, and the most proteolytically active lactobacilli are *Lacticaseibacillus casei*, *Lactobacillus delbrueckii* subsp. *bulgaricus*, *Lb. helveticus*, and *Lactobacillus acidophilus*.[21]

5.3.4 Adjunct LAB Enzymes

Adjunct LAB cultures are not required for ripening, but cheesemakers select them based on their enzyme profiles for the purpose of accelerating ripening or generating distinctive flavours and other characteristics. Adjuncts include *Propionibacterium freudenreichii* in Swiss-type cheeses, citrate-positive strains of *Lactococcus* and *Leuconostoc* spp. in Dutch-type cheeses, *Penicillium roqueforti* in blue cheeses, *Penicillium camemberti* in surface mould-ripened cheeses, and *Brevibacterium linens* and yeasts in surface smear-ripened cheeses.[28]

5.3.5 NSLAB Enzymes

The enzymes in NSLAB also play a role in ripening. Artisanal cheesemakers who use raw milk frequently do not add starter cultures, instead relying on naturally occurring microorganisms in the milk and environment. Even with starters and pasteurized milk, some NSLAB will find their way into the curd. Starter cultures along with residual chymosin and plasmin play a greater role than NSLAB during the early stages of proteolysis,[29] but the number of NSLAB in Cheddar increases from 100 to 1000 g^{-1} after pressing to 10^7–10^8 cfu per g within about 2 months.[28] The most common NSLAB are the mesophiles *Lactobacillus casei* and *Lactobacillus paracasei*. Other NSLAB lactobacilli include *Lactobacillus farciminis*, *Lb. plantarum*, *Lactobacillus pentosus*, *Lactobacillus curvatus*, *Lactobacillus rhamnosus*, *Lactobacillus fermentum*, *Lactobacillus buchneri*, *Lactobacillus parabuchneri*, and *Lactobacillus brevis.* Common non-lactobacillus species of NSLAB include *Pediococcus acidilactici*, *Pediococcus pentosaceus*, *Enterococcus durans*, *Enterococcus faecalis*, and *Enterococcus faecium*, along with some *Leuconostoc* species often used as starter cultures.[30]

5.3.6 Proteolytic Pathways

Once amino acids break away from caseins, transaminases (also called aminotransferases) convert them into α-keto acids that undergo further degradation *via* different pathways to produce volatile organic compounds[31] (Figure 5.4). In transamination, the amine (NH_2) group on an amino acid switches with the keto (=O) group on a keto acid. The primary amino acid sources for flavour production by LAB are branched-chain, aromatic, and sulfur-containing amino acids.[32] Table 5.1 shows α-keto acids and some flavour compounds generated from selected amino acids. Note that some pathways are purely chemical and do not involve transaminases.

Leucine, isoleucine, and valine, the branched chain amino acids, undergo decarboxylation into amines that carry strong unpleasant aromas and

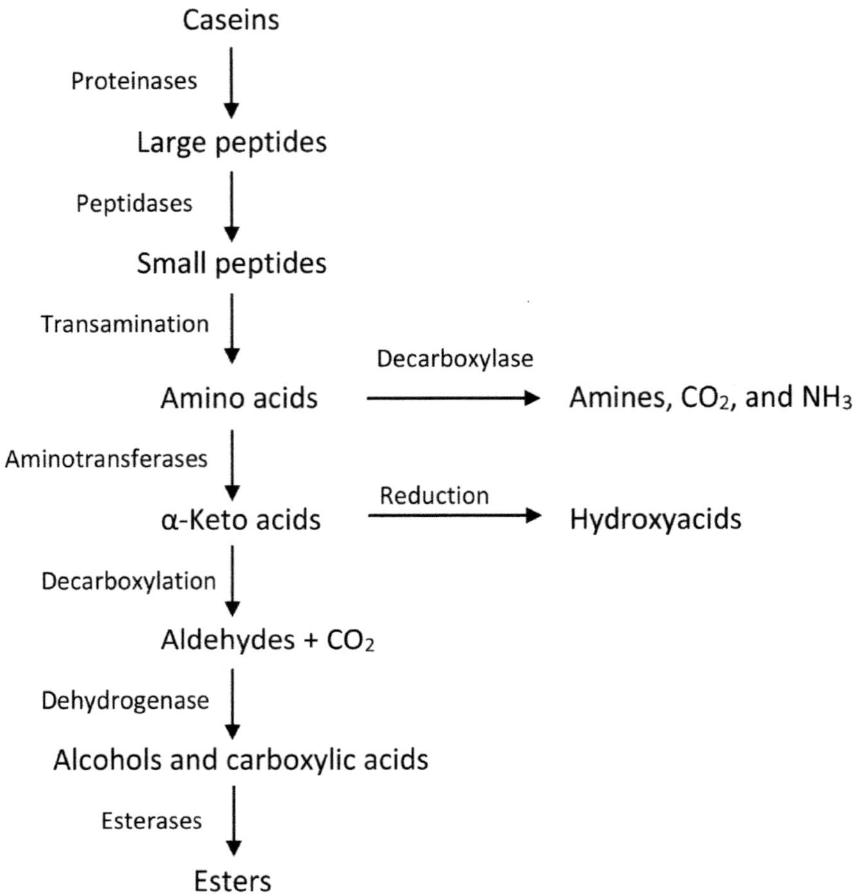

Figure 5.4 Chemical pathways for conversion of casein into important flavour compounds in cheese.

deaminate to produce ammonia and carboxylic acids such as rancid-flavoured butyric acid.[32] Figure 5.5 shows the pathway for breakdown of leucine.

The aromatic amino acids also degrade into flavour compounds: tyrosine breaks down to *p*-cresol (barny, unclean flavours),[33] phenylalanine to benzaldehyde (bitter almond flavour),[34] and tryptophan to skatole (fecal, putrid, and unclean flavours)[35] (Figure 5.6). Transamination converts the amino acids to aromatic pyruvates, which degrade to lactates, aldehydes, and acetates.[36] Tyrosine also forms tyramine, a biogenic amine identified as a trigger for migraine headaches, through decarboxylation.[37] The aromatic rings stay intact throughout these pathways.

Casein contains low levels of sulfur-containing amino acids. Cystine and methionine comprise only 3.02% of the amino acids in Cheddar.[38] These amino acids degrade to methanethiol and other volatiles that give rise to boiled cabbage, potato, and garlic flavours.[39] Figure 5.7 shows the pathway for breakdown of methionine in cheese.

Table 5.1 Flavour compounds generated from amino acids and α-keto acids *via* enzymatic (transaminase) and non-enzymatic (chemical) pathways.[13,32]

Amino acid	α-Keto acid	Organic acid	Aldehyde	Alcohol	Ester (example)
Leucine[a]	α-Ketoisohexanoic acid	3-Methyl butyric acid	3-Methylbutanal	3-Methylbutanol	Ethyl-3-methyl butanoate
Leucine[b]	α-Ketoisohexanoic acid	2-Methyl propanoic acid	2-Methylpropanal	2-Methylpropanol	
Isoleucine[a]	α-Keto-3-methyl-pentanoic acid	2-Methyl butyric acid	2-Methylbutanal	2-Methylbutanol	
Methionine[a]	α-Ketomethylthio butyrate	Methylthio butyric acid	Methional	Methionol	Ethyl-3-methylthio propionate
Methionine[b]	α-Ketomethylthio butyrate		2-(Methylthio)ethanal	Methanethiol	Methylthio acetate
Phenylalanine[a]	Phenyl pyruvate	Phenylacetic acid	Phenylacetaldehyde	1-Phenylethanol	Phenylethyl acetate
Phenylalanine[b]	Phenyl pyruvate	Benzoic acid	Benzaldehyde	Benzyl alcohol	Ethyl benzoate
Tryptophan[a]	Indole-3-pyruvate	Indole-3-acetic acid	Indole-3-acetaldehyde		
Valine[a]	α-Ketoisovaleric acid	2-Methyl propanoic acid	2-Methylpropanal	2-Methylpropanol	Ethyl isobutanoate

[a] Enzymatic pathway.
[b] Non-enzymatic pathway.

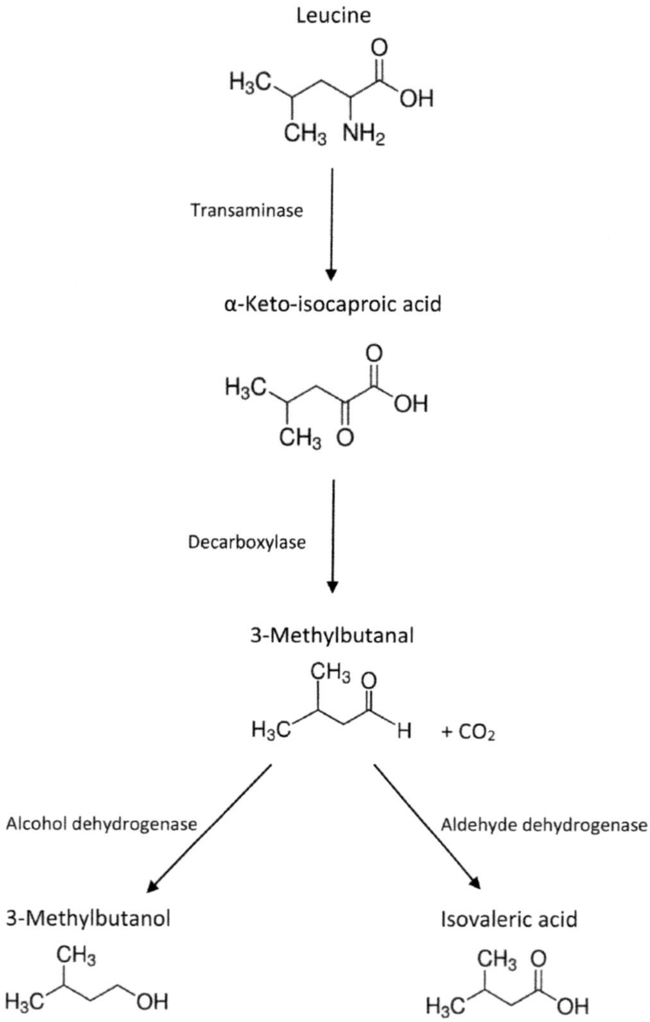

Figure 5.5 Pathways for conversion of leucine into important flavour compounds in cheese.

5.4 Lipolysis

Around 98% of the lipids in milk are in the form of triacylglycerols,[40] often called triglycerides or triacylglycerides. Triacylglycerols consist of a glycerol backbone with fatty acids attached at the sn-2 (centre) and sn-1 and sn-3 (end) positions. Lipolysis occurs when enzymatic hydrolysis of triacylglycerols releases fatty acids from the backbone and degrades them into other compounds (Figure 5.8).

Lipids in cheese undergo little oxidative degradation since the oxidation/reduction potential (E_h) is negative. The E_h of milk is around +150 mV, indicating an oxidized state, but the E_h of cheeses such as Camembert,

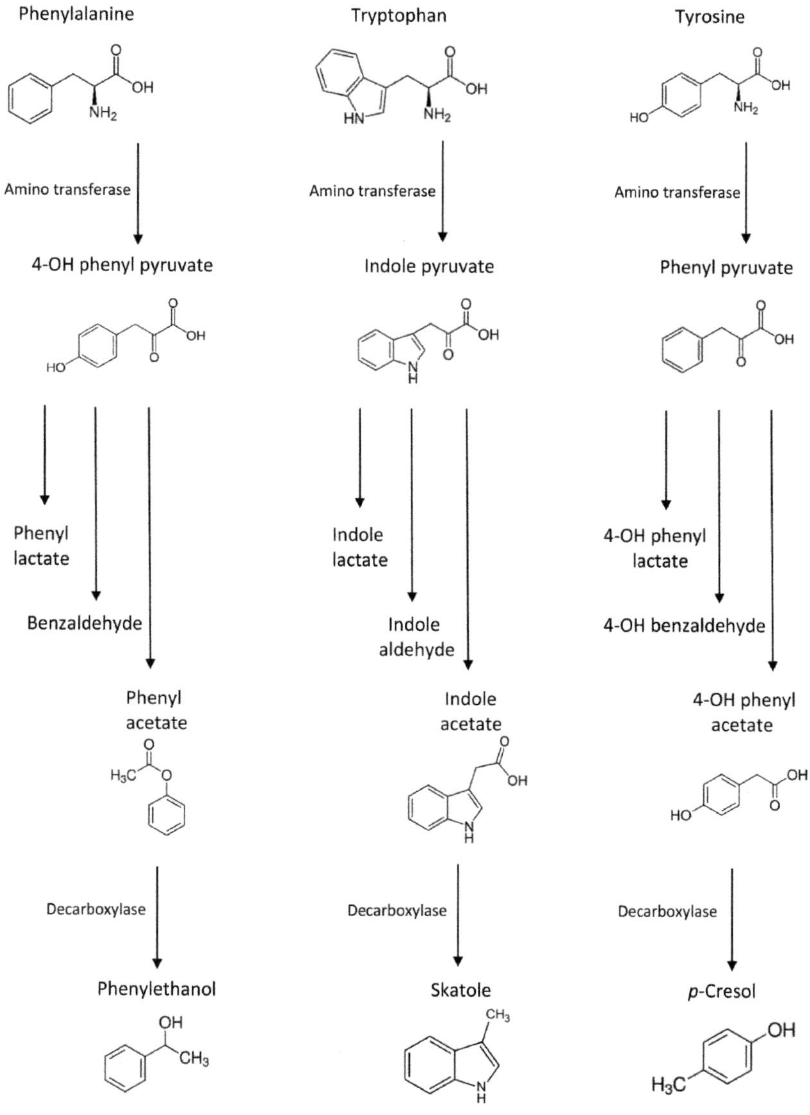

Figure 5.6 Pathways for conversion of aromatic amino acids into important flavour compounds in cheese.

Cheddar, Edam, and Swiss is in the −120 to −300 mV range, indicating a reduced state.[41] However, endogenous and added lipases cause triacylglycerols in cheese to undergo enzymatic hydrolysis, resulting in characteristic flavours in many varieties. Milk contains relatively high levels of short-chain (up to C6) and medium-chain (C6–C12) fatty acids, which have low flavour thresholds and contribute greatly to the flavour of many cheeses when they break away from triacylglycerols. Low levels of lipolysis apparently occur in all cheese and high levels result in rancidity.

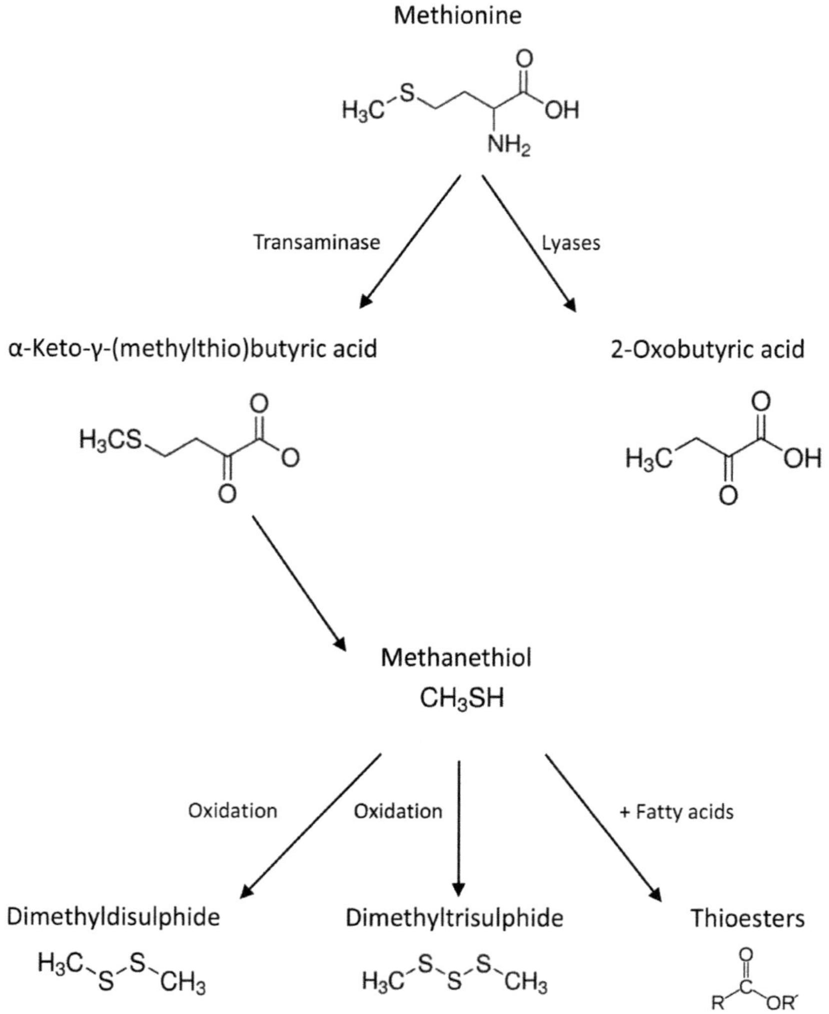

Figure 5.7 Pathways for conversion of methionine into important flavour compounds in cheese.

Lipolytic enzymes are either esterases or lipases. Esterases hydrolyse soluble C2–C8 acyl ester chains in aqueous solutions, whereas lipases hydrolyse emulsified acyl ester chains of 10 or more carbon atoms.[42] Lipase and esterase activities come from the milk, LAB, and coagulant.

5.4.1 Milk Lipases

Lipoprotein lipase (EC 3.1.1.34; LPL) serves to hydrolyse triacylglycerols in the blood and is the enzyme responsible for lipolysis in raw milk. Lipids in milk come in the form of emulsified fat globules surrounded by a protective milk fat globule membrane that prevents interaction with LPL, which is

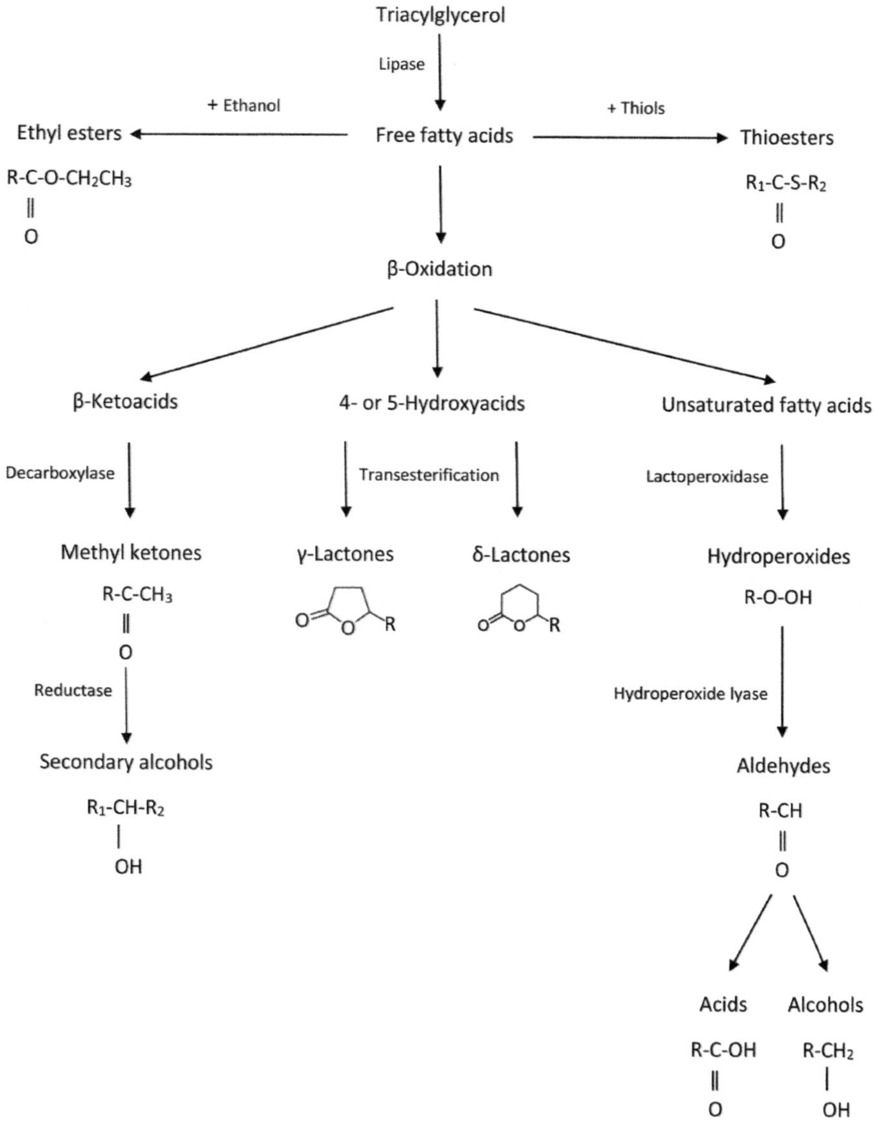

Figure 5.8 Pathways for lipolysis in cheese. R indicates a hydrocarbon chain.[10,22,46]

associated with casein micelles.[43] Agitation and homogenization damage the membrane, exposing the fat substrate and opening it to contact by LPL.[44] LPL tends to liberate medium-chain fatty acids from the sn-1 and sn-3 positions.[43]

5.4.2 Coagulant Lipases

Rennet, an extract from the fourth stomach chamber (the abomasum) of a calf, kid, or lamb, normally does not contain lipolytic enzymes, but manufacturers of hard Italian cheeses such as Romano use rennet pastes containing pregastric esterase (EC 3.1.1.3). Rennet paste comes from aged and macerated abomasum tissues.[45] Pregastric esterase is quite specific for short-chain fatty acids at the sn-3 position.[46]

5.4.3 Starter LAB Lipases

The esterases and lipases in LAB are contained inside the cells and released when the cells lyse. These LAB enzymes appear to be primarily responsible for lipolysis in Cheddar and Dutch-type cheeses.[47] Although *Lactobacillus* spp. and *Lactococcus* spp. display relatively weak lipolytic activity, their high numbers in cheese likely produce large numbers of free fatty acids.[43]

5.4.4 Fungal Lipases

Yeasts and moulds are also responsible for lipolysis. *Brevibacterium linens* and *Penicillium* spp. are important for lipolysis in surface-ripened and mould-ripened cheeses, respectively.[43] *Propionibacterium freudenreichii* is the main source of lipolysis in Swiss-type cheeses, though the activity is quite dependent on the strain.[48] *Debaryomyces hansenii*, *Geotrichum candidum*, *Kluyveromyces marxianus*, and *Yarrowia lipolytica* exhibit varying degrees of lipolytic activity in addition to proteolytic activity.[49]

5.5 Summary

Cheese acquires its unique mixture of flavours from the breakdown of carbohydrates, proteins, and lipids. Cheese is not a stable and unvarying product, but instead undergoes constant changes that affect its texture and flavour profile.

References

1. S. Bowen, *Rural Sociol.*, 2010, **75**, 209.
2. N. Silanikove, G. Leitner and U. Merin, *Nutrients*, 2015, **7**, 7312.
3. P. C. Garnsworthy, L. L. Masson, A. L. Lock and T. T. Mottram, *J. Dairy Sci.*, 2006, **89**, 1604.
4. P. L. H. McSweeney and P. F. Fox, in *Cheese: Chemistry, Physics and Microbiology*, P. F. Fox, P. L. H. McSweeney, T. M. Cogan and T. P. Guinee, Elsevier, Amsterdam, 2004, vol. 1, pp. 361–371.
5. M. T. Fröhlich-Wyder, W. Bisig, D. Guggisberg, E. Jakob, M. Turgay and D. Wechsler, in *Cheese*, Academic Press, Cambridge, MA, 2017, p. 889.
6. E. Parente and T. M. Cogan, in *Cheese: Chemistry, Physics and Microbiology*, P. F. Fox, P. L. H. McSweeney, T. M. Cogan and T. P. Guinee, Elsevier, Amsterdam, 2004, vol. 1, p. 123.
7. T. Beresford, N. Fitszimons, N. Brennan and T. Cogan, *Int. Dairy J.*, 2001, **11**, 259.

8. P. L. H. McSweeney and P. F. Fox, in *Cheese: Chemistry, Physics and Microbiology*, ed. P. L. H. McSweeney, P. F. Fox, P. Cotter and D. Everett, Academic Press, London, 4th edn, 2017, p. 411.
9. S. T. Dybing, J. A. Wiegand, S. A. Brudvig, E. A. Huang and R. C. Chandan, *J. Dairy Sci.*, 1988, **71**, 1701.
10. P. Piveteau, *Lait*, 1999, **79**, 23.
11. C. Steffen, E. Eberhard, J. O. Bosset and M. Ruegg, in *Cheese: Chemistry, Physics and Microbiology, Vol. 2: Major Cheese Groups*, ed. P. F. Fox, Chapman & Hall, London, 2nd edn, 1993, p. 83.
12. R. Richoux, E. Faivre and J. R. Kerjean, *Lait*, 1998, **78**, 319.
13. P. L. H. McSweeney, *Int. J. Dairy Technol.*, 2004, **57**, 127.
14. H. Drici, C. Gilbert, M. Kihal and D. Atlan, *J. Appl. Microbiol.*, 2010, **108**, 647.
15. J. Blaya, Z. Barzideh and G. LaPointe, *J. Dairy Sci.*, 2018, **101**, 3611.
16. M. I. Torino, M. P. Taranto and G. F. De Valdez, *Appl. Microbiol. Biotechnol.*, 2005, **69**, 79.
17. C. J. Zhao, A. Schieber and M. G. Gänzle, *Food Res. Int.*, 2016, **89**, 39.
18. M. A. Vélez, C. V. Bergamini, M. B. Ramonda, M. C. Candioti, E. R. Hynes and M. C. Perotti, *LWT-Food Sci. Technol.*, 2015, **64**, 282.
19. M. H. Tunick, E. L. Malin, P. W. Smith, J. J. Shieh, B. C. Sullivan, K. L. Mackey and V. H. Holsinger, *J. Dairy Sci.*, 1993, **76**, 3621.
20. R. C. Lawrence, L. K. Creamer and J. Gilles, *J. Dairy Sci.*, 1987, **70**, 1748.
21. P. F. Fox, T. Uniacke-Lowe, P. L. H. McSweeney and J. A. O'Mahony, in *Dairy Chemistry and Biochemistry*, Springer, New York, 2nd edn, 2015, p. 377.
22. N. Bansal, P. F. Fox and P. L. H. McSweeney, *J. Agric. Food Chem.*, 2007, **55**, 9219.
23. V. K. Upadhyay, P. L. H. McSweeney, A. A. A. Magboul and P. F. Fox, in *Cheese: Chemistry, Physics and Microbiology, Vol. 1: General Aspects*, ed. P. F. Fox, P. L. H. McSweeney, T. M. Cogan and T. P. Guinee, Elsevier, London, 3rd edn, 2004, p. 391.
24. M. Kieliszek, K. Pobiega, K. Piwowarek and A. M. Kot, *Molecules*, 2021, **26**, 1858.
25. M. Liu, J. R. Bayjanov, B. Renckens, A. Nauta and R. J. Siezen, *BMC Genomics*, 2010, **11**, 1.
26. S. Sieuwerts, F. A. de Bok, J. Hugenholtz and J. E. van Hylckama Vlieg, *Appl. Environ. Microbiol.*, 2008, **74**, 4997.
27. V. R. Sridhar, J. E. Hughes, D. L. Welker, J. R. Broadbent and J. L. Steele, *Appl. Environ. Microbiol.*, 2005, **71**, 3025.
28. M. El Soda, S. A. Madkor and P. S. Tong, *J. Dairy Sci.*, 2000, **83**, 609.
29. M. Gobbetti, M. De Angelis, R. Di Cagno, L. Mancini and P. F. Fox, *Trends Food Sci. Technol.*, 2015, **45**, 167.
30. L. Settanni and G. Moschetti, *Food Microbiol.*, 2010, **27**, 691.
31. A. R. Khattab, H. A. Guirguis, S. M. Tawfik and M. A. Farag, *Trends Food Sci. Technol.*, 2019, **88**, 343.

32. G. Smit, B. A. Smit and W. J. Engels, *FEMS Microbiol. Rev.*, 2005, **29**, 591.
33. O. Suriyaphan, M. A. Drake, X. Q. Chen and K. R. Cadwallader, *J. Agric. Food Chem.*, 2001, **49**, 1382.
34. F. J. Hidalgo and R. Zamora, *Food Chem.: X*, 2019, **2**, 100037.
35. S. Gummalla and J. R. Broadbent, *J. Dairy Sci.*, 1999, **82**, 2070.
36. S. Gao, D. H. Oh, J. R. Broadbent, M. E. Johnson, B. C. Weimer and J. L. Steele, *Lait*, 1997, **77**, 371.
37. G. D'Andrea and A. Leon, *Neurol. Sci.*, 2010, **31**, 1.
38. US Department of Agriculture, National Nutrient Database for Standard Reference, 2016, http://www.ars.usda.gov/nea/bhnrc/mafcl (Accessed August 2022).
39. Y. Ardö, *Biotechnol. Adv.*, 2006, **24**, 238.
40. R. G. Jensen and D. S. Newberg, in *Handbook of Milk Composition*, ed. R. G. Jensen, Academic Press, San Diego, CA, 1995, p. 543.
41. V. Caldeo and P. L. H. McSweeney, *Int. Dairy J.*, 2012, **25**, 16.
42. I. V. Wolf, C. A. Meinardi and C. A. Zalazar, *Protein Peptide Lett.*, 2009, **16**, 1235.
43. Y. F. Collins, P. L. H. McSweeney and M. G. Wilkinson, *Int. Dairy J.*, 2003, **13**, 841.
44. D. K. Hickey, K. N. Kilcawley, T. P. Beresford and M. G. Wilkinson, *J. Dairy Sci.*, 2007, **90**, 47.
45. M. Harboe, M. L. Broe and K. B. Qvist, in *Technology of Cheesemaking*, ed. A. Y. Tamine and B. A. Law, Wiley, Hoboken, NJ, 2010, p. 98.
46. P. F. Fox and L. Stepaniak, *Int. Dairy J.*, 1993, **3**, 509.
47. P. L. H. McSweeney and M. J. Sousa, *Lait*, 2000, **80**, 293.
48. A. Thierry, F. Valence, S.-M. Deutsch, S. Even, H. Falentin, Y. Le Loir, G. Jan and V. Gagnaire, *Dairy Sci. Technol.*, 2015, **95**, 895.
49. M. T. Fröhlich-Wyder, E. Arias-Roth and E. Jakob, *Yeast*, 2019, **36**, 129.

CHAPTER 6

Cheese Ripening: An Overview of Technological Strategies Towards Process Acceleration

M. A. VÉLEZ,* C. BERGAMINI, I. V. WOLF, G. H. PERALTA AND M. C. PEROTTI

Instituto de Lactología Industrial (INLAIN, Universidad Nacional del Litoral/CONICET), Facultad de Ingeniería Química, Santiago del Estero 2829, S3000AOM, Santa Fe, Argentina
*Email: mvelez@fiq.unl.edu.ar

6.1 Introduction

Cheese is considered to be a biocomplex ecosystem colonized by a diverse group of microorganisms provided by raw milk and starter and non-starter microorganisms. This microflora is the major contributor to the perceived sensory attributes in the different cheese types owing to its complex interaction with the milk components (proteins, carbohydrates, and fats) that mainly occur in a key technological step during cheese manufacturing, known as 'ripening'.[1]

During ripening, flavour and texture of cheese are developed. This process is governed by three main metabolic routes: (1) metabolism of residual lactose, lactate, and citrate, (2) proteolysis and amino acid (AA) catabolism, and (3) lipolysis and free fatty acid (FFA) metabolism.[2] Cheese is a biochemically dynamic product containing up to 100 billon bacteria per gram, all of which metabolize carbohydrate, proteins, and/or lipids to create a

Food Chemistry, Function and Analysis No. 40
Handbook of Cheese Chemistry
Edited by Michael H. Tunick

Published by the Royal Society of Chemistry, www.rsc.org

myriad of aromatic and sapid compounds that contribute to flavour to a greater or lesser extent depending on the cheese variety.[3]

The degree and extent of these biochemical events are affected by many factors including the type and amount of residual coagulant, presence of starter and non-starter microorganisms, addition of exogenous enzymes, technology of manufacture (temperature/time, curd washing, curd salting and pressing, *etc.*), and environmental conditions during manufacturing and ripening. Many cheese varieties are essentially similar at the end of manufacturing stage in terms of chemical composition and texture; however, a number of changes occur during ripening based on ripening conditions (temperature, time, environmental) and these influence the flavour and texture of cheese, as mentioned above.[4]

Proteolysis is a complex primary event that occurs in most cheese varieties. Breakdown of intact caseins, especially α_{S1}- and β-caseins, is mostly caused by non-microbial proteases added or indigenous such as coagulant and plasmin, respectively. Subsequent changes on casein-derived peptides are mainly attributed to the microbiota which make significant contribution to taste. The main role of secondary proteolysis is to provide the precursors for aroma development, that is, free amino acids (FAA). These compounds and small peptides are then metabolized *via* AA catabolism, a transformation also caused by lactic microbiota. Aminotransferases (AT) along with other enzymes actively participate in this process.[5,6] In bacterially ripened cheeses more than 50% of flavour compounds are produced *via* AA catabolism.[7]

Lipolysis is the enzymatic hydrolysis of triacylglycerols (TAG) resulting in the release of FFA as short-, medium- and long-chain fatty acids, glycerol, and mono- and diacylglycerides. This step is carried out by indigenous milk lipase, lipases and esterases from starter lactic acid bacteria and non-starter lactic acid bacteria (NSLAB) or exogenous lipases. FFA can be metabolized to flavour compounds such as methyl ketones, secondary alcohols, lactones, *etc.*[8] or can be involved in esterification reactions yielding esters; the main mechanism in cheeses seems to be alcoholysis (esterification of alcohols and acids catalysed by esterases) and transesterification (reaction of glycerides and alcohols catalysed by alcohol acyltransferases).[9]

Depending on the cheese variety, microflora, manufacture, and ripening conditions, lactose and citrate are metabolized by several different pathways towards various compounds; acetoin, diacetyl, acetate, acetaldehyde, and ethanol are the most common products resulting from the enzymatic activities of lactic acid bacteria (LAB).[10] Ripening is a slow process, especially for low moisture cheese varieties, and expensive in terms of controlled atmosphere storage and stocks, that is not fully predictable or controllable.[11] Long maturation time is an economic issue in developing countries as energy supplies in these countries are either very expensive or insufficient.[12] Cheese ripening acceleration is of major interest in the cheese industry, mainly in semi-hard and hard cheeses due to the high costs associated with long storage periods, providing significant economic contributions to producers such as net savings in refrigeration, labour and inventory costs,

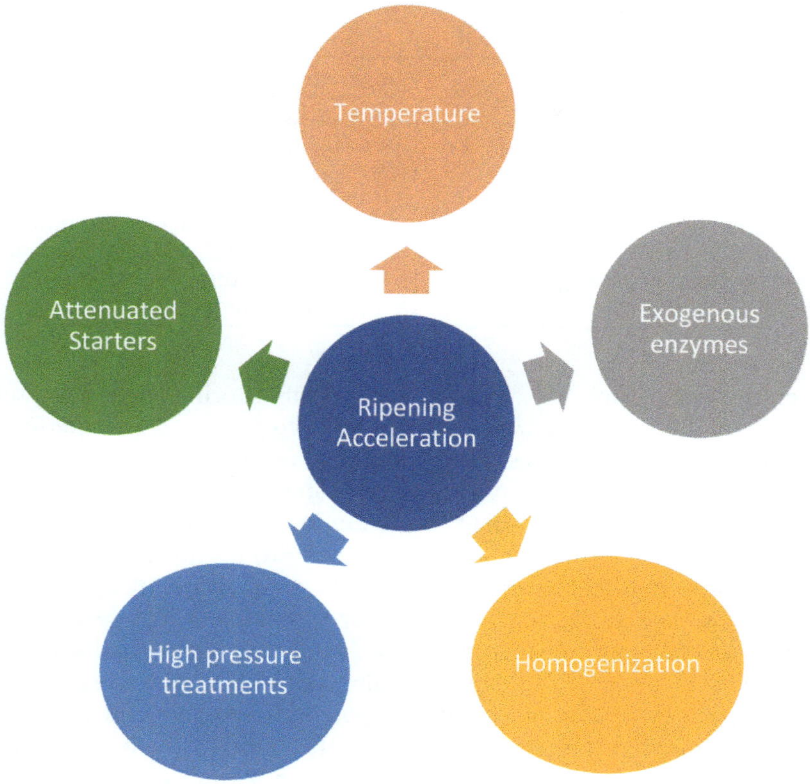

Figure 6.1 Most studied strategies to accelerate ripening.

as well as increased cheese production when storage facilities are limited.[13,14] In contrast, poor texture or flavour defects (bitterness, rancidity) are the main drawbacks that may occur, resulting in a diminution in the shelf life and quality of cheese.[15–17]

Factors that affect mainly the process of proteolysis and lipolysis can be manipulated with the potential view of accelerating ripening, without affecting shelf life and cheese quality. In this sense, various approaches have received considerable attention in the literature; even though this is relatively easy to achieve it is very difficult to control; Figure 6.1 shows the more studied strategies.[1,14,15,17,18]

6.2 Physical Treatments

6.2.1 Elevated Ripening Temperature

The rate of ripening and cheese quality are strongly influenced by ripening temperature.[2] Typical temperatures and times of ripening for different cheese varieties are shown in Table 6.1.[13,19,20]

Table 6.1 Typical ripening temperatures and times of different cheese varieties.[13,19,20]

Cheese varieties	Temperature and time of ripening
Cheddar	7–11 °C/9–12 months
Dutch type (Gouda)	10–14 °C/3 months
Swiss type (Emmental)	19–24 °C/30–70 days (warm room) and 11–14 °C/4–12 months
Blue (Roquefort)	8–10 °C/15–21 days (first maturation) and −2, − 4 °C/3 months (second maturation)
Grana type (Parmigiano Reggiano)	18 °C/20–24 months
Reggianito Argentino	12–14 °C/6–12 months

The implementation of elevated temperature during cheese ripening emerged in the 1980s. It is the most technologically simple method to accelerate cheese ripening as the biochemical reactions by the starter and non-starter microorganisms in the cheese system are promoted, thus leading to a fast flavour development. Refrigeration costs and storage requirements are reduced, providing an overall saving to the producer. However, cheese quality could be affected; the elevated temperature of ripening may accelerate the loss of starter bacteria prior to complete lactose utilization and also promote undesirable reactions. Potential problems include the development of unpredictable and unbalanced off-flavours, negative changes in texture, fat exudation, growth of potential spoilage and pathogens, and contaminating substances, making the product non-consumable. Therefore, attention should be paid to milk quality and the hygienic conditions used during cheesemaking.[2,15,17]

Several studies have documented the microbial and biochemical changes and development of accelerated flavour induced by elevated ripening temperatures for Cheddar cheese.[21–29] In particular, Folkertsma *et al.*[25] studied the effect of accelerated ripening of Cheddar cheese exposed to 12 and 16 °C, comparative to the conventional condition (8 °C) for 9 months, on NSLAB numbers, proteolysis, lipolysis, and flavour development; in addition, different cooling conditions of freshly made cheeses were applied: cooled rapidly (15 h) to ripening temperature and cooled slowly over 8 days until reaching 16 °C. Results demonstrated that the rate of ripening was increased or decreased by changing the temperature. Specifically, patterns of proteolysis and lipolysis were sped up with increasing ripening temperature and slow cooling of cheese. The increase of ripening temperature had little influence on the number of NSLAB, although rapid cooling of cheeses freshly made and ripening at 8 °C significantly reduced the growth rate of these adventitious bacteria. The flavour score of cheese ripened at 16 °C was higher in the early stages, but the texture deteriorated and fat exudation occurred after prolonged ripening at 16 °C. Hence, ripening at 12 °C was recommended for the commercial/large-scale production of Cheddar cheese. According to Rehman *et al.*,[26] ripening temperature increased all the aroma attributes (creamy/milky, fruity/sweet, acid/sharp, and pungent), except musty and unclean aroma, for Cheddar cheeses ripened at 8 °C compared to those matured at 1 °C, but all cheeses were acceptable. Cheeses ripened at 8 °C for 6 months had the highest concentrations of most of the volatile compounds quantified (fatty acids, ketones, aldehydes, esters, alcohols, lactones, and methional). Hannon *et al.*[27] analysed the effect of various combinations of temperature/time in the initial stages of ripening of Cheddar cheese on the proteolysis, microbial counts of starter lactic acid bacteria and NSLAB, and sensory descriptors. They observed an initial acceleration in the three types of experimental cheeses: ripened at 12 °C for 6 weeks followed by 8 °C up to 8 months, 20 °C for 1 week followed by 8 °C up to 8 months, and 30 °C for 1 week followed by 8 °C up to 8 months. Mature flavour attributes were detected in these experimental cheeses compared to the control counterpart (ripened at 8 °C for 8 months). They concluded that three experimental conditions helped in reducing the ripening time for 2 months as well as maintaining the desirable properties of taste and flavour of cheese, and they noticed proteolysis differences depending on the exposure time of ripening at higher temperatures. Immediately after pressing, starter counts in all cheeses were high (approx. 10^9 cfu per g) and remained at elevated levels throughout ripening, irrespective of the ripening conditions used. However, ripening temperature had a significant effect on NSLAB populations; levels of $\sim 10^2$ cfu per g at day 1 were detected in all cheeses, and thereafter NSLAB increased at a rate dependent on the ripening temperature. In contrast, increasing the ripening temperature from 4 °C to 8 °C affected the counts of starter lactococci in Cheddar cheeses but did not affect the viability of the probiotic microorganisms (*B. longum* 1941, *B. animalis* LAFT B94, *Lb. casei* 279, *Lb. casei* LAFT IL26, *Lb. acidophilus* 4962, *Lb. acidophilus* LAFTI L10) added to cheeses. Probiotic cheeses had a higher concentration of lactic acid, especially in those with *Lb. casei* 279 and *Lb. casei* L26 ripened at 8 °C for 24 weeks; this shows that careful selection of probiotic microorganisms is required when choosing higher temperature to accelerate the probiotic cheese ripening.[28] According to Batool *et al.*,[29] lipolysis and organic acids (formic, pyruvic, lactic, acetic) after 120 days of ripening increased for Cheddar cheeses ripened at 12 °C compared to samples aged at 4 °C.

Temperature-driven changes on the complex microbial consortia of cheeses was confirmed by De Filipis *et al.*[30] for a traditional Italian medium-ripened pasta-filata cheese (Caciocavallo Silano) using metatranscriptomics. An increase in the ripening temperature (from 16 °C to 20 °C) promoted the expression of genes related to proteolysis, lipolysis and AA/lipid catabolism and significantly increased the cheese maturation rate. Moreover, temperature-promoted microbial metabolism processes were consistent with the metabolomic profiles of proteins and volatile organic compounds in the cheese.

On the other hand, the impact of elevated ripening temperatures on cheese microstructure, structure, and functionality is poorly understood. In this sense, modern image analysis techniques are suitable for this purpose.

According to Soodam *et al.*,[31] an increase in ripening temperature (15 °C and 2 °C compared to 8 °C) altered the microstructure of the Cheddar cheese protein network after only a few weeks of ripening. Ripening at 20 °C significantly reduced branching within the protein network, leading to thicker protein strands and larger pores after 33 days. More granules, possibly containing calcium salts, appeared in the cheeses ripened at higher temperatures; these structures could have occurred due to the higher whey migration resulting in greater movement and concentration of salts and granule formation. Elevated ripening temperatures may also increase the rate of calcium lactate deposition due to lactate racemization by cheese microflora.[32] The presence of free whey and calcium lactate deposits is a negative aspect for the industrial application of high temperatures for cheese ripening acceleration. The structural changes observed by Soodam *et al.*[31] coincided with increased proteolysis, consistent with solubilization of the protein network, and led to a softer, less chewy, and less cohesive cheese. However, fat distribution within the cheese matrix did not change during ripening, irrespective of the ripening temperature.

In relation to the impact of elevated ripening temperature on the sensory aspects of cheeses, Walsh *et al.*[33] demonstrated that aging white Cheddar for 8 months at 10 °C produced cheese with similar consumer acceptance to that aged at 7 °C for 12 months. Trained panel results showed that 2 month cheeses were described by milk fat flavour and sweet taste, 5 month cheeses were described by nutty aroma and white color, and 8, 10, 11, and 12 month cheeses developed aged characteristics, such as umami and bitter tastes, brothy aroma, and aged flavour. Consumer panel results showed similar overall liking scores among cheeses aged at 7, 10 or 12 °C. In addition, the electronic tongue served as a valid and rapid method for predicting the age of cheese.

Sheehan *et al.*[34] found higher levels of primary and secondary proteolysis, in semi-hard cheeses with propionic bacteria ripened at 12 °C, relative to cheeses matured at 9 °C. In addition, the increase of ripening temperature promoted the growth and metabolic activity of propionic bacteria, as the levels of acetate and propionate were increased. Ripening temperature did not affect the number of *Streptococcus thermophilus* but affected that of NSLAB; growth rates of NSLAB increased at higher ripening temperatures (12 °C *vs.* 9 °C) although final NSLAB counts were similar for all cheeses. Manchego cheese matured at 16 °C had higher lipolysis and proteolysis than counterparts ripened at 12 °C and 8 °C. High ripening temperatures did not improve flavour quality, although they enhanced flavour intensity.[35] In a similar way, the elevation of ripening temperature at 16 °C for Dutch type cheeses produced two times higher contents of FAA than those obtained for cheeses ripened at the current standard temperature (10 °C); central parts of accelerated cheeses contained higher levels of FAA than edge parts. Cheeses ripened under both conditions had acceptable flavour and no off-flavour; however, the hardness of accelerated cheeses dropped faster. So the aforementioned increase of temperature could reduce the ripening time to approximately one half compared to the standard ripening temperature. Nevertheless, the elevated ripening temperature caused greater production of biogenic amines (tyramine, putrescine, and cadaverine), which represents a risk for consumers.[36]

The results of Zhao *et al.*[37] demonstrated that proteolysis, lipolysis, and volatile compounds of hard cheese were significantly increased by increasing temperature (from 4 °C to 20 °C). Compared with cheese ripened at 4 °C for 90 days, the cheeses ripened at 10 °C for 60 days and 20 °C for 30 days displayed the same or more proteolysis, lipolysis, and volatile compounds; furthermore, tiny differences in cheese microstructures were found among the three conditions. Although 20 °C accelerated cheese ripening, undesirable effects were found on the volatile compounds and texture of the cheese. Therefore, the best suitable temperature for elevating hard cheese ripening was 10 °C. For Zamorano cheese, a traditional Spanish cheese made from ewes' milk, Ferrazza *et al.*[38] concluded that the effect of increased ripening temperature was noted more particularly in the changes undergone by the population of lactobacillus NSLAB; the lactobacillus counts were 3 log units higher in cheeses ripened at 15 °C/180 days and 15 °C/60 days, followed by 10 °C/120 days, with respect to the remainder of the cheese (10 °C/180 days; 10 °C/60 days followed by 15 °C/120 days). Sihufe *et al.*[39–42] and Ceruti *et al.*[43,44] investigated the effect of elevated ripening temperatures on the ripening index, volatiles, and sensory characteristics of traditional Argentinean hard-cooked cheese (Reggianito). The elevated ripening temperature (18 °C) increased the α_{s1}- and β-casein degradation, which correlated with higher contents of AA; the total AA amount in cheeses ripened for 123 days at 18 °C was higher than that found in cheeses ripened for 186 days at 12 °C. Besides, the increase of ripening temperature accelerated the salt redistribution.[41] Lipolysis increased until 4 months of ripening in cheese stored at 12 °C, remaining almost constant during the last 2 months of ripening, while in cheese ripened at 18 °C the increase sustained over the 6 months, demonstrating the temperature effect on the lipolytic enzymes present in cheese.[40] The attribute scores of cheeses ripened at 18 °C were different from those of cheeses ripened at 12 °C but they did not result in atypical Reggianito sensory characteristics.[40] Principal component analysis (PCA) conveniently summarized the information obtained from physicochemical, proteolysis, lipolysis, and sensory analyses, and concluded that the optimal time for cheeses ripened at 18 °C was between 2 and 3 months and had similar values to those for cheeses ripened for 6 months at 12 °C.[39] Ceruti *et al.*[43] showed that Reggianito cheeses ripened at 20 °C for 2 or 4 weeks and then at 12 °C for 6 months had a higher level of proteolysis than cheeses stored at 12 °C for 6 months. The more notable variations in the volatile fraction of cheeses occurred during the first 2 months of ripening and were mostly explained by changes observed for the predominant acid fraction. The different temperature–time combinations affected the quantities of some of the volatile compounds, but the balance between the different groups of compounds remained roughly the same.[44]

6.2.2 Cheese Milk with Low Homogenization

In many cheeses lipolysis is undesirable, but in mould ripened cheeses (blue veined, Brie, Roquefort) and hard cheeses such as Italian hard cheeses (Parmigiano Reggiano, Grana Padano) or Reggianito cheeses, lipolysis is important as it contributes to the development of genuine flavour.[45,46] In these types of cheeses, the lipolytic agents are the lipoprotein lipase (LPL) and the enzymes from lactic starters and NSLAB and moulds. In raw milk, lipolysis does not occur spontaneously since lipolytic enzymes are separated from substrate fat: TAG are inside fat globules surrounded by the milk fat globule membrane (MFGM) and LPL is associated with caseins.[45] However, physical treatments applied to milk before cheesemaking (agitation, pumping, homogenization) diminish the protection of fat by the MFGM and enhance lipolysis.[47]

Homogenization was developed in 1899 by Gaulin; it is a widely used method to delay creaming in milk and dairy products during storage.[48] This is achieved by the reduction of milk fat globules, which retards coalescence and separation of the fat layer.[49] Disruption of the fat globules produces a five- to tenfold larger surface area of the fat emulsion which is stabilized by casein micelles and whey proteins.[50] Homogenization is almost always combined with a heat treatment, followed or preceded by pasteurization, to destroy pathogenic bacteria and to inactivate LPL which could cause excessive lipolysis.[51] In raw milk, the fat globule size commonly ranges from ~0.2 to 15 μm, and homogenization generally aims to reduce the maximum to <1 μm. For this purpose, two-stage valve homogenizers are commonly used. The first stage is usually operated at ~15 MPa and the second stage at 3/4 MPa.[52]

In cheesemaking, homogenization of milk can be of interest for the purpose of reducing fat losses in the whey, controlling development of free fat in the cheese, or when it is necessary to improve the sensory qualities (*e.g.* to enhance flavour in blue cheeses or to develop a correct texture in fresh cheeses and in reduced-fat cheeses giving rise to products with higher moisture and a softer texture).[48,53,54] However, it has various effects on technology and cheese parameters, including decreased rennet coagulation time,[55] decreased syneresis,[56] a weaker rennet coagulum and lower curd tension,[56] increased cheese yield,[57] compositional changes in the cheese,[58] and excessive FFA production.

In blue cheeses, the homogenization of the cream is carried out to promote fat hydrolysis, a desirable reaction for this variety, combined with the formation of volatile compounds derived from the catabolism of FFA, such as methyl ketones, enhancing flavour in the final product. In addition, it increases the whiteness and opacity of the cheese paste, which is highly desirable in blue cheeses to increase the contrast between the blue veins and the whiteness of the matrix.[59]

There are few studies aimed at accelerating cheese ripening through the application of low homogenization by means of enhancing the formation of FFA and associated flavour compounds. Deegan and McSweeney[60] studied lipolysis in Cheddar cheeses made at a pilot scale with raw milk homogenized at different pressures (0, 5, and 10 MPa) and pre-incubated (1 h at 37 °C). Trial effects were found, but in general cheeses made from milk homogenized at 10 MPa contained the highest level of FFA initially and throughout ripening. As for proteolysis, breakdown of α_{s1}-casein was more extensive in cheese made from non-homogenized milks and breakdown of β-casein was more extensive in homogenized milk cheeses.

Other authors tested the option of homogenizing only the cream portion of milk to avoid drawbacks during cheesemaking. Cao *et al.*[61] studied the influence of cream homogenization (14% fat, 13.8 MPa) in FFA release and methyl ketone formation in blue cheeses with different lipase treatments. They found a strong and positive effect of homogenization as FFA increased >10-fold (except octanoic acid). Although the premise is that an ability to increase the pool of the FFA substrate (as affected by lipase and homogenization treatments) influences the resulting conversion into methyl ketone metabolites, no such cause and effect was readily observed. Vélez *et al.*[62] performed a pre-incubation (12 h/12 °C) of cheese-milk prepared with 20% cream homogenized at 9 MPa and elaborated hard cheeses (Reggianito type) at a laboratory scale in order to study composition, ripening index, lipolysis, and volatile profiles during ripening (90 days). The homogenization step had no impact on composition and proteolysis. A lipolysis acceleration was noticed in cheeses made with homogenized cream (H). The percentage of short-chain free fatty acids in H cheeses (C4:0–C8:0) increased from 3% at 3 days to 6.6% at 45 days and remained constant until the end of ripening, while in control cheeses the percentage increased from 3% at 3 days to 9% at 45 days and then it diminished to 6% at 90 days. Therefore, at 90 days control and experimental cheeses had similar SCFFA percentages. As for volatile compounds, the levels of several compounds derived from FFA catabolism were influenced: straight-chain aldehydes such as hexanal, heptanal, and nonanal, and methyl ketones from C5 to C9 were enhanced in H cheeses at different ripening times. Vélez *et al.*[63] studied the lipolysis, volatile compounds, proteolysis, and sensory attributes of Reggianito made at a pilot scale. Experimental cheeses were made with raw milk prepared with homogenized 20% cream pre-incubated at 12 °C, cheeses were cooked at 50 °C, and a curd-washing step was performed. Control cheeses were made with pasteurized milk and cooked at 52 °C, and neither milk homogenization and pre-incubation nor a washing step was performed. Cheeses were analysed at 45 and 60 days (end of ripening). Regardless of the ripening time, the individual concentrations of all FFA were higher in experimental cheeses and the volatile profiles were characterized by compounds derived from milk fat catabolism. The acid group dominated the profiles and the formation of some key flavour compounds derived from fat, such as 2-propanone, diacetyl, 2-pentanone, 2-heptanone, 2-nonanone, 2-pentanol, and ethyl hexanoate, reached the maximum levels at 45 days. As for aldehydes, heptanal was detected only at 60 days and no differences were found between both types of cheeses. These results were in concordance with the sensorial analysis as an increase of hot taste and an overall intensification of flavour were reported at 45 days.

In the studies mentioned above, no coagulation or rheological properties were analysed, and they may have been modified. Vigneux *et al.*[64] studied the application of homogenization treatment to a portion of milk or cream in curd models to provide the cheesemaker with additional means to better control the cheese yield, composition, and texture. They found that fat globules in homogenized milk were significantly smaller with a higher protein load than those in homogenized cream. Curd moisture content, fat retention, and actual and moisture-adjusted yields increased with the proportion of homogenized fat in cheese milk, while the curd storage modulus decreased. In this sense, cheese moisture is a key factor that can reduce cheese firmness.[65] Homogenization of milk can result in cheeses with poor body, reduced stretchability and meltability.[57] According to Johnson,[59] cheeses made from homogenized milk are smoother and excessively sticky compared with cheeses made from non-homogenized milk.

6.2.3 High Pressure Homogenization

Recently, novel homogenization devices, *e.g.*, high-pressure homogenizers and microfluidizers, that can be operated at pressures of several hundred MPa and achieve greater reductions in fat globule size have been developed.[66] Depending on the nominal pressure, the technology takes the name of high-pressure homogenization (HPH) (~150 to 200 MPa) or ultra-high pressure homogenization (UHPH) (~350 to 400 MPa). In the literature, these procedures are also called high dynamic pressure treatments. In HPH and UHPH the principle of the operation is similar to that of conventional homogenizers used in the dairy industry, except that it works at higher pressures.

HPH and UHPH are food preservation treatments that have been developed and applied as minimal processes for the production of a wide variety of safe and nutritious foods. They have been studied for the treatment of milk in order to inactivate microorganisms, to produce very fine and stable emulsions, and to modify the rheological and/or coagulation properties of milk and dairy products. For these purposes, HPH and UHPH have been applied mostly in fresh cheese to increase shelf life and improve texture. In ripened cheeses, Kheadr *et al.*[67] evaluated the rheological and microstructural quality of Cheddar cheese made from treated milk by homogenization at 200 MPa, Ti (inlet temperature) = 28 °C and 5 cycles of 1 min, compared to cheeses made with pasteurized milk. They tested (a) pressurizing whole milk, (b) pressurizing skim milk which was standardized with pasteurized cream, (c) whole milk pasteurization at 72 °C/15 s, and (d) pressurizing 2% fat milk. The results showed that milk homogenized cheeses had higher moisture content, protein and yield compared to cheeses made with pasteurized milk. Also, (a) and (d) presented a firmer, more elastic texture that was more cohesive and less brittle than pasteurized milk cheeses (c) due to the formation of a very compact and regular casein matrix. In addition, treatment (b) resulted in cheese with a textural profile and structural characteristics close to those of pasteurized milk cheese.

High pressure has not been applied with the main objective of ripening acceleration. However, Guerzoni *et al.*,[68] Lanciotti *et al.*,[69] and Vannini *et al.*[70] found an increase in lipolysis in cheeses from highly homogenized goat, cow, and ewe milk (100 MPa) compared to those from heat-treated milk, which was attributed to the smaller and highly emulsified fat globules produced by the treatment, and activation of the indigenous lipase. When treatment pressure increases, different results have been obtained. It has been reported that UHPH could activate or inactivate enzymes according to the conditions reached with the treatment. The effects of UHPH on milk are due to the combination of cavitation, turbulence, shear forces and heating to which the milk is exposed, with greater inactivation occurring as the outlet temperature increases.[71] Datta *et al.*[71] found total inactivation of the LPL in UHPH-treated milk treated at 200 MPa at outlet temperatures >71 °C. Juan *et al.*[72,73] evaluated the proteolysis and lipolysis of processed goat cheeses from milk treated by UHPH (200 MPa, Ti = 30 °C) compared to those made from pasteurized–homogenized (72 °C/15 s, 18 + 2 MPa) and pasteurized (72 °C/15 s) milk, respectively. Lipolysis in cheeses made with UHPH-treated milk and that in those made with pasteurized milk were similar. However, conventional homogenization produced an extensive lipolysis in cheeses due to the reduced size of the fat globule, increasing the area for the activity of the lipase. Differences in the level of lipolysis of the cheeses made from milk treated with UHPH and conventional homogenization could be related to the intensity of UHPH as the maximum temperature with UHPH was around 82 °C, which was maintained for a short time (0.7 s), so little or no LPL activity was expected. As for proteolysis, cheeses made from UHPH-treated milk showed a greater hydrolysis of α_{s2}-, β-, and κ-caseins than those made with pasteurized milk, a fact that was attributed to the increased substrate exposure to proteolytic enzymes. UHPH treatment of milk significantly increased the levels of soluble nitrogen, and hydrophobic and hydrophilic peptides in cheeses. This could be due to the increased exposure of substrates to proteolytic enzymes.

6.2.4 High Pressure Treatment

High pressure treatment, also called high hydrostatic pressure (HHP) technology or high-pressure processing (HPP), is a non-thermal technique with similar benefits to heat treatment, but with the additional advantage of minimal damage to food quality.[74] It uses pressures up to 900 MPa (approximately equal to 9000 atm or 135 000 pounds per square inch) to kill many of the microorganisms found in foods, even at room temperature, without degrading vitamins, flavour, and color molecules.[75] It is often cited as a major technological innovation in food preservation and has emerged as a viable commercial alternative for the pasteurization of value-added fruits, vegetables, meat, and seafood products.[75] Traditionally it is a batch process and the vessels used have capacities of 35–350 L. The packaged food is placed in a sealed high-pressure vessel, and the pressure medium, usually water, is pumped into the vessel. Once the desired pressure is reached, the pumping is stopped, valves are closed, and the pressure is maintained. The high pressure is applied in an isostatic manner, so all regions of the food, regardless of size and geometry, have uniform pressure. The compression during the HP process is accompanied by a temperature increase (normally 3–9 °C/100 MPa) of food. During the decompression process, the temperature decreases at a lower degree than that in the come-up stage.[74,75] The commonly applied conditions cause the elimination of most pathogenic bacteria and spoilage microorganisms. Spore-forming bacteria like *Clostridium botulinum* require more severe conditions or combined strategies for their inactivation. The effect on LAB is strain-dependent. Enzymes can be inactivated by HP treatments. Proteins and lipids are also affected, producing modifications to the cheese microstructure and texture. Disruption of the casein micelle structure and denaturation of β-lactoglobulin and α-lactalbumin are induced. The fat globule size does not vary significantly with HP treatment from 100 to 600 MPa for 30 min at 20 °C, although the volume weighted mean diameter of globules is increased after 30 min at 600 MPa. Besides, the association of milk protein with fat globules by the HP treatment, with a maximum at 200 MPa, was reported.[75,76]

By the end of the 19th century, it was shown that milk shelf-life could be extended by pressure treatment. However, the lack of availability of suitable equipment slowed down further research for the first half of the 20th century. From there to the present, industrial equipment has expanded in number and processing capacity.[77] The first background of HP application to cheese originated with the patent presented by Yokohama *et al.*[78] who demonstrated that Cheddar cheese exposed to 50 MPa for 3 days at 25 °C had FAA concentrations and taste equivalent to those of 6 month-old cheese. The increased rate of proteolysis and acceleration of the cheese maturation was probably due to the higher release of the intracellular enzyme pool associated with cell lysis.[78] After that, other investigations were performed on Cheddar[79–84] and other varieties like fresh,[85] Gouda,[86] smear-ripened,[87] and hard cheeses.[88–91] However, the effectiveness of HP treatment in shortening cheese ripening seems to be limited, and it should be carefully considered before being selected for this purpose. In contrast, HP may be useful for arresting or slowing down ripening.

O'Reilly *et al.*[79] made cheeses with four *L. lactis* strains (303, 223, 227 and AM2) added individually as starter cultures and at 1 day after manufacture applied high pressure in the range of 100–400 MPa for 20 min at 25 °C. Reductions in *L. lactis* counts depending on the strain and the applied pressure to cheese were detected. Higher levels of β-caseins were observed for cheese treated at 400 MPa than controls on day 30. Cheeses with *L. lactis* 223 and treated at 400 MPa were the only ones in which increases in FAA concentrations were observed; for the other cases, HP treatments produced a decrease in FAA in relation to that found for control cheese. O'Reilly *et al.*[84] observed an enhancement of proteolysis in Cheddar cheese treated at 50 MPa for 3 days at 25 °C, although the levels of proteolysis were not as suggested by Yokoyama *et al.*;[92] in addition, any gross structural changes in the cheese matrix were revealed by confocal laser scanning microscopy. Rynne *et al.*[81] applied 400 MPa for 10 min at room temperature on one-day post-manufacture of Cheddar cheese. Starter lactococci and NSLAB counts fell after treatment for HP cheese comparative to untreated cheeses. They found little effect on primary proteolysis of HP cheese, consistent with the measured plasmin and chymosin activities. α_{s1}-Casein breakdown was more intense in HP cheese with a slight increase in pH 4.6-soluble nitrogen but not on FAA concentrations. Descriptive sensory analysis indicated that after 90 days of ripening the HP cheese had a greater intensity of some attributes compared to control cheese, but after 180 days of ripening the HP cheese was less intensely flavoured.

The application of 400 MPa for 5 min to hard caprine cheese caused significant quantitative and qualitative changes on proteolysis that persisted throughout the ripening. Proteolysis of cheeses treated at 50 MPa for 72 h was only slightly different from that of control cheese, with differences being less apparent by the end of ripening. Increased concentrations of FAA, possibly due to a higher peptidase activity, were found for cheese treated at 400 MPa; however, the peptide profiles and caseins of 400 MPa cheese were similar to those of 50 MPa cheese and untreated cheese. Plasmin activity in cheese was unaffected by pressure treatment, whereas coagulant activity was decreased at 400 MPa. 400 MPa cheese had a higher pH than untreated cheese, presumably due to the reduced acidification activity of LAB and the effect on calcium equilibrium after HP treatment. It was concluded that the conditions of high moisture and high pH of cheese favored the shortening of maturation.[90] However, sensory assessment by panelists evidenced bitter notes in HP cheeses, especially in cheese treated at 400 MPa, and pressurized cheeses were less crumbly and more elastic than control.[88] The instrumental texture of pressurized cheese at 50 MPa/72 h became softer, large deformations were achieved with less force applied, and a more difficult fracture was observed.[89] In other varieties of hard cheese, like Reggianito, Costabel *et al.*[91] reported the acceleration of proteolysis and peptidolysis by HP treatment at 400 MPa for 5 or 10 min at 20 °C. Cheese treated at 400 MPa for 10 min showed higher plasmin activity than cheese treated at 100 MPa or untreated cheese, similar residual coagulant activity, lower starter culture counts and higher contents of FAA, which were 1.89-fold higher than those of untreated cheese on day 45 and 1.24-fold higher on day 60.

According to Arqués *et al.*,[93] treatment at 300 and 400 MPa for 10 min applied to ewe raw milk La Serena cheese after 2 days of manufacture improved the formation of some volatiles (branched-chain aldehydes and secondary alcohols), but decelerated the generation of other volatiles (*n*-aldehydes, 2-methyl ketones, dihydroxy-ketones, *n*-alcohols, unsaturated alcohols, ethyl esters, propyl esters, and branched-chain esters). A negative effect on odour intensity, aroma quality, and aroma intensity was found for HP cheeses compared to controls. Delgado-Martínez *et al.*[94] suggested that 600 MPa treatment on mature Torta del Casar cheese can be used to prevent excessive bitterness and overripening getting a longer shelf-life and a better sensory quality. Better preservation of the sensory characteristics could be related to a decrease in the volatile compounds generated during storage, since an excess of acids and esters would provide off-flavours or would act as precursors for other off-flavour compounds.

Overall, the pressure effects on microbial growth and metabolism, and milk components are variable depending on HP parameters (pressure, holding time, and temperature, moment at which the treatment is applied during cheese ripening, and the type of cheese). In addition, enzymes involved in cheese ripening can be affected in variable degrees depending on the enzyme, substrate and also the parameters applied. Finally, all these aspects have an effect on proteolysis, lipolysis, and flavour compound formation.

6.3 Exogenous Enzyme Addition

Increasing the amount of the key enzymes involved in cheese ripening leads to cheese ripening acceleration. Therefore, the incorporation of exogenous proteinases, peptidases and lipases/esterases, and other enzymes (β-galactosidases, *etc.*), individually or in cocktails of enzymes, is another attempt to speed up cheese ripening and aid in producing characteristic flavours in cheeses.[2] In particular, the degree of enzyme incorporation into solid curd that occurs when enzymes are added to milk or to curd at the salting stage is an important aspect to take into account.[95] So adequate selection of the enzyme or mixture of enzymes, the proper adjustment of the amount to be added in each case, and the moment of addition into cheese milk or into curd have to be optimized for each case in particular, in order to obtain the desirable flavour and quality of cheese.[15]

6.3.1 Free Lipolytic Enzymes

Lipolytic enzymes – lipases and esterases – are ester hydrolases (EC 3.1) which catalyse the cleavage and formation of ester bonds.[96] They are ubiquitous and produced by various sources such as plants, animals, and microbes. Microbial sources from bacteria, yeasts or moulds are of high interest due to their commercial significance and are cheaper, easier to produce, and the application is more predictable and controllable.[97] Lipolysis – partial and selective hydrolysis of milk fat – is promoted by these enzymes, leading to the formation of FFA and other aromatic compounds derived from its catabolism (esters, ketones, and alcohols, among others).[98]

Rani and Jagtap[99] focused on reduction of the ripening period of Swiss cheese by adding a microbial lipase produced by *Bacillus tequilensis*; the influence on the physicochemical, textural, and sensory properties and microstructure of cheese was studied. Cheese ripening was shortened from 3 to 2 months by adding the lipase (200 U to 60 L) to milk, without compromising the desirable physicochemical and sensory properties of the final product. However, no significant change in quality and sensory properties was observed adding a higher level of lipase (800 U/60 L of milk). Changes in microstructure were revealed by scanning electron microscopy images; the number and size of the fat globules were reduced by the presence of the lipase. An increased hardness of cheese with added lipase was observed which could be due to the plasticizing effect of fat. In fact, changes in the textural properties of cheese take place because of changes in the interaction between casein micelles and the fat phase; a decrease in the plasticizing effect of fat leads to a compact structure, probably resulting in a harder texture.[100] Karaca and Güven[101] detected increased lipolysis and accelerated ripening in cheeses added with *Mucor miehei*-originated fungal esterase-lipase (Piccantase A) at two doses (0.05 and 0.10 g 100 L^{-1} of milk). However, the cheese yield, flavour, and texture were negatively affected as higher fat content in whey, bitter taste, and the softest texture were reported. It was recommended to use a lower dose of enzyme for reducing the ripening period without any adverse effect. The addition of commercial microbial lipase (Piccantase A) to Tulum cheese enhanced the production of volatile FFA concurrent with an increased lipase level. This has led to the shortening of ripening time without any flavour defect.[102] Hernández *et al.*[103] demonstrated that the utilization of animal lipase (pregastric lipase) had a clear effect on the FFA profile, increasing mainly short-chain FFA, and on the FAA accumulation during cheese ripening. Desirable attributes of Idiazabal cheese like odour intensity and butyric flavour were positively influenced by lipase addition. Particularly, increasing the amount of lipase resulted in higher scores for odour and flavour intensities, sharp odour, pungent flavour, nutty flavour, and bitterness. In contrast, low amounts of lipase resulted in increases of attributes such as milky, butter and toasty odours, usually associated with mild odour notes. Similarly, the inclusion of commercial lipases, namely, two pregastric lipases (A1 and A2) and one fungal lipase (A3), in milk during Idiazabal cheese making increased lipolysis in cheeses during ripening. Lipases A1 and A2 increased the concentrations of short-chain FFA (C4, C6, and C10), whereas C16:0 and C18:1 were the main FFA produced by the A3 lipase, comparative to cheeses made without lipase. Cheeses made with A1 and A2 lipases had higher scores in odour and flavour intensity, sharp odour and pungent flavour than cheeses made with the A3 lipase or without lipases.[104] Increases in the lipolysis of Manchego cheese, one of the most representative of the Spanish hard cheeses, were promoted by the addition of two fungal lipases (Palatase 200 from *Mucor miehei* and Palatase 750L from *Aspergillus niger*). Sensory analysis revealed an acceleration of ripening for cheeses treated with *A. niger* lipase; these cheeses possessed a typical sharp Manchego flavour. Soapy flavour was perceived in cheeses treated with *M. miehei* lipase probably because of the excessive release of long-chain FFA.[105] Cao *et al.*[61] evaluated the effect of *P. roqueforti* lipase and porcine pancreatic lipase, added at the salting stage during blue cheese making, on the generation of methyl ketones and the formation of blue cheese flavour. The effectiveness of *P. roqueforti* lipase as an agent in increasing methyl ketones and flavour notes in cheese was demonstrated in comparison to the other lipase which was very much less effective toward this end. This finding emphasizes the need for careful selection of lipolytic enzyme treatments prior to cheesemaking.

The oldest enzyme studies were on Cheddar. The addition of commercial lamb pregastric esterase, calf esterase, and *Mucor* spp. lipase to cheese milk for Cheddar accelerated the liberation of FFA during ripening.[106] Similarly, Cheddar cheese containing the fungal lipase from *Candida cylindracea* had higher volatile FFA than cheeses without added enzymes; the volatile FFA content in cheese increased with increasing lipase levels. The ripening process of the cheese made with enzyme was normal, though accelerated in chemical and cheese flavour changes. In this study, the enzyme preparation was mixed with salt and added to the milled curd.[107] Arbige *et al.*[108] compared similar FFA profiles in Cheddar cheese treated with FlavorAge to cheeses without enzymes, but with higher concentrations of individual compounds. FlavorAge is a commercial enzyme preparation obtained from *Aspergillus oryzae* and containing both proteases and lipases; according to Arbige *et al.*,[108] the lipase fraction was the major contributor to Cheddar cheese aroma. Other applications of lipases in Cheddar cheese were discussed by Azarnia *et al.*[14] FlavorAge was also assayed in buffalo milk Gouda by Kanawjia *et al.*[109] This study revealed that the flavour, body, and texture developments were more rapid in the cheeses made with the enzyme compared to control cheeses.

6.3.2 Free Proteolytic and Peptidolytic Enzymes

The unspecific action and impurity of some commercial food grade proteinase preparations can negatively affect texture development and lead to an unbalanced flavour. The hydrolysis of caseins by exogenous proteases during cheesemaking could lead to a lower cheese yield due to the loss of peptides in the whey. So it is crucial to optimize the type and dose of enzymes to be added, the moment of addition, and the composition of cheeses to avoid adverse effects.[15,110,111]

For example, the addition of an enzymatic preparation derived from *Aspergillus oryzae* (FlavorAge) was evaluated in Manchego cheese manufactured with a mixture of ewe and cow milk. Higher proteolysis was obtained and no bitterness and off-flavours were detected. In addition, there were significant differences between the estimated age and real ripening time of cheese when a higher dose was used (50 mg kg^{-1}): cheeses with enzymes had higher sensory scores.[112] Similar results were obtained by Mohedano *et al.*[113] in Manchego cheese made only with ewe milk. It was observed that the addition of a cysteine proteinase from *Micrococcus* sp. INIA 528 accelerated proteolysis: residual α_s- and β-caseins were lower in cheeses with the enzyme added, while hydrophobic and hydrophilic peptides, their ratio, and the soluble nitrogen in phosphotungstic acid were higher. The flavour intensity was increased in cheeses with a higher dose of the enzyme; values at 36 days of ripening were similar to the values obtained for control cheeses at 60 days of ripening.

Law and Wigmore[114] verified different performances of two proteolytic enzymes in Cheddar cheese. A neutral protease derived from *Bacillus subtilis* accelerated the development of the typical flavour, while an acid proteinase R derived from *Aspergillus oryzae* produced a bitter defect. The cheese flavour obtained with the bacterial proteinase at 2 months of ripening was comparable to the one of control cheeses at 4 months of ripening.

In Gouda cheese made with buffalo milk, two strategies were assayed, both individually and combined, to accelerate ripening: the addition of heat-shocked *Lactobacillus helveticus* and the incorporation of FlavorAge enzymatic preparation. In all cases, higher levels of proteolysis and flavour scores and faster body and texture development were observed; these effects were more pronounced when both strategies were used together. The flavour scores for all experimental cheeses reached the highest value after 4 months of ripening; these values were higher than those for control cheeses at 8 months.[109] However, after 4 months of ripening, the flavour scores of the cheeses treated with enzymes diminished due to the high levels of proteolysis.

A proteolytic enzymatic preparation containing peptidase R and prozyme 6 (Biocatalystics) was added to cheese milk to obtain Ossau-Iraty cheeses (ewe cheese). In this study an increased release of FAA due to the added enzymes was observed, which were more marked in the early months of ripening. This strategy also produced an adverse effect on texture (reduced elasticity and creaminess and increased brittleness) and flavour (bitterness), especially at higher concentrations of enzymes.[115]

Kilcawley *et al.*[16] assayed three commercial proteolytic enzyme preparations to accelerate Cheddar cheese ripening: Accelase AM317 and Accelase AHC50, unpurified preparations derived from *Lactococcus lactis* (contains permeabilized or dead cells) with high peptidase activities, and Accelerzyme® CPG (DSM Food Specialties), which is a purified preparation derived from *Aspergillus niger* and contains a carboxypeptidase. They observed a positive effect on the acceleration of secondary proteolysis, the production of more complex volatile profiles and the development of sensory attributes related to aged Cheddar cheese. However, the elevated levels of primary proteolysis produced by AHC50 and above all by AM317 led to negative changes of the texture during ripening. These researchers highlighted both the suitability of these commercial enzymes to accelerate cheese ripening and the importance of the optimization of the application to avoid adverse effects. One of these enzymes, Accelerzyme, has also been assayed in Reggianito cheese, the most popular hard cheese in Argentina, by Ceruti *et al.*[116] In this work, a higher degradation of β-casein and slight changes in the peptide profiles were observed due to the enzyme addition.

Recently, Alhelli *et al.*[12] optimized the acceleration of Cheddar cheese ripening by means of an exogenous protease (*Penicillium candidum* PCA1/TT01) using response surface methodology (RSM) to study the protease purification factor, protease concentration, and ripening time. The protease was mixed with salt and added to the milled curd. RSM allowed selecting a combination of factors (protease purification factor: 3.12, 0.01% protease concentration, and 0.6 months ripening at 10 °C) which led to an ideal cheese with characteristics similar to commercial Cheddar cheese. Another enzyme that demonstrated effectiveness at accelerating Cheddar cheese ripening was a recombinant aminopeptidase (PepN) of *Lactobacillus rhamnosus* S93 over-expressed in *Escherichia coli*.[117]

Another approach regarding the addition of enzymes to accelerate ripening is the use of enzymes to increase the activity of native proteolytic enzymes in milk. For example, urokinase catalyses the conversion of the inactive precursor plasminogen to plasmin, which is involved in proteolysis. Barret *et al.*[118] observed increased levels of plasmin and consequently higher proteolysis (increase of FAA, γ-casein, and water-soluble N, and changes in the peptide profiles) in Cheddar cheeses manufactured with urokinase added to the cheese milk. Similar results were observed in Swiss cheese and ultrafiltrated Havarti and Saint-Paulin cheese.[119,120]

Finally, the use of cheese slurries as a source of enzymes, microorganisms, and flavour precursors is another strategy to accelerate cheese ripening.[121] This approach was applied in Ras cheese (the most popular hard cheese of Egypt) made with ultrafiltrated (UF) milk. It has been demonstrated that the use of UF milk leads to cheeses with lower flavour, atypical texture, and longer ripening times in comparison with cheeses made with unconcentrated milk. Thus, the use of cheese slurry could improve the impaired quality due to the use of UF milk. In this way, Hamdy *et al.*[122] found that the addition of 2–3% of mature (12 months of ripening) Cheddar cheese slurry accelerated the ripening (proteolysis and lipolysis) and improved the sensory and textural properties of UF Ras cheese; ripening indices were almost two-fold higher in slurry-containing cheeses. A technologically advanced form of this approach is the use of enzyme-modified cheese slurries (EMC), which provide high concentrations of AA, peptides, FFA, and several flavour compounds; these slurries do not provide enzymes because they are heat-treated before their addition to cheeses. This strategy was satisfactorily used for the acceleration of Cheddar cheese ripening. In this way, the addition of a dried EMC to Cheddar cheese at the salting stage led to an improved development of cheese flavour and increased secondary proteolysis. The EMC was prepared by incubating a mixture composed of milled rennet curd, emulsifying salts, and water with several enzymes: Debitrase DBP20 (aminopeptidases from *Aspergillus oryzae* and *Lactococcus lactis*), Glutaminase F (from *Bacillus subtilis*), and Lipomod 338 and 229 (pancreatic lipase with proteinase).[123]

6.3.3 Enzyme-loaded Liposomes

The limitations in the use of exogenous enzymes are related to enzyme addition. Indeed, using free enzymes may attack substrates prematurely, losing peptides and FA in the whey. Thus, reduction in cheese yield and flavour defects from proteolysis or lipolysis during manufacture and in the early stages of maturation, as well as whey contamination with the added enzyme, are major obstacles in the application of this technology.[17] Besides, enzymes could be lost in the whey with a concomitant low retention in the curd, which increases costs.

In this context, the use of enzyme encapsulation has been proposed to circumvent the drawbacks described earlier. For food applications, the components chosen as encapsulation materials should be food grade, biodegradable, and capable of isolating the internal phase from the environment surrounding the vesicle.[124] Liposomal systems have attracted the attention of researchers because of their beneficial characteristics, as biocompatibility, biodegradability, low toxicity, and the capacity to encapsulate hydrophobic and hydrophilic compounds. Liposomes are self-assembled structures formed mainly by polar lipids (phospholipids) structured in the form of one or more bilayers, enclosing an aqueous medium inside. Currently, there are several methods for preparing liposomes. The most used methods for food applications avoid the use of toxic solvents. For enzyme encapsulation, proliposome and heating methods have been mostly applied for cheese ripening acceleration. The first involves a three-phase mixture (phospholipids, ethanol, and buffer), called the proliposome mixture (commercially available), which is converted into multilamellar liposomes by a two-stage dilution procedure with suitable buffers.[125] This approach avoids the use of energy-expensive processes, and it is possible to achieve high entrapment efficiencies. The heating method involves hydration of the liposome ingredients, followed by heating and stirring of the ingredients with glycerol.

As for lipolytic enzymes, Kheadr *et al.*[126] studied the addition of two types of lipase (Palatase M derived from *M. miehei* and Lipase 50 derived from *A. niger*) encapsulated in liposomes prepared by the proliposome method (VPF 012) during Cheddar cheese production (ripened for 90 days). Lipase-loaded liposomes were incorporated into cheese milk prior to the renneting at three levels (low, medium, and high). Treated cheeses had higher moisture (attributed to the water binding capacity of phospholipids) and lower protein, fat, and ash contents than control cheese and they were less firm, but more elastic and cohesive, than control cheese. Production of FFA was accelerated by the addition of encapsulated lipases. Cheeses with up to the medium concentration level had slightly better flavour intensities than control cheese. Cheeses with the highest levels of Palatase M and Lipase 50 developed a pronounced soapy off-flavour after 2 and 3 months of ripening, respectively.

The use of enzyme cocktails has been also proposed[127] as it is known that the characteristic aroma, flavour, and texture of cheeses are the result of numerous biochemical reactions mediated by different enzymes. Kheadr *et al.*[128] studied Cheddar cheeses added with Flavourzyme, a neutral bacterial protease (from *Bacillus subtilis*), an acid fungal protease (*A. oryzae*), and a lipase (Palatase M). Flavourzyme is a commercial preparation that includes a combination of fungal enzymes, consisting of exopeptidases and endoproteases.[129] All enzymes were entrapped individually in proliposomes (VPF 012). The authors focused their work on sensorial analysis and determination of bitter peptides. Flavourzyme was tested alone at three concentrations, and enzyme cocktails consisted in lipase and bacterial protease, lipase and fungal protease, or lipase and Flavourzyme. As for composition, no significant differences were observed among liposome-treated cheeses; the main differences were found between cheeses added with liposomes and control cheeses. Again, cheeses with liposomes presented higher moisture and consequently lower protein and fat than control cheeses. Among all treatments, the mixture of bacterial protease and lipase was preferred as it showed a well-balanced equilibrium between casein and milk fat hydrolysis products without producing bitter, rancid, or texture defects.

Jahadi *et al.*[130] studied the influence of Flavourzyme encapsulated in liposomes, ripening time, and curd retention in saturated brine on the proteolysis indices and sensory perception of Iranian white brined cheese by response surface methodology. The authors employed the heating method for enzyme liposome encapsulation and added the preparations along with rennet addition. The most influential factor was the ripening time, but the enzyme liposomal content and retention time were also significant. The maximum sensory characteristics were in line with maximum proteolysis. A ripening time of 30 days, a curd retention time in saturated brine of 8 h, and a content of 0.3% w/w liposomal enzyme resulted in cheeses with the maximum proteolysis and sensory evaluation score. In another work, Jahadi *et al.*[131] continued the study of cheese ripening acceleration to determine the effect of liposomal Flavourzyme and free Flavourzyme addition on Iranian white brined cheese ripening (matured for 30 days at 12 °C). The authors also employed the heating method for enzyme encapsulation. They found higher proteolysis rates in cheeses made with liposomal enzymes, which could accelerate the ripening time by 10 to 20 days.

Moreover, a commercial neutral proteinase from *Bacillus subtilis* (Neutrase) has been commonly applied in the acceleration of Saint-Paulin cheese. Picon *et al.*[132] applied liposomes prepared by a thin-film hydration method to encapsulate the neutral proteinase to increase proteolysis in Manchego cheeses (ripened for 90 days at 12 °C). The liposome-loaded enzyme was added with rennet in experimental cheeses; control cheeses without enzyme were also included. Contrary to the other reports mentioned previously, experimental cheeses showed higher dry matter and protein content, possibly because the enzyme leaking from the liposomes favored whey expulsion during pressing and salting. Proteolysis was enhanced in experimental cheeses by 24 h after manufacture. After 90 days of ripening, the percentages of soluble nitrogen at pH 4.6 in TCA and in phosphotungstic acid were higher in enzyme-treated cheeses than those in control cheeses. In addition, hydrophobic and hydrophilic peptides were also increased. No bitter flavour defect was detected in experimental cheeses, although development of flavour intensity was accelerated by the encapsulated proteinase. Picón *et al.*[133] continued the studies on Manchego cheese proteolysis acceleration.

They co-encapsulated chymosin, the neutral protease and cardoon cyprosins with phospholipase C in liposomes. Phospholipase C was added to favor enzyme release.[134] The encapsulated enzymes were added to cheese milk with rennet. Chymosin and the neutral proteinases accelerated α_s-casein degradation in comparison with the control cheese, whereas β-casein degradation was accelerated by neutral proteinase and cyprosins. Neutral proteinase yielded the highest increase in soluble nitrogen. Cheese flavour intensity was enhanced by the neutral proteinase and cyprosins but not by chymosin.

In general, liposome encapsulation helped to overcome the limitations of the free addition of enzymes. However, more research regarding release behavior, solubility, and potential toxicity is needed.

6.4 Design of Starter and Adjunct Cultures

6.4.1 Selection Criteria

The selection of strains to accelerate cheese ripening and control flavour development is mainly based on their enzymatic profiles. In this sense, proteinases and peptidases are some of the enzymes most widely used as selection criteria.[135–137] Extracellular cell envelope proteinase (CEP) and the oligopeptide transport system mediate the first steps in casein hydrolysis by LAB. The protein is hydrolysed into large peptides by the CEP which are subsequently absorbed by the cells through specific peptide transport systems for further degradation into small peptides and AA by the coordinated action of intracellular peptidases. The peptidases show general or specific cleavage for some AA and different substrate preferences such as di- and tripeptides and oligopeptides.[138] The influence of microbial proteolysis on cheese flavour is mainly related with the production of AA that are the major precursors of a wide range of aromatic compounds[139] and the ability of some CEP and peptidases to produce or degrade bitter peptides, respectively.[138] More than twenty years ago, several studies reported the increase of soluble nitrogen and FAA levels in cheese by the use of mesophilic *Lactobacillus*.[140–144] For example, Milesi *et al.*[141] reported that *Lactiplantibacillus plantarum* I91, used as an adjunct culture in Cremoso and Cheddar cheeses, caused changes consistent with the acceleration of proteolysis, including modification of peptide profiles and the increase of the levels of the total and some individual AA. Similarly, Ciocia *et al.*[142] reported changes in the secondary proteolysis of Cheddar cheese due to the use of *Lactobacillus* strains as adjunct cultures. In addition, one of the strains (*Lactiplantibacillus plantarum* P1.5) improved the preference scores. Peralta *et al.*[143] reported that an adjunct culture of *Lacticaseibacillus paracasei* 90 improved the flavour and odour intensities in Cremoso cheese by increasing the levels of diacetyl and acetoin and decreasing the pungent taste and compounds associated with rancidity. On other hand, *Lb. helveticus* strains were also assayed to accelerate cheese ripening; the strain B02 led to an increase of different proteolytic indices and acceptability in terms of the overall aroma and flavour of Prato cheese.[144]

Intracellular enzymes that are involved in amino acid catabolism are also good selection criteria for adjunct cultures.[145] The catabolism of AA in cheese is commonly initiated by AT. These enzymes catalyse the transfer of amino groups from AA to α-keto acids, such as α-ketoglutarate, oxaloacetate, and pyruvate. The resulting α-keto acids can then be further degraded into volatile flavour compounds (alcohols, aldehydes, and carboxylic acids) by means of enzymatic reactions catalysed by different enzymes such as α-ketoacid decarboxylase and α-keto acid dehydrogenase. On the other hand, α-keto acids can be converted by hydroxy acid dehydrogenases into hydroxy acids, compounds that do not contribute to flavour.[145,146] Therefore, a high level of hydroxy acid dehydrogenase activity is a criterion used to rule out a strain. In general, the transamination is the highest with α-ketoglutarate functioning as the amino group acceptor. Thus, glutamate dehydrogenase (GDH), which catalyses the restoration of α-ketoglutarate levels from glutamate, also has a key role in amino acid catabolism. Identification, purification, and characterization studies of AT and GDH were performed on *Lactococcus lactis*,[147,148] *S. thermophilus*[149] and *Lactobacillus* strains, including *Lacticaseibacillus paracasei*, *Lacticaseibacillus casei*, *Lacticaseibacillus rhamnosus* and *Lactiplantibacillus plantarum*.[136,149–151]

Besides the selection based on the enzymes involved in proteolysis and AA catabolism, other enzymes are also interesting. For example, acetoin and diacetyl are important compounds in the aroma of many types of cheese.[152] Thus, LAB with the ability to metabolize citrate with the concomitant production of both compounds could be selected as potential adjunct cultures to enhance the cheese flavour.[153] However, it was also reported that the depletion of citrate does not necessarily correlate with an increase in diacetyl.[154,155]

The autolysis ability of starter and adjunct cultures has been also reported as a selection criterion for strains to accelerate the peptidolytic reactions during cheese ripening. In this context, numerous studies confirm the positive correlation between the autolytic ability of starter culture and the increase of proteolysis during cheese ripening due to the release of endopeptidases.[156,157]

On the other hand, the selection of cultures to be used in cheeses also includes evaluating the feasibility to be produced at an industrial scale in cost-effective culture media.[158,159] The main selection criteria for starter and adjunct cultures to be used in cheeses to accelerate ripening or improve flavour are summarized in Figure 6.2.

6.4.2 Lysis and Permeabilization Methods

Over the past several decades lysis and permeabilization methods have been proposed to improve the accessibility of microbial enzymes to their substrates in the cheese matrix, releasing intracellular enzymes or increasing membrane permeability.[160,161] Some of them were known as attenuation methods because they produced LAB unable to grow and to produce a significant concentration of lactic acid but still deliver active ripening enzymes

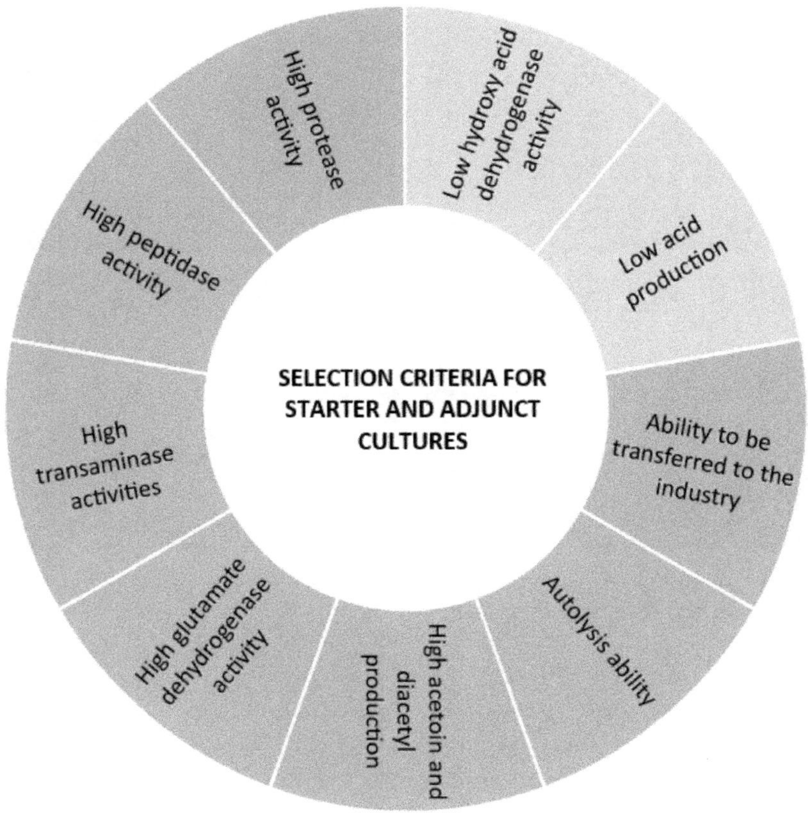

Figure 6.2 Main selection criteria for starter and adjunct culture development.

during cheese ripening.[160] The resistance of microorganisms to disruption treatments will depend on the type of microorganism, as well as on growth conditions including temperature, medium composition, and growth phase, among others.[161] The impact of a treatment on cellular envelopes (wall or membrane) also depends on the applied conditions: concentration of chemical and enzymatic treatments, intensity, and exposure time for physical and mechanical treatments. Peralta *et al.*[162] studied the effect of three attenuation methods on the cell architecture of *Streptococcus thermophilus* (Figure 6.3). It is clearly noted that ethanol treatment generated slight changes on the cell walls; meanwhile ultrasonic treatment greatly affected the cellular envelopes and mutanolysin caused wide cellular destruction.

The first attenuated starter used to accelerate proteolysis and reduce the ripening time was reported by Petterson and Sjöström[163] in semi-hard cheese. They applied a heat-shock of 59 °C and 69 °C for 15 s for mesophilic and thermophilic cultures, respectively. This study suggested for the first time that heat-shock could be a potential method for attenuated starter production. Years later, both heat-shocked and freeze-shocked cultures were used to accelerate the ripening of Gouda cheese,[164,165] Swedish hard cheese,[166] Kareish

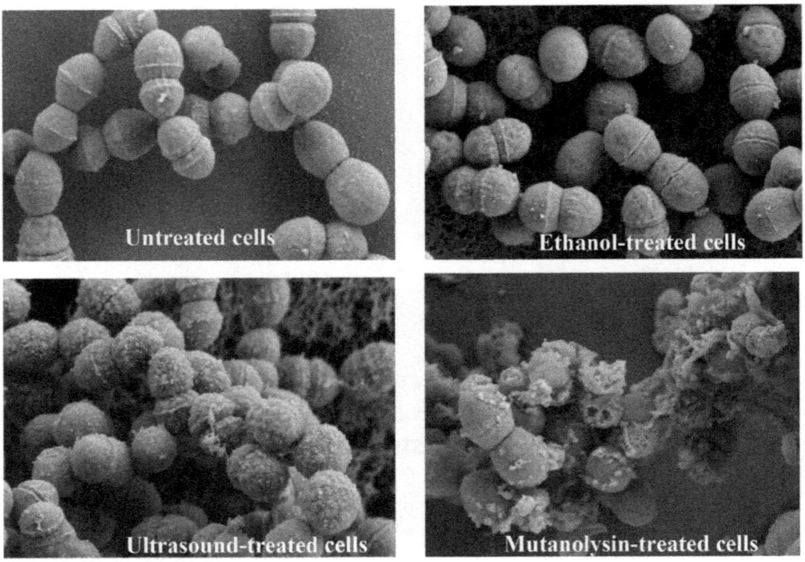

Figure 6.3 Scanning electron micrographs of *Streptococcus thermophilus* treated with three attenuation methods.

cheese,[167] Ras cheese,[168,169] and Cheddar cheese.[170] In this way, Abdek Baky[171] reported that cheeses manufactured with heat-shocked lactobacilli showed desirable flavour and consistency 1–2 months earlier than cheese made without attenuated culture. Asensio *et al.*[172] showed that early peptidolysis was induced in low-fat goat's milk cheese by addition of heat-shocked mesophilic lactobacilli, which enhanced the flavour and reduced the bitterness. In addition, heat-shocked cultures of *Leuconostoc* spp. improved the organoleptic features of Ras cheese, while freeze-shocked cultures did not show any effect.[173] In the same way, Peralta *et al.*[162] reported that both freeze-shock (−80 °C for 48 h) and heat-shock (63 °C for 20 s) were not suitable for the release of the intracellular enzymes (lactate dehydrogenase and β-galactosidase) of two strains of *Streptococcus thermophilus*. More recently, Garbowska *et al.*[174] reported that the heat-treated adjunct XT-312® culture (*L. lactis* ssp. *cremoris*, *Leuc. mesenteroides* ssp. *cremoris*, *L. lactis* ssp. *lactis biovar diacetylactis*, *L. lactis* ssp. *lactis*, *Leuc. pseudomesenteroides*) and *Lb. helveticus* LH-B01 increased primary and secondary proteolysis in a Dutch-type cheese model, which indicated cheese ripening acceleration.

Ultrasound treatment using an ultrasonic bath or an ultrasonic probe has been used as a strategy to enhance the metabolic potential of LAB in milk[175] and in cheeses.[162,176,177] The ultrasound waves produce a phenomenon of acoustic cavitation that is responsible for injuring the bacterial envelope, which results in the weakening or disruption of bacterial cells. Di Cagno *et al.*[178] showed that attenuated adjunct cultures prepared using an ultrasonic probe sonicator were suitable for accelerating the ripening of

Caciocavallo Pugliese cheese. Calasso *et al.*[179] accelerated the ripening of Pecorino-type cheese by the use of freeze-dried cell-free extracts from *Lactobacillus casei* LC01, *Weissella cibaria* 1XF5, *Hafnia alvei Moller* ATCC 51815, and *Debaryomyces hansenii* LCF-558 obtained by sonication treatment. In the same way, the peptidase activity in semi-soft Caciotta cheese was increased by the use of starter culture treated with ultrasonic waves.[176] More recently, the metabolic potential of autochthonous mesophilic lactobacilli was enhanced by ultrasonic waves.[175]

HP and HPH have also been assayed for bacterial damage. Casal and Gómez[180] reported that HP treatment increased the hydrolytic activity of lactococci and lactobacilli on the carboxyl-terminal fragment from β-casein in *in vitro* assays. Similarly, Upadhyay *et al.*[181] reported that the addition of HP-treated bacteria accelerated secondary proteolysis in Cheddar cheese.

As for HPH, cavitation and shearing forces are responsible for bacterial damage. A large number of studies have shown the effect of HPH processing on the cell envelope, affecting viability and enzyme release.[182,183] In addition, Burns *et al.*[184] and Peralta *et al.*[185] also reported changes in the proteolysis of Caciotta and Pategrás cheeses, respectively, by the incorporation of adjunct cultures treated by HPH. In the same way, the potential of microfluidisation to improve the enzymatic and metabolic performance of adjuncts in the flavour development of Gouda type cheese was also reported.[186]

Finally, chemical treatments with alcohols (ethanol, butanol, isopropanol) and detergents (sodium dodecyl sulphate (SDS), hexadecyltrimethylammonium bromide) have been reported as efficient methods for cell permeabilization.[95,162,187] In fact, the SDS application directly on the starter to obtain disrupted cultures to accelerate cheese ripening was patented by Smith *et al.*[188] In the same way, Peralta *et al.*[162] reported that *S. thermophilus* treated with SDS and used as the source of enzymes in Cremoso cheese led to a higher proteolysis compared to cheeses without attenuated culture. On the other hand, the use of lytic enzymes has also been proposed as a method to shorten the ripening time. Mutanolysin and lysozyme are the most studied enzymes.[160,162]

Abbreviations

AA	Amino acids
AT	Aminotransferases
CEP	Extracellular cell envelope proteinase
EMC	Enzyme-modified cheese slurries
FAA	Free amino acids
FFA	Free fatty acids
GDH	Glutamate dehydrogenase
HHP	High hydrostatic pressure
HPH	High-pressure homogenization
HPP	High-pressure processing
LAB	Lactic acid bacteria

LPL	Lipoprotein lipase
MFGM	Milk fat globule membrane
NSLAB	Non-starter lactic acid bacteria
PCA	Principal component analysis
RSM	Response surface methodology
SDS	Sodium dodecyl sulphate
TAG	Triacyclglycerols
UHPH	Ultra-high-pressure homogenization

References

1. A. R. Khattab, H. A. Guirguis, S. M. Tawfik and M. A. Farag, *Trends Food Sci. Technol.*, 2019, **88**, 343.
2. P. F. Fox and P. L. H. McSweeney, in *Cheese: Chemistry, Physics and Microbiology*, ed. P. F. Fox, P. L. H. McSweeney, T. M. Cogan and T. P. Guinee, Elsevier, Amsterdam, 2017, vol. 1, p. 1.
3. K. Kilcawley and M. O'Sullivan, in *Global Cheesemaking Technology: Cheese Quality and Characteristics*, ed. P. Papademas and T. Bintsis, Wiley, Hoboken, NJ, 2017, p. 45.
4. A. A. Hayaloglu and P. L. H. McSweeney, Primary biochemical events during cheese ripening, in *Dairy Microbiology and Biochemistry*, ed. B. Ozer and G. Akdemir-Evrendilek, CRC Press, Boca Raton, FL, 2014, p. 134.
5. M. M. Milesi, I. V. Wolf, C. V. Bergamini and E. R. Hynes, *J. Dairy Sci.*, 2010, **93**, 5020.
6. M. A. Vélez, C. V. Bergamini, M. B. Ramonda, M. C. Candioti, E. R. Hynes and M. C. Perotti, *LWT–Food Sci. Technol.*, 2015, **64**, 282.
7. M. Yvon, *Aust. J. Dairy Technol.*, 2006, **61**, 88.
8. Y. F. Collins, P. L. H. McSweeney and M. G. Wilkinson, *Int. Dairy J.*, 2003, **13**, 841.
9. R. Holland, S.-Q. Liu, V. L. Crow, M.-L. Delabre, M. Lubbers, M. Bennett and G. Norris, *Int. Dairy J.*, 2005, **15**, 711.
10. P. L. H. McSweeney and M. J. Sousa, *Lait*, 2000, **80**, 293.
11. P. F. Fox, J. M. Wallace, S. Morgan, C. M. Lynch, E. J. Niland and J. Tobin, *Antonie van Leeuwenhoek*, 1996, **70**, 271.
12. A. M. Alhelli, N. K. Mohammed, E. S. Khalil and A. S. M. Hussin, *AMB Express*, 2021, **11**, 45.
13. M. El Soda, *Int. Dairy J.*, 1993, **3**, 531.
14. S. Azarnia, N. Robert and B. H. Lee, *Crit. Rev. Biotechnol.*, 2006, **26**, 121.
15. B. A. Law, *Int. Dairy J.*, 2001, **11**, 383.
16. K. N. Kilcawley, A. B. Nongonierma, J. A. Hannon, I. A. Doolan and M. G. Wilkinson, *Int. Dairy J.*, 2012, **26**, 50.
17. M. El Soda and S. Awad, in *Encyclopedia of Dairy Sciences*, ed. P. L. H. McSweeney and J. P. McNamara, Elsevier, Amsterdam, 2022, vol. 3, p. 1.

18. M. El Soda, S. Awad, in *Encyclopedia of Dairy Sciences*, ed. J. W. Fuquay, P. F. Fox and P. L. H. McSweeney, Elsevier Academic Press, London, 2011, vol. 1, p. 795.
19. P. L. H. McSweeney, P. F. Fox, P. Clotter and D. Everett, *Cheese Chemistry, Physics & Microbiology*, Elsevier, London, 2017.
20. P. Papademas and T. Bintsis, *Global Cheesemaking Technology*, Wiley & Sons, Ltd., Hoboken, NJ, 2018.
21. B. A. Law, Z. D. Hosking and H. R. Chapman, *Int. J. Dairy Technol.*, 1979, **32**, 87.
22. J. W. Aston, I. A. Fedrick, I. G. Durward and J. R. Dulley, *N. Z. J. Dairy Sci. Technol.*, 1983, **18**, 143.
23. J. W. Aston, P. A. Grieve, I. G. Durward and J. R. Dulley, *Aust. J. Dairy Technol.*, 1983, 59.
24. J. W. Aston, J. E. Giles, I. G. Durward and J. R. Dulley, *J. Dairy Res.*, 1985, **52**, 595.
25. B. Folkertsma, P. F. Fox and P. L. H. McSweeney, *Int. Dairy J.*, 1996, **6**, 1117.
26. S.-U. Rehman, J. M. Banks, E. Y. Brechany, D. D. Muir, P. L. H. McSweeney and P. F. Fox, *Int. Dairy J.*, 2000, **10**, 55.
27. J. A. Hannon, M. G. Wilkinson, C. M. Delahunty, J. M. Wallace, P. A. Morrissey and T. P. Beresford, *Int. Dairy J.*, 2005, **15**, 263.
28. L. Ong and N. P. Shah, *LWT-Food Sci. Technol.*, 2009, **42**, 1260.
29. M. Batool, M. Nadeem, M. Imran, I. T. Khan, J. A. Bhatti and M. Ayaz, *Lipids Health Dis.*, 2018, **17**, 2.
30. F. de Filippis, A. Genovese, P. Ferranti, J. A. Gilbert and D. Ercolini, *Sci. Rep.*, 2016, **6**, 21871.
31. K. Soodam, L. Ong, I. B. Powell, S. E. Kentish and S. L. Gras, *Food Struct.*, 2017, **14**, 8.
32. Y.-E. Chou, C. G. Edwards, L. O. Luedecke, M. P. Bates and S. Clark, *J. Dairy Sci.*, 2003, **86**, 2516.
33. E. A. Walsh, C. Diako, D. M. Smith and C. F. Ross, *J. Food Sci.*, 2020, **85**, 268.
34. J. J. Sheehan, M. G. Wilkinson and P. L. H. McSweeney, *Int. Dairy J.*, 2008, **18**, 905.
35. P. Gaya, M. Medina, M. A. Rodriguez-Marin and M. Nuñez, *J. Dairy Sci.*, 1990, **73**, 26.
36. V. Pachlová, F. Buňka, R. Flasarová, P. Válková and L. Buňková, *Food Chem.*, 2012, **132**, 1846.
37. Z. Zhao, W. Yang, Y. Xu, D.-J. Zhang and Y.-X. Zhao, *Mod. Food Sci. Technol.*, 2013, **29**, 2586.
38. R. E. Ferrazza, J. M. Fresno, J. I. Ribeiro, M. E. Tornadijo and M. M. Furtado, *Food Res. Int.*, 2004, **37**, 149.
39. G. A. Sihufe, S. E. Zorrilla, N. G. Sabbag, S. C. Costa and A. C. Rubiolo, *J. Sens. Stud.*, 2010, **25**, 94.
40. G. A. Sihufe, S. E. Zorrilla, D. J. Mercanti, M. C. Perotti, C. A. Zalazar and A. C. Rubiolo, *Food Res. Int.*, 2007, **40**, 1220.

41. G. A. Sihufe, S. E. Zorrilla, M. C. Perotti, I. V. Wolf, C. A. Zalazar, N. G. Sabbag, S. C. Costa and A. C. Rubiolo, *Food Chem.*, 2010, **119**, 101.
42. G. A. Sihufe, S. E. Zorrilla and A. C. Rubiolo, *LWT–Food Sci. Technol.*, 2010, **43**, 247.
43. R. J. Ceruti, S. E. Zorrilla and G. A. Sihufe, *Food Res. Int.*, 2012, **48**, 34.
44. R. J. Ceruti, S. E. Zorrilla and G. A. Sihufe, *Eur. Food Res. Technol.*, 2016, **242**, 1369.
45. Y. F. Collins, P. L. H. McSweeney and M. G. Wilkinson, *Int. Dairy J.*, 2003, **13**, 841.
46. M. C. Candioti, E. Hynes, A. Quiberoni, S. B. Palma, N. Sabbag and C. A. Zalazar, *Int. Dairy J.*, 2002, **12**, 923.
47. J. M. Evers, *Int. Dairy J.*, 2004, **14**, 661.
48. A. L. Kelly, T. Huppertz and J. J. Sheehan, *Dairy Sci. Technol.*, 2008, **88**, 549.
49. M. G. Hayes and A. L. Kelly, *J. Dairy Res.*, 2003, **70**, 297.
50. M.-C. Michalski, F. Michel and C. Geneste, *Lait*, 2002, **82**, 193.
51. H. C. Deeth, *Int. Dairy J.*, 2006, **16**, 555.
52. H. D. Goff, in *Handbook of Farm, Dairy and Food Machinery Engineering*, ed. M. Kutz, Academic Press, Delmar, NY, 2013, p. 199.
53. B. S. Oommen, V. V. Mistry and M. G. Nair, *Lait*, 2000, **80**, 77.
54. V. V. Mistry, *Int. Dairy J.*, 2001, **11**, 413.
55. E. W. Robson and D. G. Dalgleish, *J. Dairy Res.*, 1984, **51**, 417.
56. M. L. Green, R. J. Marshall and F. A. Glover, *J. Dairy Res.*, 1983, **50**, 341.
57. K. Muthukumarappan and G. J. Swamy, in *Advances in Food Rheology and its applications*, ed. J. Ahmed, Elsevier, Amsterdam, 2017, p. 245.
58. A. Jana and K. G. Upadhyay, *Aust. J. Dairy Technol.*, 1992, **42**, 72.
59. M. E. Johnson, in *Encyclopedia of Dairy Sciences*, ed. J. W. Fuquay, P. F. Fox and P. L. H. McSweeney, Elsevier Academic Press, London, 2011, vol. 1, p. 544.
60. K. C. Deegan and P. L. H. McSweeney, *Dairy Sci. Technol.*, 2013, **93**, 641.
61. M. Cao, L. M. Fonseca, T. C. Schoenfuss and S. A. Rankin, *J. Agric. Food Chem.*, 2014, **62**, 5726.
62. M. A. Vélez, E. R. Hynes, C. A. Meinardi, V. I. Wolf and M. C. Perotti, *Food Res. Int.*, 2017, **96**, 215.
63. M. A. Vélez, E. R. Hynes, G. Rodriguez, L. Garitta, I. V. Wolf and M. C. Perotti, *LWT–Food Sci. Technol.*, 2019, **112**, 108241.
64. M.-P. B. Vigneux, W. Villeneuve, Y. Pouliot and M. Britten, *Int. Dairy J.*, 2022, **126**, 105254.
65. J. A. Lucey, M. E. Johnson and D. S. Horne, *J. Dairy Sci.*, 2003, **86**, 2725.
66. J. Mayta-Hancco, A.-J. Trujillo and B. Juan, *Rev. de Investig. Vet. del Perú*, 2020, **31**, e17934.
67. E. E. Kheadr, J. F. Vachon, P. Paquin and I. Fliss, *Int. Dairy J.*, 2002, **12**, 435.
68. M. E. Guerzoni, L. Vannini, C. Chaves Lopez, R. Lanciotti, G. Suzzi and A. Gianotti, *J. Dairy Sci.*, 1999, **82**, 851.

69. R. Lanciotti, C. Chaves-López, F. Patrignani, A. Paparella, M. E. Guerzoni, A. Serio and G. Suzzi, *Int. J. Dairy Technol.*, 2004, **57**, 19.
70. L. Vannini, F. Patrignani, L. Iucci, M. Ndagijimana, M. Vallicelli, R. Lanciotti and M. E. Guerzoni, *Int. J. Food Microbiol.*, 2008, **128**, 329.
71. N. Datta, M. G. Hayes, H. C. Deeth and A. L. Kelly, *J. Dairy Res.*, 2005, **72**, 393.
72. B. Juan, J. M. Quevedo, A. Zamora, B. Guamis and A.-J. Trujillo, *LWT–Food Sci. Technol.*, 2015, **60**, 1034.
73. B. Juan, A. Zamora, J. M. Quevedo and A.-J. Trujillo, *LWT–Food Sci. Technol.*, 2016, **69**, 17.
74. Q. Xia, Y. Zheng, Z. Liu, J. Cao, X. Chen, L. Liu, H. Yu, F. J. Barba and D. Pan, *Trends Food Sci. Technol.*, 2020, **106**, 365.
75. M. F. Patterson, M. Linton and C. J. Doona, in *High Pressure Processing of Foods*, ed. C. J. Donna and F. E. Feeherry, IFT Press, Blackwell Publishing, Ames, IA, 2007, p. 1.
76. T. Huppertz, A. L. Kelly and P. F. Fox, *Int. Dairy J.*, 2002, **12**, 571.
77. M. Nuñez, J. Calzada and A. D. Olmo, *Int. Dairy J.*, 2020, **100**, 104558.
78. C. Wick, U. Nienaber, O. Anggraeni, T. H. Shellhammer and P. D. Courtney, *J. Dairy Res.*, 2004, **71**, 107.
79. C. E. O'Reilly, P. M. O'Connor, P. M. Murphy, A. L. Kelly and T. P. Beresford, *Int. Dairy J.*, 2002, **12**, 915.
80. C. E. O'Reilly, A. L. Kelly, J. C. Oliveira, P. M. Murphy, M. A. E. Auty and T. P. Beresford, *Innovative Food Sci. Emerging Technol.*, 2003, **4**, 277.
81. N. M. Rynne, T. P. Beresford, T. P. Guinee, E. Sheehan, C. M. Delahunty and A. L. Kelly, *Food Sci. Emerging Technol.*, 2008, **9**, 429.
82. C. E. O'Reilly, A. L. Kelly, P. M. Murphy and T. P. Beresford, *Trends Food Sci. Technol.*, 2001, **12**, 51.
83. U. Nienaber, T. H. Shellhammer, W. J. Harper and P. D. Courtney, *J. Dairy Sci.*, 2000, **83**(Suppl. 1), 95.
84. C. E. O'Reilly, P. M. O'Connor, P. M. Murphy, A. L. Kelly and T. P. Beresford, *Innovative Food Sci. Emerging Technol.*, 2000, **1**, 109.
85. M. Capellas, M. Mor-Mur, E. Sendra and B. Guamis, *Int. Dairy J.*, 2001, **11**, 165.
86. W. Messens, K. Dewettinck, J. Van Camp and A. Huyghebaert, *Lebensm.-Wiss. Technol.*, 1998, **31**, 552.
87. W. Messens, I. Foubert, K. Dewettinck and A. Huyghebaert, *Milchwissenschaft*, 2000, **55**, 328.
88. J. Saldo, E. Sendra and B. Guamis, *J. Food Sci.*, 2000, **65**, 636.
89. J. Saldo, E. Sendra and B. Guamis, *Lait*, 2001, **81**, 625.
90. J. Saldo, P. L. H. McSweeney, E. Sendra, A. L. Kelly and B. Guamis, *Int. Dairy J.*, 2002, **12**, 35.
91. L. M. Costabel, C. Bergamini, S. R. Vaudagna, A. L. Cuatrin, G. Audero and E. Hynes, *J. Dairy Sci.*, 2016, **99**, 4220.
92. H. Yokohama, *Eur. Pat.*, EP469857-A, 1992.
93. J. L. Arqués, S. Garde, E. Fernández-García, P. Gaya and M. Nuñez, *J. Dairy Sci.*, 2007, **90**, 3627.

94. F. J. Delgado-Martínez, A. I. Carrapiso, R. Contador and M. R. Ramírez, *Innovative Food Sci. Emerging Technol.*, 2019, **52**, 34.
95. I. A. Doolan and M. G. Wilkinson, *Int. Dairy J.*, 2009, **19**, 215.
96. U. T. Bornscheuer, *FEMS Microbiol. Rev.*, 2002, **26**, 73–81.
97. F. Hasan, A. A. Shah and A. Hameed, *Enzyme Microb. Technol.*, 2006, **39**, 235.
98. A. Trani, P. Loizzo, A. Cassone and M. Faccia, in *Advances in Dairy Products*, John Wiley & Sons Ltd, Chichester, UK, 2017, p. 163.
99. S. Rani and S. Jagtap, *J. Food Sci. Technol.*, 2019, **56**, 497.
100. M. Karami, M. R. Ehsani, S. M. Mousavi, K. Rezaei and M. Safari, *Food Chem.*, 2009, **113**, 424.
101. O. B. Karaca and M. Güven, *Foods*, 2018, **7**, 0125.
102. G. Yilmaz, A. Ayar and N. Akin, *J. Food Eng.*, 2005, **69**, 269–274.
103. I. Hernández, L. J. R. Barrón, M. Virto, F. J. Pérez-Elortondo, C. Flanagan, U. Rozas, A. I. Nájera, M. Albisu, M. S. Vicente and M. de Renobales, *Food Chem.*, 2009, **116**, 158.
104. I. Hernández, M. de Renobales, M. Virto, F. J. Pérez-Elortondo, L. J. R. Barron, C. Flanagan and M. Albisu, *Enzyme Microb. Technol.*, 2005, **36**, 870.
105. E. Fernandez-Garcia, R. Lopez-Fandiño, L. Alonso and M. Ramos, *J. Dairy Sci.*, 1994, **77**, 2139.
106. B. A. Law and A. S. Wigmore, *Int. J. Dairy Technol.*, 1985, **38**, 86.
107. V. K. Sood and F. V. Kosikowski, *J. Dairy Sci.*, 1979, **62**, 1865.
108. M. V. Arbige, P. R. Freund, S. C. Silver and J. T. Zelko, *Food Technol.*, 1986, **40**, 91.
109. S. K. Kanawjia, P. Rajesh, L. Sabikhi and S. Singh, *LWT–Food Sci. Technol.*, 1995, **28**, 577.
110. V. K. Upadhyay and P. L. H. McSweeney, in *Dairy Processing. Improving Quality*, ed. G. Smit, Woodhead Publishing Ltd., Cambridge, UK, 2003, p. 417.
111. K. N. Kilcawley, A. B. Nongonierma, J. A. Hannon, I. A. Doolan and M. G. Wilkinson, *Int. Dairy J.*, 2012, **26**, 50.
112. E. Fernandez-Garcia, A. Olano, D. Cabezudo, P. J. Martin-Alvarez and M. Ramos, *Enzyme Microb. Technol.*, 1993, **15**, 519.
113. A. F. Mohedano, J. Fernández, S. Garde, M. Medina, P. Gaya and M. Nuñez, *Enzyme Microb. Technol.*, 1998, **22**, 391.
114. B. A. Law and A. Wigmore, *J. Dairy Res.*, 1982, **49**, 137.
115. J. M Izco, A. Irigoyen, P. Torre and Y. Barciná, *J. Chromatogr.*, 2000, **881**, 69.
116. R. J. Ceruti, M. B. Pirola, E. Ramos, L. Robert, A. C. Rubiolo and G. A. Sihufe, *Czech J. Food Sci.*, 2016, **34**, 445.
117. S. Azarnia, B. Lee, D. St-Gelais, K. Kilcawley and E. Noroozi, *LWT–Food Sci. Technol.*, 2011, **44**, 570.
118. F. M. Barrett, A. L. Kelly, P. L. H. Mcsweeney and P. F. Fox, *Int. Dairy J.*, 1999, **9**, 421.
119. E. D. Bastian, C. G. Lo and K. M. M. David, *J. Dairy Sci.*, 1997, **80**, 245.

120. E. D. Bastian, K. G. Hansen and R. J. Brown, *J. Dairy Sci.*, 1991, **74**, 3669–3676.
121. J. R. Dulley, *Aust. J. Dairy Technol.*, 1976, **31**, 143.
122. S. M. Hamdy, H. S. Mahmoud and K. A. H. Abbas, *Int. J. Dairy Sci.*, 2017, **12**, 318.
123. J. A. Hannon, K. N. Kilcawley, M. G. Wilkinson, C. M. Delahunty and T. P. Beresford, *J. Dairy Sci.*, 2006, **89**, 3749.
124. M. Ayelén Vélez, M. Cristina Perotti, L. Santiago, A. María Gennaro and E. Hynes, in *Nutrient Delivery*, Elsevier, Amsterdam, 2017, p. 221.
125. P. Dufour, J. C. Vuillemard, E. Laloy and R. E. Simard, *J. Microencapsulation*, 1996, **13**, 185.
126. E. E. Kheadr, J.-C. Vuillemard and S. A. El-Deeb, *J. Food Sci.*, 2002, **67**, 485.
127. M. el Soda and S. Pandian, *J. Dairy Sci.*, 1991, **74**, 2317.
128. E. E. Kheadr, J. C. Vuillemard and S. A. El-Deeb, *Food Res. Int.*, 2003, **36**, 241.
129. R. Mohammadi, M. Mahmoudzade, M. Atefi, K. Khosravi-Darani and M. R. Mozafari, *Int. J. Dairy Technol.*, 2015, **68**, 11.
130. M. Jahadi, K. Khosravi-Darani, M. R. Ehsani, M. R. Mozafari, A. A. Saboury, A. Zoghi and M. Mohammadi, *Int. J. Dairy Technol.*, 2016, **69**, 57.
131. M. Jahadi, K. Khosravi-Darani, M. R. Ehsani, T. C. Pimentel, A. G. da Cruz and M. R. Mozafari, *Biointerface Res. Appl. Chem.*, 2020, **10**, 4966.
132. A. Picon, P. Gaya, M. Medina and M. Nuñez, *J. Dairy Sci.*, 1995, **78**, 1238.
133. A. Picón, P. Gaya, M. Medina and M. Nuñez, *Biotechnol. Lett.*, 1997, **19**, 345.
134. A. Picón, P. Gaya, M. Medina and M. Nuñez, *Biotechnol. Lett.*, 1995, **17**, 1051.
135. Y. Ardo and L. Jonsson, *J. Dairy Res.*, 1994, **61**, 573.
136. E. Stefanovic, K. N. Kilcawley, M. C. Rea, G. F. Fitzgerald and O. McAuliffe, *J. Appl. Microbiol.*, 2017, **122**, 1245.
137. L. Solieri, L. Sola, A. Vaccalluzzo, C. L. Randazzo, S. Martini and D. Tagliazucchi, *Biology*, 2022, **11**, 139.
138. K. Savijoki, H. Ingmer and P. Varmanen, *Appl. Microbiol. Biotechnol.*, 2006, **71**, 394.
139. G. Smit, B. A. Smit and W. J. M. Engels, *FEMS Microbiol. Rev.*, 2005, **29**, 591.
140. E. R. Hynes, J.-C. Ogier and A. Delacroix-Buchet, *Int. Dairy J.*, 2001, **11**, 581.
141. M. M. Milesi, P. L. H. McSweeney and E. R. Hynes, *J. Appl. Microbiol.*, 2008, **105**, 884.
142. F. Ciocia, P. L. H. McSweeney, P. Piraino and E. Parente, *Dairy Sci. Technol.*, 2013, **93**, 623.
143. G. H. Peralta, C. V. Bergamini, L. Costabel, G. Audero, E. C. Ale, A. G. Binetti, I. V. Wolf, A. Cuatrín, I. Nieto, M. C. Perotti and E. R. Hynes, *Int. Dairy J.*, 2020, **109**, 104779.

144. N. C. de Azambuja, I. Moreno, D. A. Gallina, L. M. Spadoti, E. M. P. Motta, M. T. B. Pacheco, A. L. M. de Queiroz and A. E. C. Antunes, *Food Nutr. Sci.*, 2017, **08**, 512.
145. R. van Kranenburg, M. Kleerebezem, J. van Hylckama Vlieg, B. Orn, M. Ursing, J. Boekhorst, B. A. Smit, E. H. E. Ayad, G. Smit and R. J. Siezen, *Int. Dairy J.*, 2002, **12**, 111.
146. M. Yvon and L. Rijnen, *Int. Dairy J.*, 2001, **11**, 185.
147. M. Yvon, E. Chambellon, A. Bolotin and F. Roudot-Algaron, *Appl. Environ. Microbiol.*, 2000, **66**, 571.
148. L. Rijnen, S. Bonneau and M. Yvon, *Appl. Environ. Microbiol.*, 1999, **65**, 4873.
149. G. H. Peralta, C. V. Bergamini and E. R. Hynes, *Braz. J. Microbiol.*, 2016, **47**, 741.
150. E. Stefanovic, K. N. Kilcawley, C. Roces, M. C. Rea, M. O'Sullivan, J. J. Sheehan and O. McAuliffe, *Front. Microbiol.*, 2018, **9**, 1506.
151. E. Stefanovic, A. Thierry, M. B. Maillard, A. Bertuzzi, M. C. Rea, G. Fitzgerald, O. McAuliffe and K. N. Kilcawley, *J. Dairy Sci.*, 2017, **100**, 6918.
152. N. García Quintans, V. Blancato, G. Repizo, C. Magni and P. López, in *Molecular Aspects of Lactic Acid Bacteria for Traditional and New Applications*, ed. B. Mayo, P. López and G. Pérez-Martínez, Research Signpost, Kerala, India, 2008, p. 1.
153. G. Pérez, E. Cardell and V. Zá, *Int. J. Food Sci.*, 2003, **38**, 537.
154. N. A. Kocaoglu-Vurma, W. J. Harper, M. A. Drake and P. D. Courtney, *J. Dairy Sci.*, 2008, **91**, 2947.
155. S. Skeie, A. Kieronczyk, S. Eidet, M. Reitan, K. Olsen and H. Østlie, *Int. Dairy J.*, 2008, **18**, 169.
156. J. A. Hannon, M. G. Wilkinson, C. M. Delahunty, J. M. Wallace, P. A. Morrissey and T. P. Beresford, *Int. Dairy J.*, 2003, **13**, 313.
157. M. Bourdat-Deschamps, D. le Bars, M. Yvon and M. P. Chapot-Chartier, *Int. Dairy J.*, 2004, **14**, 791.
158. M. V. Beret, G. H. Peralta, L. Vera-Candioti, I. V. Wolf, R. Sánchez, E. R. Hynes and C. V. Bergamini, *Antonie van Leeuwenhoek*, 2021, **114**, 2047.
159. S. A. Hayek, R. Gyawali, S. O. Aljaloud, A. Krastanov and S. A. Ibrahim, *J. Dairy Res.*, 2019, **86**, 490.
160. N. Klein, S. Lortal and N. Klein, *Int. Dairy J.*, 1999, **9**, 751.
161. J. Geciova, D. Bury and P. Jelen, *Int. Dairy J.*, 2002, **12**, 541.
162. G. H. Peralta, C. v. Bergamini and E. R. Hynes, *Int. Dairy J.*, 2019, **92**, 11.
163. H.-E. Pettersson and G. Sjöström, *J. Dairy Res.*, 1975, **42**, 313.
164. H. J. Bartels, M. E. Johnson and N. F. dc Olson, *Milchwissenschaft*, 1987, **42**, 83.
165. H. Bartels, M. Johnson and N. Olson, *Milchwissenschaft*, 1987, **42**, 139.
166. B. Y. Ardo and H. Pettersson, *J. Dairy Res.*, 1988, **55**, 239.
167. K. M. K. Kebary, M. Salem, A. I. Hamed and A. S. El-Sisi, *Food Res. Int.*, 1991, **30**, 265.

168. N. Ezzat and H. El-Shafei, *Egypt. J. Dairy Sci.*, 1991, **19**, 347.
169. M. A. El Soda, A. A. Hantira, N. I. Ezzat and H. K. Ei-Shafei, *Food Chem.*, 1992, **44**, 179.
170. J. A. C. Johnson, M. R. Etzel, C. M. Chen and M. E. Johnson, *J. Dairy Sci.*, 1995, **78**, 769.
171. A. A. A. Baky, A. A. El-Neshawy, A. M. Rabie and M. M. Ashour, *Food Chem.*, 1986, **21**, 301.
172. C. Asensio, L. Parra, C. Peláez and R. Gómez, *Agric. Food Chem.*, 1996, **44**, 2919.
173. H. El-Shafei, *Nahrung*, 1994, **38**, 599.
174. M. Garbowska, A. Berthold-Pluta, L. Stasiak-Różańska and A. Pluta, *Animals*, 2021, **11**, 2699.
175. G. H. Peralta, M. D. Milagros Bürgi, L. J. Martínez, V. H. Albarracín, I. V. Wolf, A. A. Perez, L. Santiago, E. R. Hynes and C. V. Bergamini, *Int. Dairy J.*, 2022, 105371.
176. M. Calasso, F. Minervini, F. de Filippis, D. Ercolini, M. de Angelis and M. Gobbetti, *Appl. Environ. Microbiol.*, 2020, **99**, 571.
177. R. di Cagno, I. de Pasquale, M. de Angelis, S. Buchin, M. Calasso, P. F. Fox and M. Gobbetti, *Int. Dairy J.*, 2011, **21**, 254.
178. R. di Cagno, I. de Pasquale, M. de Angelis and M. Gobbetti, *J. Dairy Sci.*, 2012, **95**, 4784.
179. M. Calasso, L. Mancini, R. di Cagno, G. Cardinali and M. Gobbetti, *J. Dairy Sci.*, 2015, **98**, 5874.
180. V. Casal and R. Gómez, *J. Dairy Sci.*, 1999, **82**, 1092.
181. V. K. Upadhyay, T. Huppertz, A. L. Kelly and P. L. H. McSweeney, *Innovative Food Sci. Emerging Technol.*, 2007, **8**, 485.
182. D. Bury, P. Jelen and M. Kalab, *Innovative Food Sci. Emerging Technol.*, 2001, **2**, 23.
183. R. Lanciotti, F. Patrignani, L. Iucci, P. Saracino and M. E. Guerzoni, *Food Chem.*, 2007, **102**, 542.
184. P. G. Burns, F. Patrignani, G. Tabanelli, G. C. Vinderola, L. Siroli, J. A. Reinheimer, F. Gardini and R. Lanciotti, *J. Funct. Foods*, 2015, **13**, 126.
185. G. H. Peralta, A. E. Aguirre, H. G. Menzella, M. D. M. Bürgi, L. J. Martínez, V. H. Albarracín, E. R. Hynes and C. V. Bergamini, *6° Simposio Argentino de Procesos Biotecnológicos*, Posadas, Universidad Nacional de Misiones, Garupá, Argentina, 2021.
186. A. B. Yarlagadda, M. G. Wilkinson, M. G. O'Sullivan and K. N. Kilcawley, *Int. Dairy J.*, 2014, **38**, 124.
187. G. A. Somkuti, M. E. Dominiecki and D. H. Steinberg, *Curr. Microbiol.*, 1998, **36**, 202.
188. C. D. Galer, J. Moran, B. E. Dias, M. Doyle, O. Herrera, S. L. Anderson and M. A. McIlroy, *US Pat.*, US-6649200-B2, 2003.

CHAPTER 7

Flavour Development in Cheese

FREDDY MAURICIO LEMUS MUÑOZ AND MICHAEL QIAN*

Department of Food Science and Technology, Oregon State University, Corvallis, OR 97331, USA
*Email: michael.qian@oregonstate.edu

7.1 Flavour Formation in Cheese

Flavour compounds in cheese arise from biochemical reactions occurring during the ripening stage. These reactions are mainly the degradation of proteins (caseins), lipids, lactose, and citrate in milk and subsequent catabolic reactions. They are grouped into

- Glycolysis: metabolism of lactose and citrate
- Lipolysis: liberation of free fatty acids (FFA) from triacylglycerols and subsequent metabolism to volatile compounds
- Proteolysis: degradation of the casein matrix into peptides and ultimately free amino acids, followed by the catabolism of these amino acids to produce flavour compounds, such as carboxylic acids and sulphur compounds

7.1.1 Glycolysis

Although lactose metabolism does not contribute directly to cheese flavour, it is significant in determining the texture of the cheese and therefore the rate of liberation of sapid compounds.

During the syneresis stage most of the lactose is lost in the whey that is drained. However, 0.8–1.5% of the total lactose remains in the cheese curd

Food Chemistry, Function and Analysis No. 40
Handbook of Cheese Chemistry
Edited by Michael H. Tunick

Published by the Royal Society of Chemistry, www.rsc.org

and acts as a substrate for the starter lactic acid bacteria (LAB).[1] In the case of Cheddar cheese, lactose is mostly fermented to lactic acid before salting and moulding[2] because LAB cultures are stimulated by low levels of NaCl and are strongly inhibited over 2.5% NaCl.[3] Salting controls the metabolism of lactose by decreasing the starter activity and lactic acid production.

When LAB activity has declined, shortly after the addition of salt, nonstarter lactic acid bacteria (NSLAB) metabolize the remaining lactose to DL-lactate and then racemizes it to L-lactate.[2] This last reaction is not significant for the flavour of cheese, but it does help determine its structure and texture since racemization supports the development of calcium-D-lactate crystals, which are incorrectly interpreted by consumers as spoilage, resulting in product rejection. Moreover, lactate can be metabolized by LAB and NSLAB to CO_2 and acetate, which at high concentrations may be perceived as an off-flavour.[4] Alternatively, lactate can undergo anaerobic fermentation to butyrate, H_2 and CO_2.[5]

Citrate on the other hand is an important precursor of flavour compounds, and even though 90% of the citrate in milk is lost in the whey, cheese such as Cheddar contains 0.2 to 0.5%. Citrate is metabolized by LAB and NSLAB to diacetyl, acetate, acetoin, 2,3-butanediol, and CO_2.[2]

7.1.2 Lipolysis

Fat content has an important role in cheese flavour, and in addition to providing precursors for the development of flavour compounds and being a solvent for hydrophobic compounds, it provides a fat–water–protein interface for catabolic reactions.[6] Indeed, hydrolysis of lipids could be thought of as the second most important metabolic reaction toward the generation of flavour compounds in cheese.

Lipolytic activity is specific of the outer ester linkage of tri- and/or diacylglycerides.[7] Esterases and lipases hydrolyse acyl ester chains between 2 and 8 carbons, and chains of 10 or more carbons, respectively. The relative proportion of FFA in cheese from C6:0 to C18:3 is similar to that in milk fat, but it is higher for free C4:0, evidencing that it is selectively released or synthesized by cheese microflora.[8–10] Lipolytic agents are indigenous, endogenous, and/or exogenous enzymes from milk, rennet, starter culture, NSLAB, and adjunct cultures.

Milk contains a potent lipoprotein lipase (LPL) that, in the presence of an apolipoprotein activator (C-II apolipoprotein-glutamic acid), releases sufficient FFA to give a rancid flavour to milk. However, LPL rarely reaches full activity due to its compartmentalization along with fat and casein micelles, being surrounded by a lipoprotein membrane, which once damaged may later promote off-flavours in cheese. LPL is relatively not specific for fatty acids but is for the sn-1 and sn-3 sites of mono-, di-, and triacylglycerides, being selective for short and medium chain triacylglycerides.[11] Due to LPL sensitivity to temperature, its contribution to flavour generation is more important in raw-milk cheeses. Indeed, 73–95% of the LPL activity is inactivated after pasteurization of milk.[12]

Commercial rennets are usually free from lipase activity. However, Greek and Italian varieties prepared with rennet paste contain a potent lipase, pregastric esterase, which is highly specific for short chain acids at the sn-3 position.[13] Although lipolysis occurs in the majority of cheeses, it is more noticeable in varieties made with rennet paste because of longer ripening times and/or development of secondary microflora. In Cheddar cheese the esterolytic/lipolytic enzymes of LAB are the main lipolytic agents during cheese ripening. These enzymes are intracellular and can hydrolyse esters of fatty acids and tri-, di-, and monoglycerides.[14–17] Although *Lactococcus* and *Lactobacillus* spp. are weakly lipolytic, they become significant factors for the final levels of FFA due to their high concentrations during long ripening periods. Moreover, it has been reported that esterase activity is higher than lipase activity for many *Lactobacillus* and *Lactococcus* strains, such as *Lactobacillus helveticus*, *Lactobacillus delbrueckii* subsp. *bulgaricus*, *Lb. delbrueckii* subsp. *lactis*, *Lactobacillus acidophilus*, *Lactococcus lactis* subsp. *lactis*, and *Lc. lactis* subsp. *cremoris*.[18] LAB enzymes are intracellular and are therefore released by cell autolysis. It has been reported that in Cheddar cheese the use of autolytic strains such as *Lc. lactis* subsp. *cremoris* AM2 results in higher levels of FFA as well as higher levels of secondary proteolysis than those for the strain *Lc. lactis* subsp. *cremoris* HP.[12,19]

FFA are precursor molecules for many catabolic reactions resulting in flavour and aroma compounds such as methyl ketones, lactones, esters, alkanes, and secondary alcohols.[10,16,20] Methyl ketones are important flavour compounds generated from β-oxidation of fatty acids followed by decarboxylation supported by a low redox potential and micro-aerophilic conditions. They are key compounds for blue cheese flavour, and they are identified as impact compounds in other cheeses.[21,22] However, in full fat Cheddar cheese it has been reported that the levels of heptan-2-one, non-2-one, and undecan-2-one increase for 14 weeks and then decrease. In low-fat milk Cheddar cheese, methyl ketone levels correspond to 25% of those in full-fat cheeses.[23,24]

Esters are other products of FFA catabolism. They are highly flavoured and arise from the reaction between short to medium fatty acids and alcohols derived from the fermentation of lactose. In Cheddar cheese some LAB cultures hydrolyse milk fat and esterify certain short fatty acids with ethanol, resulting in esters with fruity flavour notes such as ethyl butanoate,[25] which is considered a defect in Cheddar.[10] It has been reported that in Cheddar cheese esters of FFA are ethyl derivatives.[26]

Secondary alcohols are products of lipolysis and result specifically from enzymatic reduction of methyl ketones.[27] In 1993 Urbach reported that, in Cheddar cheese, 2-propanol is produced from acetone and 2-butanol from butanone.[24] The odours can be described as fruity, green, fuel oil like, and earthy, but the real contribution to the overall cheese flavour is limited because they have high odour thresholds.

Lactones are cyclic esters formed by esterification of hydroxy fatty acids. They confer a nutty, coconut, and buttery-type character to cheese.[10,28,29] In addition, γ-lactones and δ-lactones are present in milk and consequently in all types of cheese. Their concentrations are correlated to the extension of lipolysis, and in Cheddar cheese it has been reported that the levels of lactones reach concentrations above their threshold during early ripening.[12] However, high levels of lactones with high molecular mass have been associated with rancid Cheddar cheese.[30,31] δ-Lactones have higher detection thresholds than γ-lactones and contribute to fruity notes such as coconut, peach, and apricot.

It has been proposed that cheese flavour is the balance of the contribution of many compounds present at certain levels. However, some varieties have specific major contributors, like methyl ketones in mould ripened cheeses or FFA in hard Italian cheeses. However, for many other cheese varieties, the exact role of individual compounds and the right balance among them are still the subject of research. Furthermore, in order to relate the flavour of cheese to the levels of FFA, research has been based on the determination of individual levels of FFA, addition of plain bases to determine if they can be produced or improved, selective removal of FFA to detect any alteration in perceptible flavour, and manufacture of reduced fat cheese or cheese with vegetable fat as a substitute for milk fat.[4,22,32]

7.1.3 Proteolysis

Casein metabolism is the consequence of bacteria using proteins as substrates once lactose is exhausted. This process contributes directly and indirectly to the development of flavour and off-flavour compounds in cheese. After protein breakdown into peptides and free amino acids, catabolic reactions such as transamination, deamination, decarboxylation, desulphurization and catabolism of aromatic amino acids result in aroma compounds such as volatile sulphur compounds, aldehydes, ketones, esters, and thioesters. Proteolysis is the biochemical process that dominates the later phase of ripening.[33,34] The decomposition of the casein network occurs due to the action of enzymes from the coagulant, milk, starter bacteria, NSLAB, and secondary cultures.

Traditionally, cheese is manufactured by using rennet, an enzymatic coagulant extracted from the abomasa of milk-fed calves, and only up to 15% of the coagulant activity remains after whey drainage.[35] Chymosin (EC 3.4.23.4) and bovine pepsin (EC 3.4.23.1) are the major proteolytic enzymes in the coagulant and have a clotting activity of 88–94% and 6–12%, respectively. Specifically, chymosin coagulates milk by the rupture of the bond Phe_{105}–Met_{106} in κ-casein.[13,36] Chymosin then mostly acts on α_{s1}-casein by hydrolyzing the Phe_{23}–Phe_{24} bonds to form the peptides α_{s1}-CN f24–199, α_{s1}-CN f1–23 and α_{s1}-CN f102–109.[37,38] Hydrolysis of α_{s1}-casein makes the texture of curd smoother and homogeneous, and it has been acknowledged that increasing the salt-in-moisture level inhibits the subsequent hydrolysis of the peptide α_{s1}-CN f24–199.[39] However, the chymosin activity on β-casein is extremely inhibited due to the presence of NaCl, though the presence of the peptides β-CN f1–192 and β-CN f193–209 is evidence of some enzymatic activity.[40,41] No major activity has been reported in α_{s2}-casein and/or *para*-κ-casein.[38] Therefore, with high ionic strength and low water activity, breakdown of α_{s1}-casein is faster than that of β-casein.[42]

Among the most representative indigenous milk proteinases are plasmin (the principal and most studied one), cathepsin D, cathepsin B, and other proteolytic enzymes in the lysosomes of somatic cells such as the serine proteinase elastase. Plasmin acts mainly on β-casein, and due to the complex system of activators and inhibitors of the precursor plasminogen, its enzymatic activity differs between cheese varieties. Plasmin hydrolyses β-casein at the bonds Lys_{28}–Lys_{29}, Lys_{105}–His_{106} and Lys_{107}–Glu_{108}, resulting in the formation of γ1-[β-CN f29–209], γ2-[β-CN f106–209], and γ3-[β-CN f108–209] caseins, along with proteose peptones. The γ-caseins accumulate during ripening, while proteose peptones are hydrolyzed by starter bacterial peptidases which yield small peptides and free amino acids.[43]

Cathepsin D is an acid protease, with an optimum pH of 4.0, which produces the glycomacropeptide κ-CN f106–169.[44] However, despite the specificity of cathepsin D being similar to that of chymosin, cathepsin D is not significant in milk coagulation since its milk clotting potential is fairly low.[44,45] Cathepsin D cleavage sites on α_{s1}-casein and β-casein are similar to those on chymosin, but significantly different from those on α_{s2}-casein.[45] In addition, due to cathepsin D being a heat labile enzyme, it is important in proteolysis of dairy products made with pasteurized milk and in proteolysis of rennet free cheeses. The activity of cathepsin D during proteolysis is difficult to quantify.[46,47]

Cathepsin B is a lysosomal cysteine proteinase that works to degrade proteins in the cell. It is activated by dithiothreitol, has a pH optimum of 6.0, partially survives the pasteurization processes, and is capable of extensively degrading α_{s1}-casein and β-casein.[48,49] The proteolytic activity of cathepsin B is of great interest since the activity of lysosomal enzymes has been related to the poor quality of dairy products.[50,51]

Elastase is a neutral serine proteinase, and its essential physiological function is the degradation of elastin. However, it has a broad specificity on α_{s1}-casein and β-casein, with a preferred specificity for bonds involving uncharged non-aromatic amino acids,[52] cleaving 25 and 19 sites, respectively.[53] On β-casein, some of the elastase cleavage sites are identical or nearly identical to those cleaved by plasmin, chymosin, or cell envelope-associated proteinases of several strains of *Lactococcus.*[54]

The principal starter cultures used in cheese manufacturing are the mesophilic *Lactococcus* and *Leuconostoc* species and the thermophilic *Lactobacillus* and *Streptococcus thermophilus* species. Their main role is to decrease the pH by producing lactic acid from lactose. LAB possess a complex proteinase/peptidase system that is very important during secondary proteolysis (see Table 7.1). These peptidases can be classified into endopeptidases, relevant for the degradation of oligopeptides to shorter peptides, and exopeptidases such as carboxypeptidases or aminopeptidases, which release free amino acids from short peptides. The most important

Table 7.1 Proteinases and peptidases involved in cheese proteolysis.[10]

Category	Enzymes
Cell envelope proteinase	Lactocepin
Intracellular oligoendopeptidases	PepO, PepF
General aminopeptidases	PepN, PepC, PepG
Glutamyl aminopeptidase	PepA
Pyrrolidone carboxyl peptidase	PCP
Leucyl aminopeptidase	PepL
Proline-specific peptidases	PepI, PepP, PepQ, PepR, PepX
General dipeptidases	PepV, PepD, PepDA
General tripeptidase	PepT

enzyme is lactocepin, a serine proteinase that degrades intermediate size peptides produced from chymosin and plasmin activity.[35] However, since caseins are rich in proline and because of its unique structure, proline specific enzymes have significant contributions to the proteolysis of cheese, by making released peptides accessible to other peptidases.

Although the role of NSLAB in the development of the flavour of cheese is not totally understood, it has been used to manipulate the final sensory characteristics of the product. Indeed, when the LAB population declines, the NSLAB population becomes the dominant microflora in the maturation of cheese.[55–57] NSLAB essentially consist of homo- and hetero-fermentative species of *Lactobacilli* and are factors during secondary proteolysis. However, certain hetero-fermentative *Lactobacillus* strains have been associated with off-flavours in Cheddar cheese,[58] and it has been reported that thermophilic *Lactobacilli* do not influence the development of Cheddar flavour.[59] It has been suggested that the addition of selected adjunct strains of *Lactobacillus* spp. positively influences the quality of cheese[60–63] and accelerates the ripening of standard and reduced-fat Cheddar cheese.[64,65] Research has been done to clarify the proteolytic and lipolytic systems of NSLAB which contribute to cheese maturation.

7.2 Factors Affecting the Development of Volatile Compounds

Cheese flavour development and most of its chemical and physical properties are related to microbiological and biochemical events that depend on the physicochemical parameters, established during the early stages of cheese manufacture. Factors such as pH, water activity, buffer capacity, oxidation–reduction potential (redox), the physical history of the milk, and the size and geometry of curds are set during manufacture of cheese, which as mentioned earlier is a dehydration process in which fat and caseins are concentrated 6- to 12-fold through the following steps:

- Preparation of milk (pasteurization)
- Acidification (by LAB)

- Coagulation (by rennet)
- Syneresis of whey
- Pressing and shaping the curd
- Salting
- Additional operations

However, other physical parameters such as time and temperature are relevant dynamic factors during the ripening stage which are of great interest in the discussion of cheese flavour.

7.2.1 Milk Preparation: Selection, Pasteurization, and Standardization

After refrigeration of raw milk, the microflora is dominated by psychrotrophs which produce heat stable lipases and proteinases, and at concentrations over 10^6 colony forming units per mL, they reduce yield or can cause development of off-flavours during ripening.[66]

Mainly because of heat-induced changes in cheese microflora, cheese made from raw milk ripens faster and develops stronger flavours than those made from pasteurized milk.[9] Heat treatments of milk, such as pasteurization or heat shock, cause inactivation of indigenous enzymes (mainly lipoprotein lipase), resulting in cheeses with weaker flavour and slower ripening.[9,67] On the other hand, standardisation is a process through which it is possible to control the ratio of casein to fat, influencing in this way parameters such as moisture and moisture in non-fat substances, improving cheese yield.

Some of the different methods to pre-determine the composition of milk to assure a homogenous production during the year involve the use of membrane concentration techniques or vacuum concentration techniques. Membrane concentration techniques, including reverse osmosis, nanofiltration, microfiltration, and ultrafiltration, are common in cheese production. Indeed, low concentration or protein standardization by ultrafiltration is a popular method to secure the uniformity of milk composition, lower casein loss through a firmer curd, more efficiency, and better yield.[68] Additionally, different dairy products such as milk powder, milk protein, and milk permeate can be used to standardize components in cheese milk.[69] Much of the research work on full fat and reduced fat Cheddar cheese is related to vacuum condensation, from which industrial and academic results show that the concentration cannot exceed 1.8 : 1; at higher concentration sweetness and saltiness problems occur due to excessive lactose and minerals.[70]

7.2.2 Acidification

In addition to initiating the fermentation of lactose to lactate during acidification, starter cultures provide enzymes for the ripening stage, as well as promoting the prevention of spoilage because of the reduction of the redox potential of the system. LAB cannot synthesize some amino acids and have proteinase and peptidase systems to liberate these amino acids from caseins. Also, they possess intracellular metabolic enzymes which catalyse the catabolism of amino acids and contribute to the formation of volatile flavour compounds.[5,67,71] Consequently the reduction of pH due to the fermentation of lactose to lactate and lactic acid determines the following:

- Retention of coagulant activity in the curd
- Rate of syneresis
- Growth of added and native microorganisms
- Enzymatic activity during ripening

The LAB used for many cheese varieties is a mesophilic culture with optimum temperature for growth at 20–30 °C. When the cooking temperature is higher than the optimum, the rate of syneresis is increased but the acid development decreases.

7.2.3 Coagulation

As mentioned earlier, coagulation occurs *via* limited proteolysis of κ-caseins at or near the Phe_{105}–Met_{106} bond, followed by the aggregation of the Ca^{2+} altered micelles at temperatures over 18 °C.[72]

The traditional rennet is a brine of the abomasa of milk-fed calves (or other young dairy animals), which contains mainly chymosin and low levels of pepsin. The retention of the activity of rennet added to milk varies from 0 to 15% depending on factors such as the type of enzyme, pH at whey drainage, cooking temperature, and moisture content of the curd.[35] Thus, later on during the ripening stage, the parameters for coagulation will influence the degree of hydrolysis of α_{S1}-casein, the generation of flavour compounds, the texture of curds and cheeses, and subsequently the liberation of sapid compounds.

The pH determines the characteristics of the casein micelle matrix, and along with other factors such as heating, protein concentration in milk, and addition of Ca^{2+}, it establishes the equilibrium between colloidal and dissolved calcium phosphate and between dissolved calcium phosphate and other ions. For example, acidity reduces the negative charge on the micelles (the isoelectric point of casein is 4.6) and increases the solubility of citrates (which are completely soluble at pH 5.5) and colloidal calcium phosphate (completely soluble at pH 5). Heating reduces dissolved calcium and phosphates while promoting coagulation by association of colloidal phosphates with the casein micelles. Milk concentration increases buffer capacity and colloidal calcium phosphate and its association to the micelles. Therefore, it is possible to observe how coagulation depends on dynamic properties like pH and the solubility of calcium salts, and the cleavage of κ-caseins by rennet enzymes. Defects such as sour and/or bitter flavour and soft and pasty body are associated with excessive acidity and a pH under 5.0.

7.2.4 Syneresis

The gel formed due to the rennet induced coagulation is stable, but once it is broken, most of the liquid entrapped in the gel is expelled readily as whey. Syneresis, and consequently the moisture content of cheese, is controlled by the milk composition, size of the curd particles, cooking temperature, time, rate of acidification, and rate of stirring of the curd–whey mixture.[5] Therefore, these parameters and the operations involved in syneresis have an impact on ripening (*e.g.*, high moisture cheese ripens faster than low moisture cheese). In addition to promoting syneresis, high cooking temperatures inactivate the remaining chymosin and increase the levels of plasmin due to denaturation of plasmin inhibitors and inhibitors of plasminogen activators.[73]

High temperatures during rennet gelation increase the initial porosity of the gel and consequently the initial rate of syneresis after the curd is cut. The rate of syneresis is also increased by smaller curd size at cutting and vigorous agitation.[74]

7.2.5 Cheddaring

In cheddaring, a step used in making some cheeses, the curd is piled into the centre of the vat and cut into slabs, which are periodically turned and stacked. The pH of curds decreases during cheddaring, causing dissolution of colloidal calcium phosphate, which modifies the texture of the curd by altering the Ca : protein ratio. The objectives of the cheddaring process are to remove a small amount of whey to allow a better acid development, to have moisture control, to develop a proper texture, and to repress the growth of gas forming spoilage organisms.

7.2.6 Salting

Salting can be achieved by immersion in brine, mixing dry salt with the milled curd, or by application of salt on the surface after moulding. The brine option is intended for cheeses with a high level of salt, and the rate at which salt-in-moisture increases is slow since NaCl should diffuse from the surface. In contrast, NaCl uptake is very fast in the dry salted varieties.[75] This process influences the flavour profiles of cheese by controlling microbial growth and affecting enzyme and water activity. Inhibition of acid production occurs at NaCl concentrations over 1.5%, which has been used in some cheese varieties to stop acidification and to fix the pH.[75] Microbial enzymes can be inhibited or stimulated by moderately high NaCl levels, particularly at low pH, depending on the type of enzyme. Approximately 2 kg of H_2O is lost by absorption of 1 kg of NaCl, which means a reduction of the water activity of the system. For example, plasmin is stimulated by 2% NaCl but inhibited at high concentrations,[73] but for chymosin the addition of NaCl increases the ionic strength, promoting interactions between the hydrophobic C-terminals of β-caseins, inhibiting the access of the enzyme to cleavage sites. Thus decreasing NaCl levels facilitates chymosin action on β-caseins and production of hydrophobic peptides.

In the case of Cheddar cheese, most of the acid is developed before salting and moulding, so cooking time and temperature along with culture activity determine curd pH, buffer capacity, and rennet retention at draining.

7.2.7 Pressing and Moulding

Curds for brine salted cheeses are moulded prior to salting, while those for dry-salted ones are moulded and pressed after addition of salt. Cheese size is determined not only due to aesthetic reasons, but also because many biochemical events occur at the surface of the cheese during the ripening stage. Therefore, controlling the ratio of surface area to volume is key assuring a homogenous ripening and, depending on the variety, it allows the development of certain characteristics such as eye formation due to sufficient partial pressure of CO_2. In the case of Cheddar, the time and temperature of pressing and the temperature during the first days of ripening influence the extent of acid development and the minimum pH.

Another important factor involved in flavour development is the oxidation/reduction system of milk, which includes Fe^{2+}/Fe^{3+}, Cu^{+}/Cu^{2+}, dehydroascorbate, riboflavin, and lactate/pyruvate.[6] By decreasing the pH and increasing the temperature, free sulphydryl groups are produced in proteins and amino acids, lowering the redox potential. Usually, the redox potential decreases during setting, rises during cutting, cooking and draining, decreases again during cheddaring, increases again during milling, and finally decreases during pressing and ripening. Maintaining a low redox potential (-150 to -300 mV) promotes the production and stability of sulphur volatile compounds.[76,77]

Therefore, after the early stages of cheesemaking, the main factors that determine the structure and flavour of most varieties are the extent of acid production in the vat, the amount of starter culture, the residual plasmin and rennet, the pH of curds at draining, the residual lactose, and the mineral content of the curd.

7.3 Introduction to Cheese Flavour and Aroma Analysis

As a response to the need for exact and predictable results of the ripening stage, monitoring the ripening process has been of interest for cheesemakers and food scientists, especially when large-scale production is involved, and standardization of flavour is associated with high quality products by the consumer. Thus, different methods for monitoring the extension of ripening have been developed. From this perspective, ripening can be followed by determining the metabolite products from primary carbon metabolism paths such as the conversion of lactose and citrate, lipolysis, and proteolysis.

Aroma analysis relies greatly on the techniques for extraction and concentration used to isolate different aroma compounds. Since there is no single universal method to extract all kinds of aroma compounds at once, many techniques have been used to study aroma compounds of cheese, such as solvent extraction, simultaneous distillation extraction, static and dynamic head space, purge and trap, ion exchange chromatography, and solid phase microextraction. Moreover, for research intended to track the progress of ripening, the methodology selected should focus on the expected results as a consequence of the assessed agent or parameter.

7.3.1 Proteolysis

Since the degradation and conversion of caseins is the most important biochemical step for flavour formation in hard and semi-hard cheeses, tracking the progress of proteolysis is an excellent indicator of the ripening process.

There are different analytical techniques to monitor patterns of proteolysis, such as a the fractionation scheme to extract nitrogen compounds based on pH. Choosing the scheme and methodology depends on the availability of equipment and resources, the cheese variety and its characteristics, and the objective of the study.[78,79] These fractions can be used to obtain information about one or more proteolysis agents, such as the action of a certain enzyme, comparison of coagulants, and analysis of different starters or adjunct culture. The insoluble fraction can be used to study effects on primary proteolysis by means of urea-polyacrylamide gel electrophoresis or by capillary electrophoresis. Insoluble and soluble fractions can be used to determine peptide profiles by reverse phase high performance liquid chromatography (RP-HPLC) and to perform an analysis of individual amino acids. However, it is important to remember that proteolysis is too complex to be described by a single index, and other chromatographic techniques involving mass spectrometry (MS), such as HPLC or gas chromatography (GC) with MS detection, are a more powerful approach for identifying and monitoring specific key flavour and taste compounds in a fast and accurate manner.[80,81]

7.3.2 Lipolysis

Not all metabolites should be tracked because of certain limitations related to the metabolic path. For example, the remaining lactose in ripening cheese cannot be considered a relevant parameter to monitor the ripening progress since the lactose is metabolized to L-lactic acid, ethanol, and CO_2 within the first few weeks. Ethanol can be esterified with FFA to produce ethyl esters, compounds associated with fruity notes, but they are not optimal descriptors. On the other hand, lipolysis is an important metabolic path to track despite being limited for many varieties, because of a low but steady increase in FFA; particularly, those levels for C4 to C8 FFA can be used as an indicator of the progress of ripening.[12,67] However, it is important to keep in mind that lipolysis as well as some indicators of proteolysis does not correlate directly with the production of relevant volatile compounds.

By using the methods mentioned earlier, a wide variety of compounds such as acids, esters, ketones, aldehydes, alcohols, lactones, phenols, and volatile sulphur compounds have been identified in cheese. Most volatile aroma compounds are found in the lipid content because they are hydrophobic. In the case of fatty acids, it is relevant to know the individual concentrations because each one has a different sensory threshold, aroma attributes, and contribution. They can be determined by GC without derivatisation, but they require separation from triacylglycerides and other lipids such as cholesterol and phospholipids prior to GC analysis. This is done by the use of aminopropyl weak anion exchange columns, in which 100% recoveries have been achieved and a good repeatability can be expected.[22,82,83]

7.3.3 Amino Acid Catabolism

Breakdown of key amino acids such as phenylalanine (aromatic), leucine (branched chain), and methionine and cysteine (sulphurous), particularly by LAB enzymes, has been characterized.[84–86] Volatile sulphur compounds (VSC) such as methanethiol, dimethyl sulphide, dimethyl disulphide, and dimethyl trisulphide, most of them with low odour thresholds, are important flavour compounds that can be used as tracking metabolites for monitoring the progress of the ripening stage. VSC cannot be studied by conventional techniques such as static head space and purge and trap due to the loss of analyte during the concentration stage and the potential formation of thermal artefacts, particularly in this case where the studied compounds are highly volatile and chemically reactive. A better approach is the solid phase microextraction (SPME) technique along with a sulphur pulsed photometric detector, which makes possible the analysis of VSC in food faster and without tedious sample preparation involving the use of solvents.[87–89] Nonetheless, there are many kinds of fibres, but the most appropriate type for the study of dairy products is carboxen-polydimethylsiloxane, which can readily extract highly volatile and low molecular weight compounds including VSC.

7.4 Instrumental Analysis of Cheese

Chromatographic methods are essential in understanding the aroma chemistry of cheese, and consequently they offer a thorough picture of the impact of manufacturing variables during the production process on the ready-to-sell product. Some of the preferred methods are RP-HPLC and GC, including the use of detectors such as flame ionization detectors (FID) and pulsed flame photometric detectors, among others. However, the cheese matrix is complex, samples cannot be injected directly into the injection port of the devices, and consequently the isolation of volatile, non-volatile, and/or sapid compounds is required in order to protect the instruments and to avoid degradation of samples and formation of artefacts due to the high temperatures in the GC injection ports. Thus, to achieve meaningful and reproducible results, isolation and concentration of aroma compounds is mandatory, and depending on the chemical composition of the target analytes, different approaches can be used to perform their separation from the matrix. Therefore, procedures involving pH, solubility, polarity, and volatility are often used in the aroma analysis of cheese.

7.4.1 Gas Chromatography

The separation, identification, quantification, and analysis of volatiles is generally complicated due to their concentration levels, which could be as low as parts per billion (ppb) in the non-volatile matrix. The most common isolation methods used for volatiles are solvent extraction and distillation, and headspace techniques, including static headspace analysis, dynamic headspace analysis and headspace SPME.

7.4.2 Extraction–Distillation

Cheese can be low in moisture and high in fat and protein, and in spite of that the difficult matrix solvent extraction became a common procedure to separate volatiles. It involves the use of organic solvents such as acetonitrile, dichloromethane, and diethyl ether, among which diethyl ether is considered the most suitable for cheese analysis due to its low density, low boiling point and high selectivity for aroma compounds. Moreover, the purity of solvents is critical, which implies an obligatory pre-distillation/purification process before its use. In the case of hard and semi-hard cheeses, the samples are usually frozen employing liquid nitrogen, then grated or ground, and finally the analytes are extracted with the selected solvent,[22,90–95] or alternatively an aqueous extract is prepared followed by solvent extraction.[96] However, because cheese is not a free-fat product, an additional extraction is required and usually done by dialysis[97] or by a low-temperature high vacuum distillation[92] to separate non-polar and non-volatile lipids, and other non-volatile constituents, followed by the separation of analytes into neutral, basic, and acidic fractions if necessary.

The approach of direct solvent extraction is effective to separate semi-volatiles such as lactones, FFA, and phenolics, but has some disadvantages such as a slow process, evaporation of large quantities of solvent, and adsorption of components of the membrane surface and catalysis of acetone condensation in the case of dialysis.[97]

High vacuum distillation and steam distillation techniques are suitable for purifying solvent extracts when the sample is to be analysed by GC with a capillary column. The most popular techniques are simultaneous steam distillation extraction (SDE)[98] and solvent assisted flavour evaporation (SAFE) distillation.[99] SDE has been used for the analysis of aqueous slurries, of Cheddar, Gouda, Edam, Swiss, and Parmesan cheese, with a Likens–Nickerson extractor using diethyl ether, following a procedure described by Parliment.[100] However, this technique can lead to artefact formation since it is performed at high temperatures. SAFE has been reported to result in better recovery than the traditional high vacuum transfer, and it was effective for most volatiles from the solvent extracts of cheese.[95,101–104]

7.4.3 Headspace Methods

Headspace methods include static headspace analysis (SHA), dynamic headspace analysis (DHA), and SPME, which are limited to the equilibrium of volatile compounds into the gas phase, a reason why they are considered non-destructive and require minimal sample preparation.

SHA is used when analysis of major components is satisfactory. In this technique, the sample is contained in a closed vessel and volatiles are allowed to reach equilibrium by partition into the headspace and the matrix, a process influenced by temperature, vessel size, the ratio of sample to headspace volumes, addition of salt, and agitation. An aliquot of the headspace is taken and injected into the GC column. SHA has been used for the analysis of highly volatile sulphur compounds such as hydrogen sulphide, methanethiol and dimethyl sulphide,[105] along with acetaldehyde and other low molecular weight Strecker aldehydes such as methylpropanal, 2-methylbutanal, and 3-methylbutanal,[106] which are important contributors to the flavour of Cheddar and other cheeses. However, this technique lacks sensitivity and is quite difficult to standardize.

By using a flow of carrier gas and an intermediate adsorption or cryogenic step before injection into the GC, larger amounts of volatiles can be collected, improving the efficiency of the headspace analysis, turning it into DHA or purge-and-trap analysis. Usually, the carrier gas is He or N_2, while the adsorbent might be a polymeric material such as Tenax (poly-2,6-diphenyl-*p*-phenylene oxide) or activated charcoal. Adsorbent trapping is more common than cryogenic focusing because it prevents damage to the GC column. Once the volatiles are isolated, they are desorbed by heating, possibly concentrated by cryo-focusing, and then transferred to the GC column by thermal desorption. Some of the advantages are the low risk of artefact formation, the minimal sample preparation, and the fact that it is solvent-free. However, the principal limitation is that the method is not efficient for semi-volatiles. Extracts by this technique have been analysed by multidimensional GC/FID/MS and in total five aldehydes, six ketones, eight alcohols, three esters, 11 hydrocarbons, three halides and three sulphur compounds were identified.[22,26,107–114]

Another solvent-free isolation method is SPME, which can concentrate volatiles from various matrices in a single step.[115] It reduces the time required for sampling and the cost of analysis, and it has become the most common method for analysis of volatiles in the last decade. It is based on the partition of the analytes in the headspace and the polymer coated fibre. The method comes with great selectivity and specificity due to the several adsorbent phases and film thicknesses available. The adsorption depends on temperature, vessel size, the ratio of sample to headspace volumes, addition of salt, agitation, the nature of the coating, and exposure time, for which short times (1–5 min) are recommended for highly volatile compounds and longer times (5–30 min) for semi-volatiles.[116] Volatiles are transferred to the GC column by thermal desorption in the splitless injection mode. Cheddar cheese was initially analysed using fibres coated with non-polar polydimethylsiloxane (PDMS) and polar polyacrylate, showing better results for the polar coating.[117] Volatile components such as fatty acids and δ-lactones were found, and differences in profiles were observed between varieties. Later, the technique was consolidated for the analysis of Cheddar cheese aroma, by a comparison of five types of fibre coatings, namely, PDMS, polydimethylsiloxane/divinylbenzene (PDMS–DVB), polyacrylate, carboxen/PDMS and carbowax/DVB. The study revealed that bipolar coatings PDMS–DVB and carboxen/PDMS showed better selectivity, measured by the number of peaks absorbed, besides confirming the relevance of the adsorption temperature and exposition time.[118] Initially it was suggested that, unlike solvent extraction techniques or DHA, quantification of volatiles was not possible; however, it was demonstrated that it is possible to perform a quantification of volatiles extracted by SPME.

7.4.4 Separation

The injection technique and the analytical column stationary phase could be considered as the most critical parameters influencing the results of the GC analysis. Regarding injection, the programmable temperature vaporizer and on-column injectors are the best options for analysis of aroma extracts. They provide versatility, which makes them suitable for working with cryogenic focusing from DHA, and ramped heated injections in the splitless or split mode. They also help avoid thermal degradation in the cool on-column injection for extracts from direct solvent extraction coupled with high vacuum distillation. On the other hand, the polarity of the stationary phase determines the type of analyte that could be identified. DBWAX and FFAP phases constitute adequate options for analysis of polar compounds, while the DB-5 phase is the right choice for non-polar compounds.

For studying aroma compounds that contribute to the odour of a food, the different approaches include gas chromatography–olfactometry (GCO), odour activity value (OAV) calculations, and sensory analysis. In GCO the most often used techniques are dilution analysis, where the most popular choices are aroma extract dilution analysis (AEDA),[119] the AEDA variation known as aroma extract concentration analysis (AECA),[120] and GCO-headspace dilution analysis. In AEDA a dilution series, or a concentration series in the case of AECA, of an aroma extract is evaluated by GCO, where compounds are ranked according to potency based on the highest dilution or concentration that can be perceived, which is defined as the flavour dilution (FD) factor. However, the FD factor does not account for highly volatile compounds lost during extraction and concentration, underestimating important compounds. GCO-headspace dilution analysis provides a complementary evaluation of the aroma composition. In these methods, dilutions are achieved by decreasing the headspace volume, or the purge gas volume in the case of DHA.

Retention indices are used for confirming a proper identification of compounds contributing to the aroma of a food. These are based on the retention times of target compounds compared to those of certain standards, usually hydrocarbons. In addition, OAV relates the analytical data to sensory characteristics of a sample, and essentially it establishes a relation between the concentration of a target compound in the food to its odour threshold, allowing accurate discrimination of the compounds that really contribute to the specific aroma of a food.

The methodologies mentioned earlier, coupled with the separation efficiency of GC and the excellent identification potential of MS, offer to scientists and engineers the possibility to understand, describe, and improve production processes towards the satisfaction of customers by maintaining constancy in the quality of a product, in this case the traditional aroma of cheese.

7.4.5 RP-HPLC

The assessment of proteolysis through peptide profiling of cheese extracts by RP-HPLC is widely used to study the quality and authenticity of samples. It is based on the principle that caseins might not be soluble in certain solvents, but peptides resolved from their degradation can be. The amounts of intermediate and small peptides increase as the proteolysis of caseins takes place during ripening of cheese.[121,122]

Most of the HPLC separation processes are done in the reverse-phase mode, using a non-polar stationary phase and a polar mobile phase, where solutes are mainly retained due to hydrophobic interaction with the non-polar one and are eluted in order of decreasing polarity. Indeed, solute retention decreases by increasing the organic solvent content of the mobile phase. As in the case of GC, samples cannot be directly injected into the instrument, meaning that isolation of analytes is required.

The most common way to perform the extraction of peptides from cheese is following a fractionation scheme that involves the use of solvents such as water, buffers at pH 4.6, trichloroacetic acid (TCA), phosphotungstic acid, and ethanol.[78] These, in addition to being a sort of sample preparation, allows measuring proteolytic activity through the analysis of the nitrogen content of resolved fractions, and RP-HPLC analysis of water and pH 4.6 soluble extracts. In fact, fractionation with water is a method used for mature Cheddar cheese due to its low and relatively constant pH.[78] Furthermore, water soluble nitrogen (WSN) extracts have been used as the index of ripening, which is only employed when there is no variation of pH during ripening or between samples. Bansal and others[121] have suggested that pH 4.6 soluble nitrogen should be used instead of WSN as the index of primary proteolysis. However, it is important to keep in mind that the results from this method are somewhat smaller and it is more difficult to perform but is easier to standardize. On the other hand, cheese is a dynamic system, meaning that the results of analysis depend on the age of the samples.[79] RP-HPLC analysis of water and/or pH 4.6 soluble extracts can be used as a discriminant technique for the characterization of secondary proteolysis.

TCA has been used to precipitate peptides from water and pH 4.6 soluble extracts at concentrations ranging from 2 to 12%, which depends on the degree of fractionation required.[121] It has been reported that, during fractionation with TCA, parameters such as extraction time and extraction temperature have little or no effect on the amount of nitrogen obtained. In contrast, the amount and type of peptides that can be extracted vary according to parameters such as the cheese-to-water ratio, pH, NaCl content of cheese and the type of previous fraction. Ethanol, ranging in concentration from 30 to 80%, is currently being used as an alternative to TCA because it offers similar precipitation levels and can be easily evaporated for further analysis of peptides in this fraction. Phosphotungstic acid is a very discriminating protein precipitant that is employed in the range from 1 to 6.5%, where only free amino acids (expect lysine and arginine) and peptides under 600 Da are soluble. However, there are other techniques to perform the fractionation based on the molecular mass of peptides rather than on solubility such as dialysis, ultrafiltration, and size exclusion chromatography, which are useful for taste panel work since they are solvent free.[123]

Separation and characterization of peptides has been performed by RP-HPLC analysis of WSN extract, pH 4.6 soluble and insoluble extracts, 10 kDa ultrafiltration permeate, 70% soluble and insoluble extracts, and fractions from gel permeating chromatography. The elution of analytes takes place in a Nucleosil RP-8 (250×4.6 mm^2, 5 μm particle size, 300 Å pore size) analytical column and a guard column (4.6×10 mm^2) and is usually done in the gradient mode using a system such as water/acetonitrile or water/methanol;[124] however, it is possible to perform it by isocratic elution using a phosphate buffer as the mobile phase.[125] The detection of analytes is performed by monitoring the carbonyl group in the peptide bonds using a UV detector at wavelengths ranging from 200 to 230 nm, for which the most used ion-pair reagent is trifluoroacetic acid.

Since peptide profiles are multivariate in nature, the identification of analytes from raw data has been done by visual matching[126,127] or by division of chromatograms in classes of retention time and peaks in each class,[128] followed by MS. Data analysis has been performed using descriptive techniques such as principal component analysis[122,126,127] or using descriptive and inferential techniques such as linear discriminant analysis on principal component scores,[129,130] Partial least squares regression and/or partial least squares discriminant analysis.

7.4.6 Electrophoresis

Besides chromatographic methods such as GC and RP-HPLC, another analytical method for assessment of proteolysis is electrophoresis, which is a specific technique that, in addition to giving information about the extent of proteolysis and the general contribution of proteolytic agents, also resolves, isolates, and identifies peptides.[121] It is a technique applied to the study of cheese ripening and is particularly useful for the comprehension of primary proteolysis since it is limited to monitoring hydrolysis of parent caseins, where only proteins and large peptides can be visualized.

The different electrophoretic methods used in the study of cheese ripening include paper electrophoresis, free boundary electrophoresis, high voltage paper electrophoresis, starch gel electrophoresis, isoelectric focusing, capillary electrophoresis, and polyacrylamide gel electrophoresis (PAGE).[121,123,130–140] PAGE is the most common method applied and usually works with discontinuous buffer systems utilizing urea or sodium dodecyl sulphate (SDS) as the dissociating agent. However, in the case of cheese analysis where caseins have similar molecular weights, it has been reported that buffer systems using urea are more sensitive than those using SDS, allowing proteins or peptides of comparable size to be resolved. Nonetheless, other research studies using SDS as the dissociating agent suggest that valuable information can be obtained.

The most recommended method for the study of cheese involves the stacking system of Andrews[141] in alkaline gels containing 6 M urea and the direct staining procedure of Blakesley and Boezi,[142] with Coomassie blue G250. Also, it has been reported that low molecular weight peptides can be visualized using a silver staining technique involving glutaraldehyde as the fixing agent, which has not been applied yet to cheese analysis. On the other hand, the arrangement between electrophoretic methods, or the application of a separation method followed by electrophoresis, which is known as two-dimensional electrophoresis, has been widely used in the study of cheese ripening. Examples of this are works where SDS-PAGE is one dimension and isoelectric focusing is the other, or where thin layer chromatography is followed by electrophoresis. Usually, peptides resolved on PAGE are isolated and identified by excision of the bands or by electroblotting. The latter is the most utilized, and peptides are identified by N-terminal amino sequencing rather than MS since they are stained. Nonetheless, most of the major degradation products are known in most PAGE systems.

An alternative and potentially strong methodology to PAGE systems is capillary electrophoresis, which to date has been a technique not well implemented in the study of cheese proteolysis and comes with the capacity to resolve complicated mixtures of peptides, using a buffer-filled capillary and an electric field that promotes separation based on the net charge, molecular mass, and Stokes' radius.

7.5 Specific Compounds in Cheese

Different varieties of cheese obviously contain different flavour compounds, though many compounds are found in all aged cheeses. This section will cover important constituents of three types of cheese.

7.5.1 Cheddar

Cheddar is the most-studied cheese in terms of flavour, with dozens of publications covering the development of its volatile compounds. Table 7.2 shows many of the volatiles that have been detected in this variety.

7.5.2 Surface-ripened Cheese

This category of cheese ripens from the outside toward the centre due to the action of bacteria, moulds, and yeasts that are introduced to the surface after pressing. Yeasts such as *Debaryomyces hansenii* and *Geotrichum candidum* are commonly used, and aerobes such as *Arthrobacter* spp. and *Brevibacterium* spp. begin to grow later.[146] These cheeses contain high salt concentrations and low pH values and are stored at 10–20 °C and at a relative humidity above 90%. These conditions are ideal for growth of the microorganisms added to the surface, which develops a visible orange smear on the surface of the cheese. Surface-ripened cheeses are also called bloomy rind cheeses because of the "bloom" of microbes on the surface and also washed-rind cheeses since they are usually washed with a dilute brine solution to spread microbes more uniformly.[147] Camembert is a prominent surface-ripened cheese, and the most important compounds responsible for Camembert aroma are ketones and primary and secondary alcohols.[25] Table 7.2 shows the important volatile compounds in Camembert.

7.5.3 Blue Cheese

Common blue cheese varieties include Roquefort, Stilton, and Gorgonzola. Blue cheese is typically made with *Lactococcus lactis* starter culture and *Penicillium* mould such as *Penicillium roqueforti*. The cheese is ripened for at least 60 days at 10 °C, during which the cheese is pierced with skewers to let in oxygen, which allows the mould spores to grow into hyphae and leads to the creation of veins. Sometimes *Leuconostoc* spp. is added as an adjunct starter culture to create CO_2 gas that opens the structure and permits mould growth.[148] The *Penicillium* metabolizes fatty acids into methyl ketones that are central to blue cheese aroma, as shown in Table 7.2.[149]

7.6 Summary

Microbial activity and processing conditions greatly affect the development of aromas and flavours in cheese. The breakdown of carbohydrates, triacylglycerols, and proteins during aging leads to a multitude of compounds that are responsible for the characteristics of cheese. Further analytical research on techniques for detecting these compounds and microbiological research on tailoring microorganisms to improve flavour will lead to improvements in cheese production.

Table 7.2 Volatile compounds in Cheddar (Ch), Camembert (Ca), and blue (B) cheese with aroma notes (if identified).[25,143–145]

Compound	Aroma notes	Where found
Volatile sulphur compounds		
Methanethiol	Rotten cabbage, faecal	Ch, Ca, B
Methyl sulphide		Ca
Dimethyl sulphide	Boiled cabbage	Ch, Ca
Dimethyl disulphide	Cabbage, cauliflower, garlic, strong onion	Ch, Ca
Dimethyl trisulphide	Ripe cheese, garlic	Ch, B
Methional	Roast potato	Ch, Ca, B
Methionol	Garlic, sulphur	Ch, B
Dithiapentane	Garlic	Ca
Hexanethiol	Burnt fat, sulphury, meaty, fatty, garlic, roasted, burnt	Ch
Hydrogen sulphide	Rotten egg	Ch
Organic acids		
Acetic acid	Vinegar, sour, pungent	Ch, Ca, B
Propionic acid	Vinegar, pungent	Ch, Ca
Butanoic acid	Sweaty, cheesy, faecal, rancid, sharp	Ch, Ca, B
Isobutanoic acid	Sweaty, rotten apple	Ca
Pentanoic acid	Swiss cheese, meaty, rotten	Ch, B
3-Methylbutanoic acid	Swiss cheese, waxy, sweaty, old socks, faecal	Ch, Ca, B
Hexanoic acid	Goat like	Ch, Ca, B
Octanoic acid	Body odour, sweaty	Ch, Ca
4-Methyloctanoic acid	Waxy, goat, sheep	Ca
4-Ethyloctanoic acid	Goat	Ca
Decanoic acid	Rancid, waxy, soapy	Ch, Ca
Phenyl acetic acid	Flowery	Ch
Lactones		
γ-Butyrolactone	Pungent, fetid, rubber, buttery	Ca
β-Angelica lactone		Ch
γ-Octalactone	Fruity, coconut	Ca
δ-Octalactone	Fruity, peachy, sweet, coconut, wine, goat, animal	Ch, Ca
γ-Decalactone	Coconut, fruity	Ch, Ca
δ-Decalactone	Peachy, coconut	Ch, Ca, B
γ-Dodecalactone	Peachy, buttery, musky	Ca
δ-Dodecalactone	Cheesy, various fruits	Ch, Ca, B
6-(*Z*)-Dodecenyl-γ-decalactone	Soapy	Ch
Sotolon	Curry, seasoning	Ch, B
Furaneol	Sweet, caramel, burnt sugar, strawberry	Ch, B
Homofuraneol, ethyl furaneol	Caramel, burnt sugar	Ch, B
Alcohols		
Ethanol	Alcohol	Ch, Ca
Propanol	Pungent	Ch, Ca
Butanol	Floral, fruity, sweet	Ch, Ca
2-Butanol	Alcoholic	Ch
2,3-Butanediol	Fruity	Ch

Table 7.2 (*Continued*)

Compound	Aroma notes	Where found
2-Methyl-1-butanol	Wine	Ch
3-Methyl-1-butanol	Fruity, alcohol, solvent, grainy	Ch
2-Ethyl butanol		Ch
Isobutanol		Ch
2-Pentanol	Sweet, alcoholic, fruity, nutty	Ch, Ca
2,4-Pentanediol		Ch
Hexanol, 2-hexanol	Fatty, green, floral	Ch, Ca
Heptanol, 2-heptanol		Ca, Ca
2-Octanol	Mushroom, coconut, oil, rancid	Ch, Ca
1-Octen-3-ol	Mushroom	Ca
1,5-Diocten-3-ol	Soil, geranium	Ca
2-Nonanol		Ca
2-Undecanol		Ca
Guaiacol	Smoky, spicy	Ch
2-Methylisoborneol	Musty, soil	Ca
Phenol	Floral, medical	Ca
2-Methoxyphenol	Smoked, phenolic	B
2-Phenylethanol	Rosy, floral	Ch, Ca
Ethylphenol	Disinfectant	B
3-Propylphenol	Disinfectant	B
p-Cresol	Unclean, medical, cowy, barny	Ch, Ca, B
Aldehydes		
Acetaldehyde	Sweet, pungent	Ch
Propanal	Solvent	Ch
Propenal		Ch
2-Methylpropanal, 3-methylpropanal	Malt	Ch, Ca
Butanal	Pungent	Ch
2-Methylbutanal, 3-methylbutanal	Dark chocolate, malt	Ch, Ca
Pentanal	Pungent, almond like	Ch
Hexanal	Green	Ch, Ca
2-Hexenal	Almond bitter, green, fatty	Ch, Ca
Isohexanal		Ch
Heptanal	Fatty, oily, green	Ch, Ca
(*Z*)-4-Heptenal	Creamy, biscuit	Ch
Octanal	Green, fatty, soapy, fruity, orange peel	Ch, Ca
Nonanal	Green, orange	Ch, Ca
2-Nonenal	Green, fatty, cucumber	Ch, B
(*E*,*E*)-2,4-Nonadienal	Soapy	Ch
(*E*,*Z*)-2,6-Nonadienal	Melon, cucumber	Ch
Decanal	Soapy, flowery, orange	Ch, Ca
(*E*,*E*)-2,4-Decadienal, (*E*,*Z*)-2,4-decadienal	Mayonnaise, bread, fatty, tallow, fruity	Ch
trans-4,5-Epoxy-2-(*E*)-decenal	Metallic	Ch
Dodecanal	Orange	Ca
Benzaldehyde	Almond	Ch, Ce
Phenyl aldehyde	Rosy	Ch
Thiophene-2-aldehyde		Ch
Ketones		
Acetone	Solvent-like	Ch
2-Propanone	Acetone	Ca

Table 7.2 (*Continued*)

Compound	Aroma notes	Where found
2-Butanone	Etheric, acetone	Ch, Ca, B
2,3-Butanedione (diacetyl)	Buttery	Ch, Ca, B
3-Hydroxy-2-butanone (acetoin)	Buttery	Ch, Ca
3-Methyl-2-butanone	Camphor	Ch
2-Pentanone	Acetone, sweet, fruity, ketone	Ch, Ca
3-Methyl-2-pentanone		Ch
2-Hexanone	Fruity, ketone	Ch, Ca
1-Hexen-3-one	Cooked vegetable	
2-Heptanone	Blue cheese, fruity, musty, soapy, varnish	Ch, Ca, B
2-Octanone	Floral, fruity, soapy, ketone, musty, green	Ch, Ca, B
1-Octen-3-one	Mushroom, fungus, woody	Ch, Ca, B
(*Z*)-1,5-Octadien-3-one	Green metallic	Ch
2-Nonanone	Green, earthy, blue cheese, fatty, musty, varnish	Ch, Ca, B
2-Undecanone	Floral, fruity, green, musty, tallow	Ch, Ca
2-Tridecanone	Fruity, green	Ch, Ca
(*E*)-β-Damascenone	Apple sauce, woody	Ch, Ca
Acetophenone	Almond, musty, glue, orange blossom	Ch, Ca
Esters		
Methyl acetate		Ch
Ethyl acetate	Fruity, solvent, sweet, pineapple	Ch, Ca
Propyl acetate		Ch
Butyl acetate	Pear, pineapple	Ch, Ca
2-Butylacetate		Ch
Isoamyl acetate	Pear, banana	Ca
Methyl propanoate		Ch
Ethyl propanoate	Fruity, pineapple, melon	Ch, Ca, B
Isoamyl propanoate	Apricot, pineapple	Ca
Ethyl butanoate	Bubble gum, fruity, pineapple	Ch, Ca
Propyl butanoate	Pineapple, banana	Ch, Ca
Butyl butanoate		Ch
Isoamyl butanoate	Apricot, pineapple	Ca
Methyl hexanoate	Pineapple	Ca
Ethyl hexanoate	Fruity, pineapple, banana, grape, melon	Ch, Ca, B
Ethyl octanoate	Fruity, apricot, wine	Ch, Ca
2-Phenylethyl acetate	Floral, rose	Ca
2-Phenylethyl propanoate	Floral, fruity	Ca
Phenylethyl butyrate	Floral, rose, honey	Ca
Terpenes		
Geosmin	Earthy, moistened soil	Ch
Limonene	Citrus	Ch
Linalool	Sweet, floral, honey	Ch
α-Pinene	Pine	Ch
Nitrogen-containing compounds		
2-Acetylpyrazine	Popcorn	Ch
2,5-Dimethylpyrazine, 2,6-dimethylpyrazine	Toasted hazelnut, nutty, roast grain	Ca, B
2-Ethyl-5-methylpyrazine	Peanut, green	Ch, B
Trimethylpyrazine	Nutty, musty, beany	Ch, B
Dimethylethylpyrazine	Brothy	B

Table 7.2 (*Continued*)

Compound	Aroma notes	Where found
2-Acetyl-1-pyrroline	Roasted	Ch
Indole	Mothball	Ch
2-Isopropyl-3-methoxypyrazine	Earthy, soil, green, beany	Ch
2-Isobutyl-3-methoxypyrazine	Bell pepper	Ch
Skatole	Unclean, mothball, faecal	Ch
γ-Cyanotoluene	Smoked, chemical	B
Various amines (C1–C8)	Fruity, alcoholic, varnish	Ca

Abbreviations

AECA	Aroma extract concentration analysis
AEDA	Aroma extract dilution analysis
DHA	Dynamic headspace analysis
DVB	Divinylbenzene
FFA	Free fatty acids
FID	Flame ionization detector
GC	Gas chromatography
GCO	Gas chromatography–olfactometry
HPLC	High performance liquid chromatography
LAB	Lactic acid bacteria
LPL	Lipoprotein lipase
MS	Mass spectrometry
NSLAB	Nonstarter lactic acid bacteria
OAV	Odour activity value
PAGE	Polyacrylamide gel electrophoresis
PDMS	Polydimethylsiloxane
RP	Reverse phase
SAFE	Solvent assisted flavour evaporation
SDE	Simultaneous steam distillation extraction
SDS	Sodium dodecyl sulphate
SHA	Static headspace analysis
SPME	Solid phase microextraction
TCA	Trichloroacetic acid
VSC	Volatile sulphur compounds
WSN	Water soluble nitrogen

References

1. L. M. Huffman and T. Kristoffersen, *N. Z. J. Dairy Sci. Technol.*, 1984, **19**, 151.
2. P. F. Fox, J. A. Lucey and T. Cogan, *Crit. Rev. Food Sci. Nutr.*, 1990, **9**, 237.

3. K. W. Turner and T. D. Thomas, *N. Z. J. Dairy Sci. Technol.*, 1980, **15**, 265.
4. J. W. Aston and J. R. Dulley, *Aust. J. Dairy Technol.*, 1982, **37**, 59.
5. P. F. Fox and P. L. H. McSweeney, in *Cheese Chemistry, Physics and Microbiology*, ed. P. F. Fox, P. L. H. McSweeney, T. M. Cogan and T. P. Guinee, Academic Press, Cambridge, MA, 2004, vol. 1, p. 1.
6. P. Walstra, T. J. Geurts, A. Noomen, A. Jellema and M. A. J. S. van Boekel, in *Dairy Technology: Principles of Milk Properties and Processes*, CRC Press, Boca Raton, FL, 1999, p. 651.
7. H. C. Deeth and V. Touch, *Aust. J. Dairy Technol.*, 2000, **55**, 153.
8. D. D. Bills and E. A. Day, *J. Dairy Sci.*, 1964, **47**, 733.
9. P. F. Fox and P. L. H. McSweeney, in *Dairy Chemistry and Biochemistry*, Springer, New York, 1998, p. 67.
10. P. L. H. McSweeney and M. J. Sousa, *Lait*, 2000, **80**, 293.
11. T. Olivecrona, S. Vilaró and G. Olivecrona, in *Advanced Dairy Chemistry*, ed. P. F. Fox, Springer, New York, 1992, vol. 1, p. 292.
12. Y. F. Collins, P. L. H. McSweeney and M. G. Wilkinson, *Int. Dairy J.*, 2003, **13**, 841.
13. P. F. Fox, T. P. Guinee, T. M. Cogan and P. L. H. McSweeney, in *Fundamentals of Cheese Science*, Springer, New York, 2000, p. 98.
14. R. Holland and T. Coolbear, *J. Dairy Res.*, 1996, **63**, 131.
15. J.-F. Chich, K. Marchesseau and J.-C. Gripon, *Int. Dairy J.*, 1997, **7**, 169.
16. P. F. Fox, J. Law, P. L. H. McSweeney and J. M. Wallace, in *Cheese Chemistry, Physics and Microbiology*, ed. P. F. Fox, P. L. H. McSweeney, T. M. Cogan and T. P. Guinee, Academic Press, Cambridge, MA, 2004, vol. 1, p. 382.
17. S.-Q. Liu, R. Holland and V. L. Crow, *Int. Dairy J.*, 2001, **11**, 27.
18. E. Tsakalidou and G. Kalantzopoulos, *Lait*, 1992, **72**, 533.
19. M. G. Wilkinson, T. P. Guinee, D. M. O'Callaghan and P. F. Fox, *J. Dairy Res.*, 1994, **61**, 249.
20. J.-C. Gripon, V. Monnet, G. Lambert and M. J. Desmazeaud, in *Food Enzymology*, ed. P. F. Fox, Elsevier, Edinburgh, 1991, vol. 1.
21. S. Sablé and G. Cottenceau, *J. Agric. Food Chem.*, 1999, **47**, 4825.
22. M. Qian and G. Reineccius, *J. Dairy Sci.*, 2002, **85**, 1362.
23. T. Dimos, MSc thesis, Latrobe University, 1992.
24. G. Urbach, *Int. Dairy J.*, 1993, **3**, 389.
25. P. Molimard and H. E. Spinnler, *J. Dairy Sci.*, 1996, **79**, 169.
26. G. Arora, F. Cormier and B. Lee, *J. Agric. Food Chem.*, 1995, **43**, 748.
27. W. J. M. Engels, R. Dekker, C. de Jong, R. Neeter and S. Visser, *Int. Dairy J.*, 1997, **7**, 255.
28. J. M. Wallace and P. F. Fox, *Int. Dairy J.*, 1997, **7**, 157.
29. P. Dirinck and A. De Winne, *J. Chromatogr. A*, 1999, **847**, 203.
30. N. P. Wong, R. Ellis and D. E. LaCroix, *J. Dairy Sci.*, 1975, **58**, 1437.
31. R. C. Jolly and F. V. Kosikowski, *J. Agric. Food Chem.*, 1975, **23**, 1175.
32. C. Wijesundera and L. Drury, *Aust. J. Dairy Technol.*, 1999, **54**, 28.
33. J. E. Christensen, E. G. Dudley, J. A. Pederson and J. L. Steele, *Antonie van Leeuwenhoek*, 1999, **76**, 217.

34. J. D. Tammam, A. G. Williams, J. Noble and D. Lloyd, *Lett. Appl. Microbiol.*, 2000, **30**, 370.
35. V. K. Upadhyay, P. L. H. McSweeney, A. A. A. Magboul and P. F. Fox, in *Cheese: Chemistry, Physics and Microbiology*, ed. P. F. Fox, P. L. H. McSweeney, T. M. Cogan and T. P. Guinee, Academic Press, Cambridge, MA, 2004, vol. 1, p. 391.
36. D. M. Mulvihill and P. F. Fox, *J. Dairy Res.*, 1979, **46**, 641.
37. B. C. Richardson, L. K. Creamer, K. N. Pearce and R. E. Munford, *J. Dairy Res.*, 1974, **41**, 239.
38. T. K. Singh, P. F. Fox, P. Højrup and A. Healy, *Int. Dairy J.*, 1994, **4**, 111.
39. F. A. Exterkate, A. C. Alting and C. J. Slangen, *Syst. Appl. Microbiol.*, 1995, **18**, 7.
40. S. Visser, G. Hup and F. A. Exterkate, *Neth. Milk Dairy J.*, 1983, **37**, 169.
41. P. L. H. McSweeney, S. Pochet, P. F. Fox and A. Healy, *J. Dairy Res.*, 1994, **61**, 587.
42. S. Visser, *J. Dairy Sci.*, 1993, **76**, 329.
43. T. K. Singh, P. F. Fox and Á. Healy, *J. Dairy Res.*, 1995, **62**, 629.
44. P. F. Fox and P. L. H. McSweeney, *Food Rev. Int.*, 1996, **12**, 457.
45. L. B. Larsen, C. Benfeldt, L. K. Rasmussen and T. E. Petersen, *J. Dairy Res.*, 1996, **63**, 119.
46. M. J. Hurley, B. M. O'Driscoll, A. L. Kelly and P. L. H. McSweeney, *Int. Dairy J.*, 1999, **9**, 553.
47. V. Crow, B. Curry and M. Hayes, *Int. Dairy J.*, 2001, **11**, 275.
48. E. Knecht, *Cell Biol. Int.*, 1999, **23**, 590.
49. T. Considine, Á. Healy, A. L. Kelly and P. L. H. McSweeney, *Int. Dairy J.*, 2004, **14**, 117.
50. A. S. Grandison and G. D. Ford, *J. Dairy Res.*, 1986, **53**, 645.
51. R. J. Verdi and D. M. Barbano, *J. Dairy Sci.*, 1991, **74**, 2077.
52. M. A. Naughton and F. Sanger, *Biochem. J.*, 1961, **78**, 156.
53. T. Considine, Á. Healy, A. L. Kelly and P. L. H. McSweeney, *Food Chem.*, 2000, **69**, 19.
54. T. Considine, Á. Healy, A. L. Kelly and P. L. H. McSweeney, *Food Chem.*, 1999, **66**, 463.
55. S. D. Peterson and R. T. Marshall, *J. Dairy Sci.*, 1990, **73**, 1395.
56. V. L. Crow, T. Coolbear, P. K. Gopal, F. G. Martley, L. L. McKay and H. Riepe, *Int. Dairy J.*, 1995, **5**, 855.
57. P. F. Fox, P. L. H. McSweeney and C. M. Lynch, *Aust. J. Dairy Technol.*, 1998, **53**, 7.
58. R. Puchades, L. Lemieux and R. E. Simard, *J. Food Sci.*, 1989, **54**, 885.
59. G. T. Lloyd, J. F. Horwood and I. Barlow, *Aust. J. Dairy Technol.*, 1980, **35**, 137.
60. M. A. Drake, T. D. Boylston, K. D. Spence and B. G. Swanson, *Food Res. Int.*, 1996, **29**, 381.
61. C. N. Lane and P. F. Fox, *Int. Dairy J.*, 1996, **6**, 715.
62. C. M. Lynch, P. L. H. McSweeney, P. F. Fox, T. M. Cogan and F. D. Drinan, *Int. Dairy J.*, 1996, **6**, 851.

63. D. D. Muir, J. M. Banks and E. A. Hunter, *Int. Dairy J.*, 1996, **6**, 407.
64. M. A. El Soda, *FEMS Microbiol. Rev.*, 1993, **12**, 239.
65. J. E. Christensen, M. E. Johnson and J. L. Steele, *Int. Dairy J.*, 1995, **5**, 367.
66. M. J. Payne and R. G. Kroll, *Trends Food Sci. Technol.*, 1991, **2**, 315.
67. Y. F. Collins, P. L. H. McSweeney and M. G. Wilkinson, in *Cheese Chemistry, Physics and Microbiology*, ed. P. F. Fox, P. L. H. McSweeney, T. M. Cogan and T. P. Guinee, Academic Press, Cambridge, MA, 2004, vol. 1, p. 373.
68. V. V. Mistry and J.-L. Maubois, in *Cheese Chemistry, Physics and Microbiology*, ed. P. F. Fox, P. L. H. McSweeney, T. M. Cogan and T. P. Guinee, Academic Press, Cambridge, MA, 2004, vol. 1, p. 261.
69. V. V. Mistry, L. E. Metzger and J. L. Maubois, *J. Dairy Sci.*, 1996, **79**, 1137.
70. D. L. Anderson, V. V. Mistry, R. L. Brandsma and K. A. Baldwin, *J. Dairy Sci.*, 1993, **76**, 2832.
71. M. Yvon and L. Rijnen, *Int. Dairy J.*, 2001, **11**, 185.
72. D. S. Home and J. M. Banks, in *Cheese Chemistry, Physics and Microbiology*, ed. P. F. Fox, P. L. H. McSweeney, T. M. Cogan and T. P. Guinee, Academic Press, Cambridge, MA, 2004, vol. 1, p. 47.
73. N. Y. Farkye and P. F. Fox, *J. Dairy Res.*, 1990, **57**, 413.
74. P. Walstra, J. T. M. Wouters and T. J. Geurts, *Dairy Science and Technology*, CRC Press, Boca Raton, FL, 2006.
75. T. P. Guinee, *Int. J. Dairy Technol.*, 2004, **57**, 99.
76. G. Urbach, *Int. Dairy J.*, 1995, **5**, 877.
77. T. P. Beresford, N. A. Fitzsimons, N. L. Brennan and T. M. Cogan, *Int. Dairy J.*, 2001, **11**, 259.
78. Y. Ardö, *Bull. Int. Dairy Fed.*, 1999, **337**, 4.
79. M. J. Sousa, Y. Ardö and P. L. H. McSweeney, *Int. Dairy J.*, 2001, **11**, 327.
80. T. Cserháti, *Biomed. Chromatogr.*, 2002, **16**, 303.
81. M. Careri and A. Mangia, *J. Chromatogr. A*, 2003, **1000**, 609.
82. C. D. Jong and H. T. Badings, *J. High Resolut. Chromatogr.*, 1990, **13**, 94.
83. F. Chavarri, M. Virto, C. Martin, A. I. Nájera, A. Santisteban, L. J. R. Barrón and M. De Renobales, *J. Dairy Res.*, 1997, **64**, 445.
84. D. J. Manning, H. R. Chapman and Z. D. Hosking, *J. Dairy Res.*, 1976, **43**, 313.
85. B. Weimer, K. Seefeldt and B. Dias, *Antonie van Leeuwenhoek*, 1999, **76**, 247.
86. M. R. Stuart, L. S. Chou and B. C. Weimer, *Appl. Environ. Microbiol.*, 1999, **65**, 665.
87. Y. Fang and M. C. Qian, *J. Chromatogr. A*, 2005, **1080**, 177.
88. P. A. Vazquez-Landaverde, J. A. Torres and M. C. Qian, *J. Dairy Sci.*, 2006, **89**, 2919.
89. H. M. Burbank and M. C. Qian, *J. Chromatogr. A*, 2005, **1066**, 149.
90. M. Preininger and W. Grosch, *LWT–Food Sci. Technol.*, 1994, **27**, 237.
91. C. Milo and G. A. Reineccius, *J. Agric. Food Chem.*, 1997, **45**, 3590.

92. O. Suriyaphan, M. A. Drake, X. Q. Chen and K. R. Cadwallader, *J. Agric. Food Chem.*, 2001, **49**, 1382.
93. G. Zehentbauer and G. A. Reineccius, *Flavour Fragrance J.*, 2002, **17**, 300.
94. Y. K. Avsar, Y. Karagul-Yuceer, M. A. Drake, T. K. Singh, Y. Yoon and K. R. Cadwallader, *J. Dairy Sci.*, 2004, **87**, 1999.
95. M. E. Carunchia Whetstine, K. R. Cadwallader and M. A. Drake, *J. Agric. Food Chem.*, 2005, **53**, 3126.
96. L. Moio, D. Langlois, P. Etievant and F. Addeo, *J. Dairy Res.*, 1993, **60**, 215.
97. K. F. Benkler and G. A. Reineccius, *J. Food Sci.*, 1980, **45**, 1084.
98. A. Chaintreau, *Flavour Fragrance J.*, 2001, **16**, 136.
99. W. Engel, W. Bahr and P. Schieberle, *Eur. Food Res. Technol.*, 1999, **209**, 237.
100. T. H. Parliment, in *Flavor Analysis*, ed. C. J. Mussinan and M. J. Morello, ACS Books, Washington, DC, 1998, pp. 8–21.
101. P. Werkhoff, S. Brennecke, W. Bretschneider and H. J. Bertram, *Food Sci. Technol.*, 2002, **19**, 139.
102. M. E. Carunchia Whetstine, P. J. Luck, M. A. Drake, E. A. Foegeding, P. D. Gerard and D. M. Barbano, *J. Dairy Sci.*, 2007, **90**, 3091.
103. H. Schlichtherle-Cerny, R. Gauch and M. Imhof, in *Developments in Food Science*, ed. W. L. P. Bredie and M. D. Petersen, Elsevier, Oxford, UK, 2006, p. 289.
104. K. R. Cadwallader, M. A. Drake, M. E. Carunchia-Whetstine and T. K. Singh, in *Developments in Food Science*, ed. W. L. P. Bredie and M. D. Petersen, Elsevier, Oxford, UK, 2006, p. 157.
105. J. C. C. Lin and I. J. Jeon, *J. Food Sci.*, 1985, **50**, 843.
106. E. Fernández-García, *J. Agric. Food Chem.*, 1996, **44**, 1833.
107. H. C. Dunn and R. C. Lindsay, *J. Dairy Sci.*, 1985, **68**, 2853.
108. G. Barbieri, L. Bolzoni, M. Careri, A. Mangia, G. Parolari, S. Spagnoli and R. Virgili, *J. Agric. Food Chem.*, 1994, **42**, 1170.
109. A. Thierry, M. B. Maillard and L. Quéré, *Int. Dairy J.*, 1999, **9**, 453.
110. A. Thierry, M. B. Maillard, C. Hervé, R. Richoux and S. Lortal, *Food Chem.*, 2004, **87**, 439.
111. P. Larráyoz, M. Addis, R. Gauch and J. O. Bosset, *Int. Dairy J.*, 2001, **11**, 911.
112. M. Rychlik and J. O. Bosset, *Int. Dairy J.*, 2001, **11**, 903.
113. E. Valero, J. Sanz and I. Martínez-Castro, *J. Chromatogr. Sci.*, 2001, **39**, 222.
114. E. Boscaini, S. Van Ruth, F. Biasioli, F. Gasperi and T. D. Märk, *J. Agric. Food Chem.*, 2003, **51**, 1782.
115. H. Kataoka, H. L. Lord and J. Pawliszyn, *J. Chromatogr.*, 2000, **880**, 35–62.
116. D. D. Roberts and T. E. Acree, *J. Agric. Food Chem.*, 1995, **43**, 2179.
117. H. W. Chin, R. A. Bernhard and M. Rosenberg, *J. Food Sci.*, 1996, **61**, 1118.
118. J. P. Dufour, P. Delbecq and L. P. Albela, in *Gas Chromatography-Olfactometry: The State of the Art*, ed. J. V. Leland, P. Schieberle, A. Buettner and T. E. Acree, ACS Books, Washington, DC, 2001, p. 123.

119. W. Grosch, *Flavour Fragrance J.*, 1994, **9**, 147.
120. W. Grosch, R. Kerscher, J. Kubickova, T. Jagella, in *Gas Chromatography-Olfactometry: The State of the Art*, ed. J. V. Leland, P. Schieberle, A. Buettner and T. E. Acree, ACS Books, Washington, DC, 2001, p. 138.
121. N. Bansal, P. Piraino and P. L. H. McSweeney, in *Handbook of Dairy Foods Analysis*, ed. F. Toldrá and L. M. L. Nollet, CRC Press, Boca Raton, FL, 2009, p. 405.
122. P. Piraino, E. Parente and P. L. H. McSweeney, *J. Agric. Food Chem.*, 2004, **52**, 6904.
123. P. F. Fox, *J. Dairy Sci.*, 1989, **72**, 1379.
124. A. J. Cliffe, J. D. Marks and F. Mulholland, *Int. Dairy J.*, 1993, **3**, 379.
125. A. M. Pham and S. Nakai, *J. Dairy Sci.*, 1984, **67**, 1390.
126. A. H. Pripp, S. Ur-Rehman, P. L. H. McSweeney and P. F. Fox, *Int. Dairy J.*, 1999, **9**, 473.
127. A. H. Pripp, L. Stepaniak and T. Sørhaug, *Int. Dairy J.*, 2000, **10**, 249.
128. D. Barile, J. D. Coïsson, M. Arlorio and M. Rinaldi, *Food Control*, 2006, **17**, 197.
129. E. Hynes, C. Bach, G. Lamberet, J. C. Ogier, O. Son and A. Delacroix-Buchet, *Lait*, 2003, **83**, 31.
130. B. A. O'Shea, T. Uniacke-Lowe and P. F. Fox, *Int. Dairy J.*, 1996, **6**, 1135.
131. P. Trieu-Cuot and J. C. Gripon, *J. Dairy Res.*, 1982, **49**, 501.
132. F. Addeo, C. Stingo, L. Chianese, P. Petrilli, A. Scudiero and G. Anelli, *Latte*, 1983, **8**, 795.
133. F. Addeo, L. Moio, L. Chianese, C. Stingo and A. D. Luccia, *Milchwissenschaft*, 1990, **45**, 221.
134. L. Moio, A. D. Luccia and F. Addeo, *Electrophoresis*, 1989, **10**, 533.
135. L. Moio, L. Chianese, M. Rivemale and F. Addeo, *Lait*, 1992, **72**, 87.
136. L. Amigo, M. Ramos, L. Calhau and M. Barbosa, *Lait*, 1992, **72**, 95.
137. E. D. Strange, E. L. Malin, D. L. Van Hekken and J. J. Basch, *J. Chromatogr. A*, 1992, **624**, 81.
138. R. A. Ledford, A. C. O'Sullivan and K. R. Nath, *J. Dairy Sci.*, 1966, **49**, 1098.
139. S. I. Shalabi and P. F. Fox, *Ir. J. Food Sci. Technol.*, 1987, **1**, 135.
140. A. M. Gouldsworthy, J. M. Banks, A. J. Law and J. Leaver, *Milchwissenschaft*, 1999, **54**, 620.
141. A. T. Andrews, *Analytical Methods of Food Authentication*, ed. P. R. Ashurst and M. J. Dennis, Thomson Science, London, 1998, p. 204.
142. R. W. Blakesley and J. A. Boezi, *Anal. Biochem.*, 1977, **82**, 580.
143. B. Ganesan, B. C. Weimer, M. C. Qian and H. M. Burbank, in *Improving the Flavour of Cheese*, ed. B. C. Weimer, Woodhead Publishing, Cambridge, UK, 2007, p. 26.
144. T. K. Singh, M. A. Drake and K. R. Cadwallader, *Compr. Rev. Food Sci.*, 2003, **2**, 166.
145. D. C. Frank, C. M. Owen and J. Patterson, *LWT–Food Sci. Technol.*, 2004, **37**, 139.

146. J. Hugenholtz and J. E. T. van Hylckama, in *Improving the Flavour of Cheese*, ed. B. C. Weimer, Woodhead Publishing, Cambridge, UK, 2007, p. 351.
147. N. M. Brennan, T. M. Cogan, M. Loessner and S. Scherer, in *Cheese Chemistry, Physics and Microbiology*, ed. P. F. Fox, P. L. H. McSweeney, T. M. Cogan and T. P. Guinee, Academic Press, Cambridge, MA, 3rd edn, 2004, vol. 2, p. 199.
148. M. H. Tunick, in *The Science of Cheese*, Oxford University Press, New York, 2014, p. 107.
149. M. Qian, C. Nelson and S. Bloomer, *J. Am. Oil Chem. Soc.*, 2002, **79**, 663.

CHAPTER 8

Cheese Microstructure

G. SWAMY[a] AND K. MUTHUKUMARAPPAN*[b]

[a] Department of Nutrition, Food Science & Packaging, San José State University, San José, CA 95192, USA; [b] Department of Agricultural and Biosystems Engineering, South Dakota State University, Brookings, SD 57007, USA
*Email: Kas.muthukum@sdstate.edu

8.1 Introduction

Cheese is a good source of protein, vitamins, and minerals, particularly calcium and phosphorus, which are essential components in most highly consumed foods.[1] Therefore, it is of high importance to assess the cheese properties, *i.e.*, the end-product characteristics, which contribute to the quality of this dairy food. The end-product characteristics, such as flavour, physicochemical and functional properties (texture and melting properties), and quality of cheeses, are significantly affected by the microstructure.[2] Cheese microstructure is the spatial arrangement of the casein micelles that combine into clusters and chains to form a viscoelastic protein network throughout which moisture, fat globules, minerals, and bacteria are dispersed.[3] The microstructure is one of the major controlling factors of the texture (firmness, softness, cohesiveness, elasticity, pastiness, crumbliness) and functional properties of cheese; it also affects the physicochemical, transport, and nutritional cheese properties. As texture and functional properties are significant quality requirements for consumers, microstructure analysis plays an important role in the quality evaluation of the dairy products; better quality usually brings higher revenues and consumer satisfaction.

Food Chemistry, Function and Analysis No. 40
Handbook of Cheese Chemistry
Edited by Michael H. Tunick

Published by the Royal Society of Chemistry, www.rsc.org

Cheese microstructure, comprising the casein matrix in which the fat globules are entrapped, is routinely and most examined using light microscopy (LM), confocal laser scanning microscopy (CLSM), and electron microscopy such as scanning electron microscopy (SEM) and transmission electron microscopy (TEM). The main advantage of electron microscopy techniques is that they allow for a much higher resolution imaging of the components of the cheese in comparison to LM and CLSM techniques. Comprehensive reviews on cheese microstructure highlighting all historical developments and techniques, including not only microscopy but also rheology, light scattering, scanning calorimetry, and magnetic resonance instruments, have been provided by several food scientists.[4–6] These techniques are used in the study of factors affecting cheese microstructure, such as physical/thermal or chemical treatments and ripening, and its impact on functional properties.

8.2 Factors Affecting Cheese Microstructure

8.2.1 Processing Techniques

8.2.1.1 Homogenization

Homogenization of cheese milk creates smaller globules with a greater total fat–water interfacial surface area. To prevent immediate globule coalescence, the newly created interface must be coated with surface-active components present in the serum phase, usually casein micelles, micellar fragments, and whey proteins. The casein component of the new interfacial area can interact with the surrounding casein matrix to participate in the structure. The outcome is a firmer cheese with less free oil formation compared with non-homogenized cheese with the same fat content.[7] Homogenization of milk prior to the manufacture of cream cheese allows the fat globules to interact with the casein matrix and produce a more elastic cheese with less free oil formation. Microscopy, such as TEM and cryo-SEM, has shown that larger fat globules and less adsorption of casein micelles onto the globule surface enhance the spreadability of American-style cream cheese containing a minimum of 33% fat.[8] The effect of homogenization of cheese milk is to increase the whiteness of cheese by reducing the size of globules such that light scattering is more effective and to increase moisture retention in the finished cheese.[9,10] Reduced-fat Mozzarella cheese is less white due to a smaller number of colloidal-sized fat globules able to scatter light and more translucent due to less fat that would otherwise impair transmission of light.[4,11] The opacity of Mozzarella cheese is a thermally reversible process. Heating the cheese increases the whiteness level due to scattering by serum pockets and protein aggregates that are of similar size to the wavelength of light.[12,13] Cooling the cheese gives a translucent look as the light scattering intensity is decreased. This reversibility may be due to proteins aggregating and contracting at higher temperatures due to stronger hydrophobic association.[14–16] Microfluidization is a process wherein streams of a coarse emulsion collide under high pressure to form a very fine emulsion because of turbulence and cavitation. Milk that has been microfluidized comprises casein micelles with very tiny, embedded fat globules, suggesting that high-pressure homogenization disrupts the casein micelle structure.[17–19] This may be used to create different functionalities in cheese made from this type of homogenized milk.

8.2.1.2 High Pressure Processing

High pressure processing (HPP) usually implies applying pressure up to 600 MPa. Structural alterations can be caused in proteins of foods at above 200 MPa.[20,21] Above 500 MPa, a non-alterable impact can occur such as unfolding, aggregation, and formation of gel structures. High pressure treatment of skim milk results in a cheese with a dense network of casein strands, as observed by SEM, formed from partially disintegrated casein micelles.[22,23] As a result, the elastic modulus (G') is higher due to a greater density of casein chain cross-links, and syneresis is lowered due to the retention of whey in micro-pockets of serum phase. High pressure treatment of milk boosts the rate of setting of renneted milk, stable with a greater casein cross-linking density, and with an enhanced gel strength.[24,25] Treatment of milk at high pressures of around 500 MPa reduces the size of milk fat globules and disperses these more evenly in a more compact protein matrix, making a cheese that is firmer and more elastic.[26] The effect of high-pressure treatment of milk is to create a cheese that flows more easily, at least in the case of young Gouda.[27] High pressure treatment of goats' milk prior to cheese manufacture produces thinner strands of casein with more moisture retention in the final cheese.[28] The calcium mineral balance and milk fat crystallization are both affected by high pressure.[29,30] In addition to pressurization of milk, cheese can also be treated by high pressure. Pressure treatment of immature Mozzarella cheese boosted meltability and hardness and improved the swelling of the casein matrix concomitant with greater water retention.[31] Shredding of freshly made Cheddar cheese is augmented by pressurization. High pressure processing of milk or cheese can provide an opportunity to influence the functional properties of cheese.

HPP can reduce proteolysis and extend the shelf life of cheese products. This study evaluated the physicochemical, rheological, microbiological, and microstructural characteristics of HPP cheese made from ultrafiltered milk (UF-cheese). In ultrafiltered milk, the distance between casein micelles is smaller compared to the non-ultrafiltered milk.[32] Thus, the interaction between proteins is stronger due to the short distance between them. In addition, as verified from the texture results, HPP favours a higher cheese hardness. Therefore, this increased interaction of proteins in ultrafiltered milk associated with the effect of HPP may contribute to a greater compaction of proteins in cheese, as shown by the micrograph of cheese at day 1

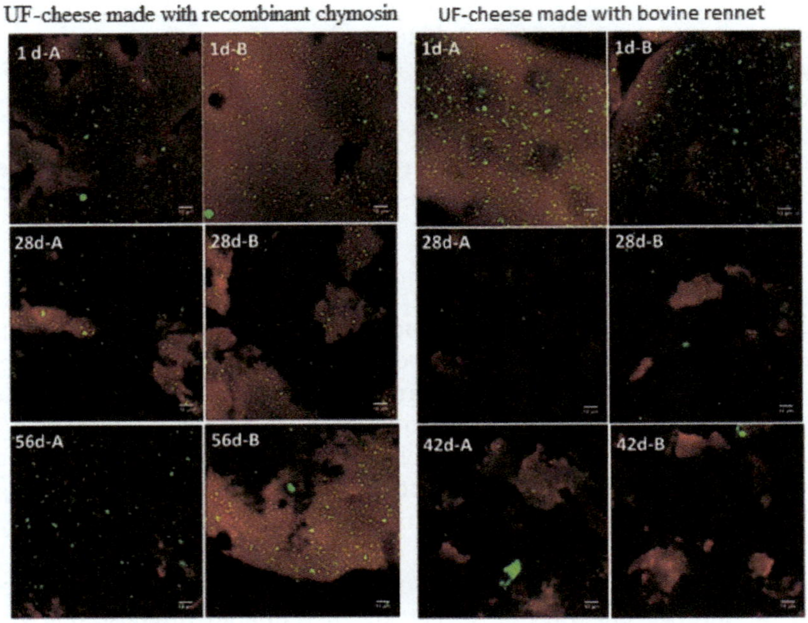

Figure 8.1 Effect of high-pressure processing on the microstructure of ultrafiltered milk-based cheese. Reproduced from ref. 32 with permission from Elsevier, Copyright 2018.

in Figure 8.1. During storage, proteolysis may make the casein network less cohesive as the gap between proteins becomes larger and there is greater syneresis, evidenced by the evolution of dark regions. However, after 56 days of storage of UF-cheese coagulated with RC, a greater aggregation of the protein network was verified in the samples processed by HPP compared to the control samples. This suggests less hydrolysis of the recombinant chymosin on the protein network due to inactivation during HPP. This result corroborates the results of higher hardness and lower proteolysis in HPP UF-cheeses.

8.2.1.3 *Freezing*

Slow freezing of cheese produces large ice crystals, whereas rapid freezing produces smaller crystals. Large crystals of ice can disrupt the cheese protein matrix, as shown by SEM, and increase the meltability of Mozzarella cheese.[6,33] Large ice crystals can create serum pockets within the protein matrix, as shown by the increase in water mobility by nuclear magnetic resonance (NMR) measurements.[34,35] Frozen Mozzarella cheese has a more porous protein structure, as shown by SEM, due to ice crystal formation.[36] The effect on frozen and thawed non-pasta filata cheese is similar to that on unfrozen pasta filata cheese in that both contain large pockets of trapped water. If the freezing rate is sufficiently fast to produce small crystals, the impact on protein matrix disruption may be minimal.[4,31] Freezing has also been shown to decrease sheep milk cheese firmness, presumably by reducing the density of casein cross-linkages and creating a more porous cheese and quantified by a smaller G' value.[33,37]

Freezing is an important process applied to cream cheese in cases where robust supply chains are needed, for example, to extend market reach, to provide food service customers with a consistent product, or to decrease waste and improve shelf life by reducing the rate of degradation during refrigerated storage. Hence, freezing is becoming a convenient tool for export or when the product is used as an ingredient for further manufacturing. However, from a physical point of view, freezing can have a negative impact on the cheese matrix: it can cause the rupture of the casein matrix because of ice crystal formation, with the creation of voids and large serum channels in the structure, the modification of the water status and distribution (*e.g.*, protein dehydration phenomena), and some rheological and sensory changes.[34]

The microstructural observations of the cheese samples in Figure 8.2 agreed with those obtained in Figure 8.3 in terms of microstructural changes caused by freezing and thawing for the protein and fat domain. The differences in the protein distribution at the interface, and the organization of the fat globule clusters, may have a role in modulating the freezing–thawing stability of the different cheese formulations; for example, a higher level of adsorbed proteins on the surface of fat globules/clusters may result in a reduction of fat globule clustering and coalescence, and improved stability. The use of confocal Raman microscopy allows one to better distinguish the effect of freezing–thawing processes of the two different cream cheeses. This was coherent with the low resolution of Raman microscopy obtained with the measurement parameters applied.

8.2.2 Cheese Composition

8.2.2.1 Water

Water exists in three main phases in cheese: bulk water in the serum channels, entrapped water near the casein matrix, or as bound water tightly associated adsorbed to the caseins and therefore unavailable as a solvent.[35,38] Heating and stretching fresh curd increase the proportion of mobile water molecules in Mozzarella cheese, and this is thought to represent the bulk water phase.[39] In the first few weeks of ripening, the proportion of entrapped water increases at the expense of bulk water, brought about by casein swelling and hydration.[35] This is confirmed by water mobility measurements using NMR. NMR water mobility measurements show that the protein phase in Mozzarella is less hydrated due to the presence of pockets of serum phase, compared with Gouda, and that the degree of casein hydration increases with age.[38,40]

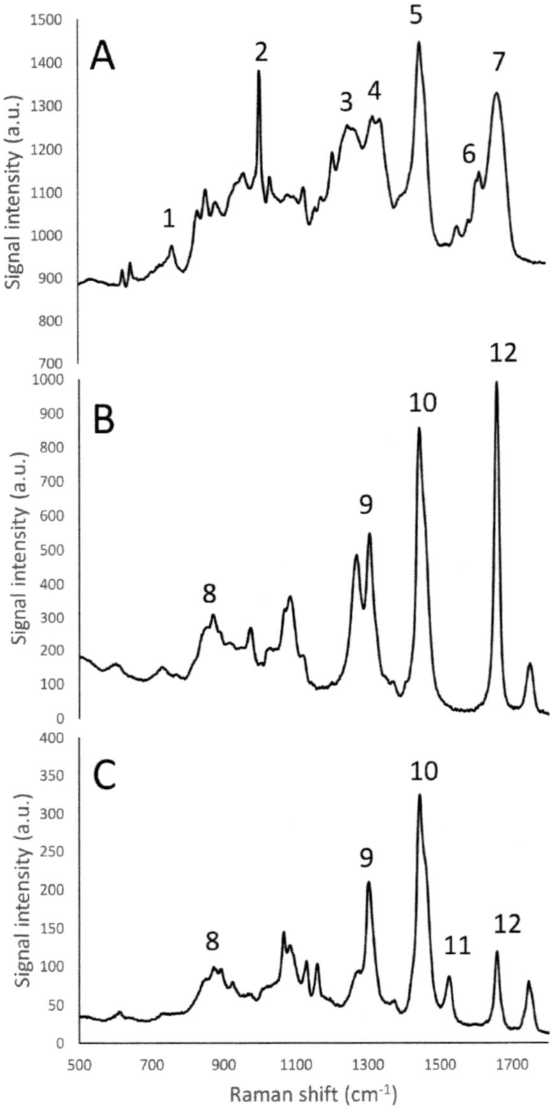

Figure 8.2 Freezing and thawing effects in the cream cheese microstructure using confocal Raman microscopy. Reproduced from ref. 74, https://doi.org/10.3390/foods9050679, under the terms of the CC BY 4.0 license, http://creativecommons.org/licenses/by/4.0/.

8.2.2.2 Calcium

The effects of calcium addition to milk and pH are not independent events. A reduction in the ratio of insoluble to soluble calcium is known to increase cheese meltability. This reduction can be facilitated by the addition of citrate, which chelates calcium ions, or by decreasing the pH.[41] Examination of

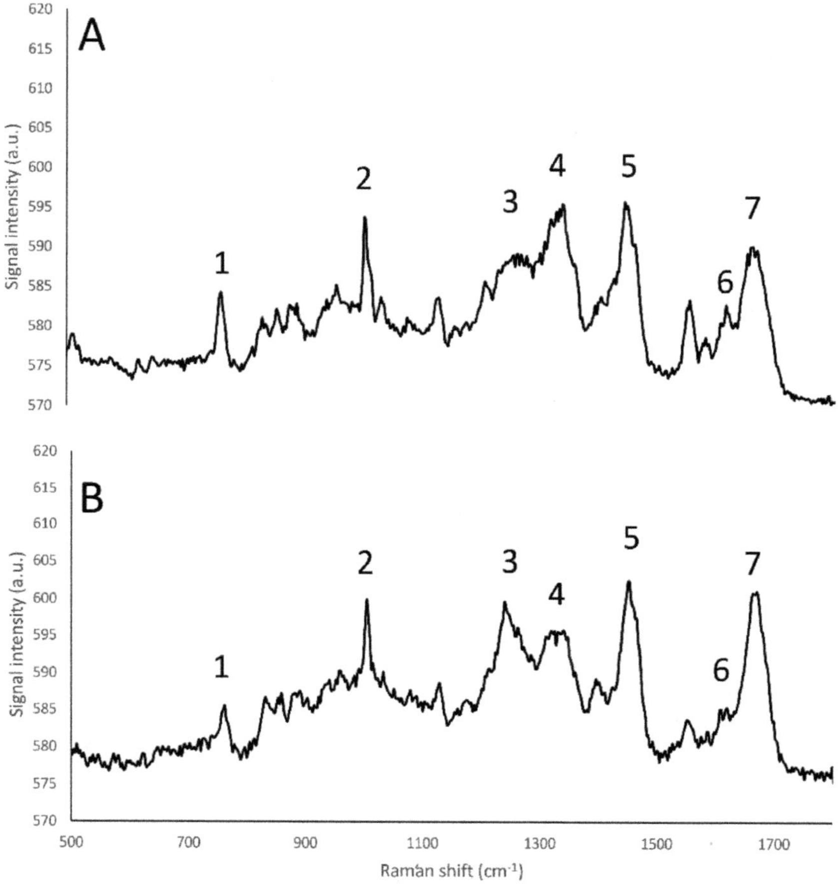

Figure 8.3 Freezing and thawing effects in the cream cheese microstructure using confocal laser scanning microscopy. Reproduced from ref. 74, https://doi.org/10.3390/foods9050679, under the terms of the CC BY 4.0 license, http://creativecommons.org/licenses/by/4.0/.

Cheddar cheese by SEM has shown that cheese with added citrate contains an expanded protein matrix, concomitant with the idea that a decrease in casein-bound calcium reduces casein–casein interactions, thus facilitating flow during melting.[42,43] As the pH of cheese is reduced toward the isoelectric point of casein (pH 4.6), there is a contraction of the casein matrix brought about by an increasing propensity for hydrophobic association of micellar aggregates. The size of protein subaggregates is reduced from 10 to 2 nm as the pH decreases from 5.3 to 4.7, illustrating well the compaction and rearrangement of the micellar structure in fresh curd as the pH is reduced over time. At a pH of around 5.6, the casein structural units interact through calcium-mediated electrostatic bonds. As the pH is further lowered, the bound calcium solubilizes, resulting in interactions between casein units by increasingly hydrophobic association. A maximum in flowability occurs at around pH 5.2–5.4. At greater pH values where calcium-mediated interactions are still prevalent, this decrease in casein subaggregate size is associated with an increase in cheese flowability. Increasing the amount of total (ionic plus insoluble) calcium in cheese enhances casein–casein interactions, thereby decreasing meltability and enhancing cheese firmness. A consequence of boosting casein interactions through insoluble (not soluble) calcium is the closing up of serum pockets, with less expressible serum. This is a consequence of caseins aggregating *via* bound calcium phosphate. Increasing pH above 5.4 reduces meltability; however, if the amount of calcium is reduced at this higher pH, protein hydration and swelling are promoted, concomitant with improved melting and stretching behaviour.

Calcium lactate crystals form when Ca^{2+} and lactate ions exceed their solubility, supersaturate the serum phase of the cheese, and then crystallize at nucleation sites within cheese matrices.[44] Figure 8.4 shows the comparison of second harmonic generation (SHG) images of the crystals, overlaid with two-photon excitation (TPE) fluorescence of the cheese matrix. Most starter bacteria ferment lactose to L-lactate; however, considerable amounts of D-lactate are formed by the activity of NSLAB during cheese ripening. The lower solubility of Ca-D-lactate then causes crystal formation. Crystallization of calcium lactate on the surface of Cheddar cheese is a common and troublesome defect. The deposits are visible to the naked eye; products containing crystal deposits are not aesthetically pleasing and can be sometimes rejected by the retailer/consumer who may confuse them with mould contamination. Calcium lactate crystals can be visualized using X-ray crystallography, but we will miss the chemical information about the placement fat and protein around the crystals. The crystalline deposits are also frequently observed by standard light microscopy, SEM or TEM in processed cheese products. Using various techniques such as electron microscopy, X-ray diffraction analysis, infrared spectroscopy, and energy dispersive X-ray spectrometry analysis (EDSA), several crystal types have been identified, *e.g.*, monoclinic calcium pyrophosphate dehydrate, disodium phosphate dodecahydrate, unreacted melting salts, tyrosine, calcium citrate,

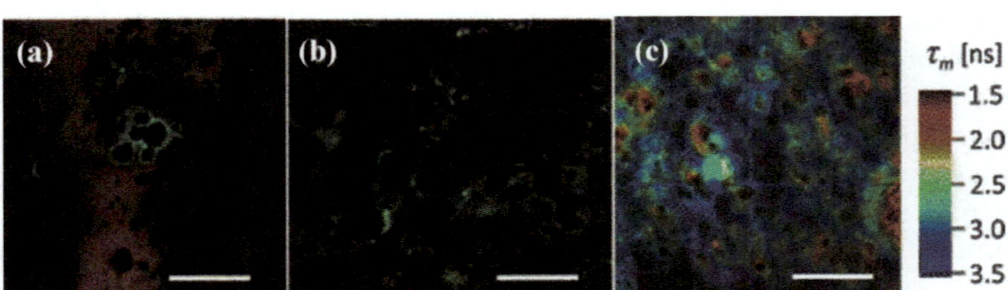

Figure 8.4 Calcium lactate crystals embedded in the Cheddar cheese matrix. Reproduced from ref. 44, https://doi.org/10.1007/s13594-015-0253-2,

fatty acids, protein, and lactose. The composition of crystalline structures in natural cheeses and PCPs was studied by SEM EDSA and X-ray diffraction. The calcium specific fluorescent stain Alizarin Red was used to identify calcium lactate crystals by optical microscopy.

8.2.2.3 Salt Content

Salt is added to freshly made curd to limit the action of starter bacteria from reducing the pH to unacceptably low values and assisting in whey expulsion immediately after curd milling (in the case of dry-salted cheeses), as well as the secondary effect of flavour enhancement. Higher levels of salt induce a swelling of the casein phase with subsequent adsorption and absorption of moisture by the casein matrix, up to at least 1.6% NaCl in Mozzarella.[4] This has been observed in Mozzarella and Muenster cheese. The protein phase in unsalted cheese will swell at a slower rate compared with salted cheese, and this explains the slower rate of reduction of expressible serum over time. Such unsalted cheese has larger pockets of serum phase. Unsalted directly acidified Mozzarella cheese contains a more porous protein matrix because of enhanced casein–casein interactions, thus impairing meltability. The application of salt will draw moisture out of a curd particle. Paradoxically, if this is done over a short period of time the water will remain entrapped in the curd structure due to the formation of a dehydrated cheese surface barrier that impedes the flow of water. To clarify this point, movement of water from Ragusano, a Sicilian brine-salted cheese, is hindered owing to the development of a dense dehydrated protein layer at the surface of the cheese, as examined by SEM as shown in Figure 8.5.[45] This effect is enhanced by higher salt levels and longer brining times.

8.2.2.4 Fat Molecule Size and Microstructure

Fat globules serve several purposes in cheese. First, the fat is necessary for flavour development. This can clearly be shown by the uncharacteristic flavour in low-fat cheese. Whether this is due to fat-soluble flavour precursors, components of the milk fat globule membrane, fatty acid flavour precursors, or perhaps the requirement of an oil–water interface for enzymes to maintain high activity and produce flavour compounds is open to debate. Likely, all these mechanisms play a role in flavour development. It should also be noted that the differences in protein structure between full-fat and non-fat cheese may have some impact on sensory perception of flavour. The second role played by fat globules is a casein matrix structure-breaker, thus reducing the firmness of cheese and providing some lubrication during mastication. A third requirement is to open the casein matrix to form serum channels. The absence of fat globules yields cheese, in many cases with a texture that is too firm and with low flavour intensity.

Fat is found in cheese either as small globules of 2 μm in size, aggregates of globules, or as large areas of fat 10–50 μm in size that are most likely pools

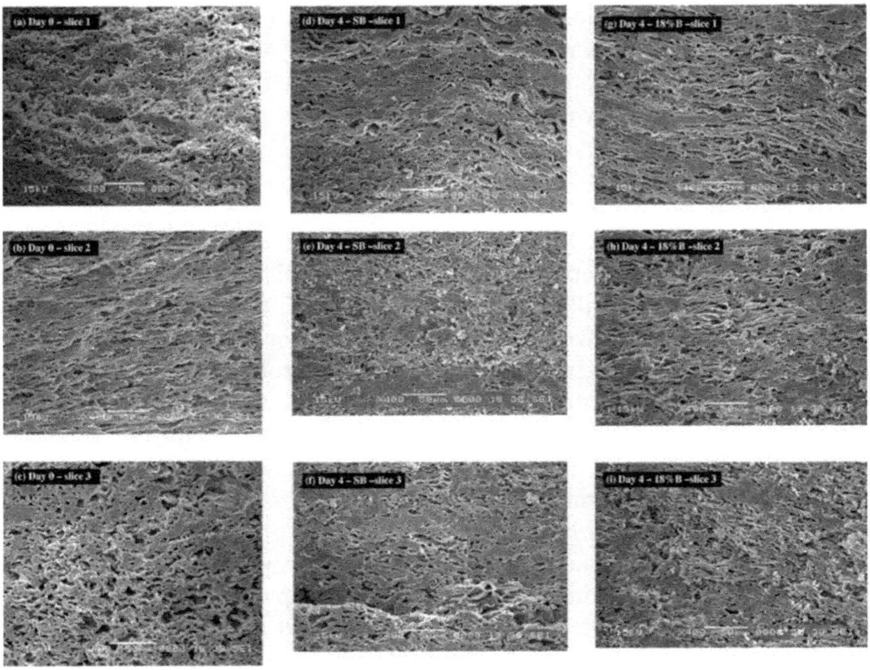

Figure 8.5 Impact of salt content on water movement in Ragusano cheese. Reproduced from ref. 45 with permission from Elsevier, Copyright 2005.

of free fat trapped within the protein matrix. Analysis of the size and shape of fat areas in cheese can be done by computerized two- and three-dimensional analysis of confocal micrographs with appropriate fluorescent staining of the fat or the aqueous protein phase. Aggregates of fat globules in Cheddar cheese are observed to align along the direction of the casein fibres, presumably due to the forces imposed by the cheddaring process. Elongation of fat globules is not observed in Gouda and Edam, in which cheddaring or stretching does not take place during manufacture, or in Cheshire. Fat globules physically interfere with the integrity of the cheese casein matrix and soften texture. This process will occur if the globules are held into place through relatively weak interactions with the surrounding casein matrix, or to a lesser extent if the globules participate as copolymers with the casein matrix, as will be the case when globules are homogenized and partially coated with casein micelle fragments. There is some debate about whether natural fat globules are inert filler materials or if there is some interaction with the casein matrix. The observation using cryo-SEM that fat globules are surrounded by serum cavities in fresh Cheddar cheese supports the idea of globules being inert filler materials.

A comparison between confocal micrographs of cheese with native fat globules and those of cheese containing globules coated with phosphatidylcholine indicated a lower amount of fat in cheese containing globules coated with phosphatidylcholine.[8] The fat globule diameter of the phosphatidylcholine-coated emulsions was standardized to 2 μm prior to addition to the cheese milk by adjusting the homogenization pressure, matching the diameter of the native fat globules. Smaller globules are more likely to be entrapped within small voids in the casein matrix and be retained in the curd structure compared with large globules. For native fat globules, binding with the casein matrix would prevent the loss of the larger globules after cutting the curd during cheese manufacture. The state of fat is important for dictating the sensory properties of dairy products. Fat globules coated with caseinate and recombined with skim milk for cheese manufacture tend to be more spherical compared with native globules, which are distorted and partially aggregated. Manufacture of cheese containing globules coated by homogenization with relatively poor emulsifiers such as α_{a2}-casein produces large pools of fat in the cheese that may explain the observed greasy and crumbly texture. Evidently large pools of fat do not integrate well into the casein matrix. The larger globules produced by homogenization of milk fat with α_{a2}-casein (compared with other caseins that are better emulsifiers) are apparently easier to rupture, producing oil that is located within voids in the casein matrix and that does not bind to the matrix. In this case, no strengthening of the cheese protein matrix takes place.

The structure of fat has a knock-on effect for the melting of cheese and for free oil formation. Fat globule rupture is likely to lubricate layers of casein and allow easier flow, thus enhancing melting. Cheese made from recombined milk containing a lower melting point fraction of anhydrous milk fat has more free oil formation and lower viscosity (improved melting), presumably as the more liquid fat phase is able to permeate among layers of casein and facilitate melting. Pools of free fat likely contribute to free oil formation on the surface of heated cheese. One application of manufacturing cheese from recombined milk containing fat globules coated with specific emulsifiers is to produce a cheese where the globules have built-in rupturability. The globules can be induced to partially rupture over time to produce a controlled small amount of free oil that may be required for specific cheese functionality, such as enhanced spreadability or meltability. The size of globules will impact upon texture. Larger globules with a smaller Laplace pressure of deformability will decrease the firmness of cheese. Conversely, small globules are much harder to deform and rupture, and are also less likely to disrupt the casein matrix, thus contributing to a firmer cheese.

Confocal microscopy is a simple technique to assess globule size. It is preferable to optically section into the cheese and take a succession of two-dimensional micrographs with subsequent assembly into a three-dimensional structure by computerized image analysis. Although this is a process that requires more computing power, it eliminates the problem of estimating the size of non-spherical globules from two-dimensional micrographs. Although analysis of a two-dimensional image can potentially underestimate the size of a fat particle by taking a section through the tip of the structure, stereology can be used to derive statistically valid data on size and shape from two-dimensional images. Fat globule aggregation is reduced by homogenization of milk prior to cheese manufacture. Flowability and stretchability of cheese are adversely affected in this case and are quantified by a reduction in tan δ (ratio of viscous modulus, G'', to G'), as measured by dynamic rheology. Aggregation is particularly evident when cheese is heated. In pasta filata cheese varieties, the fat globules begin to aggregate—as shown by SEM—and coalesce to form pools of free oil within the interstices of the casein matrix. Extensive fat globule aggregation and rupture is supposed to contribute to free oil formation. These pools of free oil have a large degree of distortion. One problem with labelling of triglycerides in cheese and observation is that there is no clear way to see if elongated fat structures are composed of aggregated spherical globules or if the entire structure is a pool of free fat trapped within the protein matrix. One solution is to label the fat phase of the cheese with fluorescent tags that selectively bind to milk fat globule lipids or use antibodies conjugated to a component of the fat globule membrane to measure the extent of membrane rupture. Often these conjugated labels are difficult to produce and expensive to purchase.

Process cheese is manufactured by heating natural cheese in the presence of emulsifying salts such as sodium salts of citrate, orthophosphate, pyrophosphate, and sodium aluminium phosphates. The structure differs from natural cheese in that there are no evident curd granules, but there are a more uniform protein matrix and less fat globule aggregation. Higher heating and mixing speeds promote fat globule homogenization and a smaller globule size. Emulsifying salts are added to process cheese to promote the solubilization of the casein matrix and allow caseins to function effectively as water-soluble emulsifiers of milk fat that are released during heating of the cheese. As a result of the addition of these salts, caseins are hydrated, calcium ions are chelated, and the pH rises. All these mechanisms promote solubilization of the casein matrix. As the degree of emulsification of the fat globule increases, meltability, free oil formation, and globule size decrease, whereas firmness increases as the globules can participate as copolymers with the surrounding casein matrix. This increase in firmness is in comparison to cheese containing the same fat level, but where the globules are not homogenized.

8.2.2.5 Protein Microstructure

A milk gel is formed by the addition of rennet at around pH 6.5 as a precursor to cheese manufacture. Lactic acid bacterium (LAB) cultures are added to reduce the pH. Casein demineralization occurs as the pH decreases, thus reducing electrostatic interactions between casein molecules. Along with this trend, the propensity for hydrophobic interaction of micelles increases as progressive charge neutralization takes place and the pH decreases toward the isoelectric point. This latter interaction occurs when sufficient glycomacropeptide has been enzymatically removed from the *para*-casein by chymosin. These interaction mechanisms can be observed using dynamic rheological methods. At high pH around 5.5–6.0, a local maximum of $\tan\delta$ indicates increased fluidity due to partial demineralization of the casein micelle, thus less interaction between casein molecules. A local minimum of $\tan\delta$ at around pH 5.0–5.5 indicates increased solid-like behaviour due to rennet-induced interactions. As the pH is further reduced, a transition takes place from a rennet to an acid gel structure. At pH 4.6 rennet gels are more permeable than acid gels, and rennet gels have thicker protein fibres, as established by electron microscopy studies. Therefore, rennet gels are more able to undergo syneresis, producing lower moisture cheese, compared with acid milk gels. Rennet gels are more open and porous at pH 6.5 than at pH 5.2. A slower rate of cooling of rennet gels produces a larger number of smaller casein micelle aggregates, more junction zones, and stronger and less permeable gels. A higher number density of casein chain cross-links will increase the value of G'. Heating of milk prior to acidification produces a less permeable and firmer acid gel due to casein cross-linking by denatured whey proteins. By contrast, unheated acid milk gels are more porous and prone to syneresis, as observed by CLSM and permeability measurements.

A characteristic of cheese ripening is the transformation of a fibrous casein matrix, formed from chains of aggregated micelles, into a more amorphous structure. Over time, the fusion of casein micelle chains creates a more homogenous structure with smaller serum cavities, providing an explanation for the observed reduction in expressible serum by centrifugation in aged cheese. During manufacture of Cottage cheese, an example of an acidified milk gel structure, the size of casein micelles increases from around 80 to 200 nm, consistent with aggregation of micelles into chains during the gelation process. The fusion of curd particles creates junction zones in the ripening cheese. These originate from two aspects of curd fusion: first, thin lines from fused curd particles, and second, thicker boundaries deficient in fat globules from fused salted milled curd fingers. Cream cheese is manufactured from milk fat standardized to a higher fat content and cultured with LAB to a pH of 4.6. A defect in some cream cheeses is a gritty mouthfeel. The grittiness may originate from large, compacted casein and denatured whey protein aggregates up to 1 mm in size, as observed using SEM. The coarseness of the casein matrix can be quantified by TEM. Faster rates of coagulation of UF milk result in a coarser casein matrix with a higher surface area-to-volume ratio of the casein, and this is positively correlated with a higher stress at fracture in Feta cheese. As an example, increasing the ratio of caprine to ovine milk in Feta cheese manufacture gave a finer casein matrix, as measured by cryo-SEM, and a firmer cheese. Camembert cheese made from reconstituted milk yields a more porous protein matrix due to incomplete fusion of casein aggregates. An increase in curd acidity, or storage at a higher temperature within a day of manufacture, also increases the degree of coarseness of the protein matrix.

Vigorous arguments have been presented for the presence of highly structured casein submicelles and for a more disordered structure of casein substructural units as the constituents of casein micelles. Models featuring a more disordered structure are the dual-binding and nanocluster models, although there is some argument that the nanocluster model is just a particular case of the dual-binding model. An increase in the size of electron-dense regions from 12 to 23 nm has been observed in Mozzarella cheese over 6 weeks of ripening using TEM, corresponding to the size of casein substructural units, and consistent with the idea of increased casein solvation over time. The spacing between these units also increases during ripening and perhaps may facilitate increased melting and softening. These units are smaller in size for lower pH cheese: 10–15 nm in Gouda cheese at pH 5.3 to 3–4 nm in Cheshire cheese at pH 4.6, which is correlated with a decrease in elasticity and an increase in the brittleness of cheese.

8.3 Cheese Microstructure Analysis

Cheese microstructure illustrates the three-dimensional arrangement of the casein micelles that link together into bands and chains. These micelles create a viscoelastic protein network in which moisture, fat globules, minerals, and microbial cells are distributed thoroughly. One of the major influencing factors of the texture and functional properties of cheese is the microstructure. It also has an impact on the physicochemical, transport, and nutritional properties of cheese.

8.3.1 Microscopy

8.3.1.1 Light Microscopy

LM is regularly used to obtain rapid, inexpensive qualitative and quantitative information in cheese microstructural analyses.[4,40] This is considered as the principal advantage of using LM techniques. The traditional LM techniques used are bright-field, polarised, and fluorescence LM. The basic microscope is a conventional bright-field microscope, to which polarising or fluorescence accessories are easily attached. In addition, confocal microscopy has also been developed. The major difference between a confocal and a conventional LM is the placement of a pinhole, at the focal plane of the image, which removes out-of-focus glare, therefore generating a clearer image.[5] In microstructural analyses of cheese, CLSM, which uses laser illumination, has frequently been used.[46,47] In all LM techniques, there is a need for sample preparation as summarised in Table 8.1 for the microscopic examination, where a wide variety of stains are used. However, sample preparation can result in production of artefacts, a major disadvantage of LM, especially if the cheese has been subjected to freezing, cooling and/or heating.

Table 8.1 Summary of microscopic techniques used for the microstructural analyses of cheeses.

Cheese	Objective	Results	Ref.
Swiss type	Investigation of curd grains in Swiss-type cheese using light microscopy and CLSM.	LM allowed direct observation of curd grain junctions and showed an anisotropic curd grain organization as they were only visible perpendicularly to the pressing axis of the cheese. Moreover, curd grains around eyes appeared highly stretched, whereas they maintained a brick-like pattern 1 mm away from eyes. CLSM showed local differences in composition between the insides of the grains and their interfaces, with fat-depleted junctions only composed of protein, which could act as a structuring element of the body of the cheese. However, it seems that protein interfaces can turn into zones of fragility inside the body of the cheese under certain stress conditions.	55
Cheddar	To study the impact of milk protein concentration during the process of ripening, including changes in proteolysis and texture development.	The milk protein concentration did not appear to have any effect on the structure of the cheese at any period of ripening when observed by CLSM.	56
Block	Block-processed Ras cheese was produced with two salt mixtures: (1) Na diphosphate + Na polyphosphate + Na tripolyphosphate in various ratios and (2) Na polyphosphate + Na citrate + Na orthophosphate + Na diphosphate in various ratios. Texture profile analysis and microstructure (LM and TEM) of resultant cheeses were studied.	Light microscopy images indicated different emulsification degrees with various emulsifying salt mixtures. These observations were confirmed with the image analysis and TEM. Among the entire treatments, mixture (1) in the ratio of 30 : 40 : 30 and mixture (2) in the ratio of 40 : 10 : 20 : 30 gave the best and closest texture to the commercial salts and can be recommended in the production of block processed cheese.	57

Table 8.1 (*Continued*)

Cheese	Objective	Results	Ref.
Australian cream	CLSM was used to compare the microstructure of six Australian commercial cream cheese products.	CLSM revealed a typical cream cheese microstructure of homogenized fat globules embedded in a non-continuous protein network. The association between fat and protein within the microstructure was influenced by product composition (fat : protein ratio, moisture content) and ingredients. The addition of an emulsifier led to a softer product with a distinct microstructure. Cryo-SEM also revealed a honeycomb-like structure, which was interpreted as a eutectic artefact formed by the addition of gum(s). Product hardness and gel strength generally correlated with high fat, low moisture content and a compact microstructure.	58
Cheddar	To investigate sample preparation parameters including the freezing method, etching time, thickness of the metal coating, acceleration voltage, and the type of detector used for SEM observation with an aim to minimise the formation of artefacts. The potential of cryo-SEM was also assessed as a technique to visualize the microstructure of the highly hydrated gel, curd and cheese samples collected during Cheddar cheese manufacture using homogenized or unhomogenized milk. A further aim was to follow the development of the microstructure during the manufacture	The results showed that cryopreservation did not alter the microstructure of the sample when it was fixed by rapid freezing in slush liquid nitrogen due to the formation of amorphous ice. The etching time, thickness of gold coating, accelerating voltage and the type of detector used for cryo-SEM observation were varied to minimise the formation of such artefacts and optimise conditions for imaging. Chains and clusters of casein micelles and fat globules were best observed in the gel and the cooked curd when the samples were freeze-fractured and etched for 30 min, coated with a	59

	of Cheddar cheese using CLSM and to compare qualitative and quantitative data obtained by cryo-SEM and CLSM.	mixture of gold and palladium alloy approximately 6 nm thick at −140 °C and observed using a backscattered electron detector at 15 kV. The structure of the gel, curd and cheese was also observed using CLSM. Spherical fat globules were mostly present in the serum pores of the gel prepared from unhomogenized milk but were found to be embedded in the aggregated chains of the casein network within the gel prepared from homogenized milk when observed using CLSM. The porosity measurements obtained using cryo-SEM were similar to those obtained using CLSM. These two complementary techniques can potentially be used to assist studies for the control of cheese texture and functionality.	
Sheep milk	To investigate the effect of processing parameters, curd grain size assessed by image analysis, and cooking temperature on the microstructure, texture, composition, and yield of cheese from sheep milk.	The rennet-induced gel had a compact structure for all cheesemaking treatments, with a mean porosity of 0.36 ± 0.08. This low porosity was expected due to the high fat and protein content of raw sheep milk.	47
Probiotic buffalo milk Ricotta	To develop a probiotic cheese and study the textural properties.	Scanning electron microscopy revealed that Ricotta cheese matrix mainly consisted of protein network. The particles of coagulated whey proteins were interconnected, and casein appeared as small clusters. Though fat is present in the matrix, it seems that fat did not interact with the protein network. In probiotic Ricotta cheese rod-shaped lactobacillus species was clearly observed in SEM analysis.	60

Table 8.1 (*Continued*)

Cheese	Objective	Results	Ref.
Kashar	To microstructurally characterise Kashar cheese using SEM.	The control Kashar cheese was characterized by a compact homogeneous structure, mostly composed of the casein matrix with dispersed fat globules of variable size and shape distributed within the protein matrix. In the control sample, fat globules were round with diameters ranging from 5 to 15 μm, and individual casein micelles could be observed. However, in the compressed samples, there were an increased number of pores, which had a more open structure, probably led by a reduced distribution of fat within the protein matrix. This was more the case for the samples compressed at further compression levels. Compression seems to have destroyed the fat globules or forced the fat to exude from the surface.	61
Processed	To investigate the emulsification process and protein interactions in more detail by high-resolution TEM augmented by LM of samples taken along the creaming reaction.	In samples taken after 15 min and 50 min of processing time and prepared for LM, fat globules of various sizes were observed as in homogenously distributed clusters within the casein matrix. Fat globules are visible as electron-dense spherical objects of varying size that predominantly clustered together. The separation of the protein matrix, on the other hand, was easily followed, and first signs were already evident after 50 min of processing (this was barely noticeable in LM images after 175 min of processing) with the appearance of slightly inhomogeneous areas, sometimes with a thin strip of higher electron density at their periphery.	62

In cheese, information on chemical composition or structural and spatial relationships is required. There is a need for sample preparation prior to the microscopic examination. The sample is sectioned to sections of ~5 µm thickness before examination. Samples are made rigid either by embedding, using a resin after dehydration by alcohol, or by freezing to allow for the sectioning. However, it must be noted that both methods might produce artefacts in the micrographs obtained due to the dissolution and shrinkage of the fat component of the cheese, respectively. Embedding of the cheese samples is done by using paraffin or a low viscosity resin. In this method, dehydration is achieved by using alcoholic formalin and acetone, followed by sample embedding and sectioning using a glass ultra-microtome. Freezing of the samples is frequently and recently used in the microstructural analyses of cheese. Small rectangular pieces of samples are frozen in liquid nitrogen and mounted on an object disk with a cryo-protective embedding agent, usually a water-based glycol such as a medium known as the optimal cutting temperature (OCT) compound. Samples are then cryosectioned in a cryostat microtome at about −30 °C. In both methods, the sections obtained are then fixed in a microstructure fixator such as formaldehyde or glutaraldehyde. Fixed sections are then stained for lipid, protein and other ingredients present such as carbohydrates or starch.

Specific stains are used to identify mainly protein, fat, and carbohydrate components within the cheese matrix and also to provide contrast of the sections obtained. The most frequently used stains for protein are fast green, acridine orange and methylene blue, whereas the stains for fat are oil red O and Sudan dyes. To identify the carbohydrates, such as starch, within the cheese matrix iodine is used in combination with the stains of fat and protein. For instance, fast green, Sudan III and iodine have been used to stain the protein, fat and starch constituents, respectively, in the LM micrographs of processed cheese.

In polarised LM, the same sample preparation as in bright field LM may be used to examine the microstructure of cheese, but generation of artefacts remains a concern. Polarised microscopy is used to examine food components within the cheese matrix that exhibit birefringence, *i.e.*, an ordered crystalline structure. When a polarised light is passing through a birefringent material, this light vibrates in different planes and travels at different speeds. The material has different refractive indices according to its thickness and this difference is termed birefringence. Therefore, birefringent materials appear bright or coloured against a dark background.

Many food components are birefringent, such as fats and carbohydrates. For instance, polarised LM has shown that the microstructure of milk fat, mainly the crystal morphology, can be modified by changing its crystallization temperature and composition. In cheese, starch that has been used as an ingredient exhibits birefringence. Processed cheese containing starch was observed using polarised LM to detect whether the starch has been gelatinized. The micrographs of the cheese did not show birefringence in the case of non-gelatinised starch, whereas birefringence was observed in microwave-heated cheese products indicating the presence of gelatinised starch. The advantage of using polarised over bright field LM is mainly that it does not require staining, but polarised LM is only suitable for the examination of birefringent materials.

8.3.1.2 Scanning Electron Microscopy

In SEM, an electron gun generates electron beams that interact with the specimen and as a result several derived radiations, called secondary electrons, are emitted. The technique operates with secondary electrons, where the technique requires that microscopic examination is under vacuum and that the specimen is completely dry to generate an image providing sample information including a topographic view. To obtain a high number of secondary electrons that allow for the imaging, it is necessary to coat the specimen with a conducting layer to prevent charging. SEM has been used to study the microstructure of cheese for many years. While observing the microstructure through SEM, a stream of electrons is fired at the surface of the sample, and the image obtained is based on electrons that are scattered back from the surface. Cheese is cut with a razor blade into rectangular strips, followed by the fixation, dehydration, and defatting by using glutaraldehyde solutions (~2%, buffered near the pH value of the cheese), ethanol, and chloroform, respectively.[48] The cheese samples are then freeze-fractured in liquid nitrogen, thawed in ethanol, and finally, at the critical point, dried with liquid carbon dioxide. The dried pieces are mounted on aluminium stubs using silver paint or 2-sided sticky tape and coated with a thin layer (20–30 nm in thickness) of gold–palladium in a sputter coater in argon medium.[49] If fat is the component of interest, cheese samples that are fixed in glutaraldehyde are post-fixed with imidazole buffered osmium tetroxide and not defatted. Furthermore, sample preparation is carried out at low temperatures to prevent any damage to the fat globules. However, the use of ethanol and freezing in sample preparation might cause a slight damage and rupture to the structure of the cheese matrix, which is considered as an artefact related to the use of SEM. In recent work on the microstructure of cheeses, bacterial cells rather than fat were the components of interest and therefore a series of concentrations of ethanol were used in the dehydration step. This step could help in the localisation of the bacterial cells within the cheese matrix without any interruption from overlapping fat globules. Finally, after sample preparation, the microstructure examination of the cheese using conventional SEM of the prepared specimens is performed immediately, at an acceleration voltage of about 5–15 kV and a magnification of 500–10 000×.

Cryo-SEM, where a cryo-chamber is attached to the conventional SEM, has facilitated the microscopic examination of dairy foods high in water, air, or fat with minimal distortion, such as yogurt, ice cream, and most types of cheeses.[46] This SEM technique is used at temperatures below −80 °C during the whole sample examination protocol. Cryo-SEM has been used for the microstructural analyses of cheese due to its advantages of minimized sample preparation and much less consumption of electron microscopy grade chemicals. For instance, cryo-SEM is more suitable for full fat cheese, especially where the fat component could be highly altered by methods of defatting or dehydration, which might result in artefacts, used in the protocol of the conventional SEM technique. However, in processed cheese, the very low temperature used in cryo-SEM during the sample preparation and microscopic examination can lead to the shrinkage of the cheese, resulting in the appearance of the protein matrix as smooth surfaced in the micrographs as shown in Figure 8.6.[50,51]

Cryo-SEM has been extensively used in the microstructural analyses of processed and cheddar cheeses. Briefly, cheese blocks of approximately 1×3×5 mm^3 size were mounted on a specimen holder, followed by cryo-fixation by rapid plunging into nitrogen slush (−210 °C). Specimens were then transferred (under vacuum at −180 °C) to the cryochamber and the interior exposed by slicing off the apex of the block using a scalpel. The samples were sublimed, sputter-coated with gold, and then introduced in the microscope chamber where it was examined at a similar acceleration voltage and magnification to the conventional SEM technique. In the cryo-SEM micrographs of processed cheese, containing fat and starch, the protein

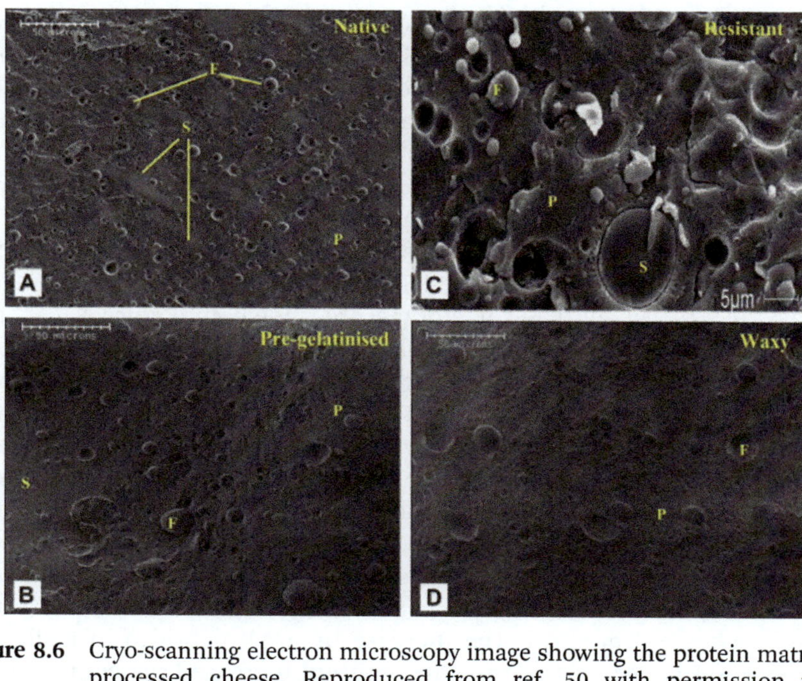

Figure 8.6 Cryo-scanning electron microscopy image showing the protein matrix of processed cheese. Reproduced from ref. 50 with permission from Elsevier, Copyright 2008.

matrix had a smooth surface in the presence of nearly spherical fat globules. The identification of the particles of starch was possible as these particles were broadly spherical but separate from the protein matrix, with bigger sizes than fat globules.

Environmental scanning electron microscopy (ESEM) differs from conventional SEM in that ESEM uses a gaseous environment in the chamber in which the sample is being microscopically examined. Compared to the examination under vacuum in the case of the conventional SEM, ESEM has the advantage of allowing the specimen to be examined at a relatively higher pressure and it is not required to have the samples completely dry.[52] This means that if water vapour is the gas used in the sample chamber of ESEM, hydrated samples can be imaged in their native state, *i.e.*, with no alterations due to sample preparation necessary. In addition, there is another major advantage of using ESEM rather than conventional SEM, in that the samples are not subjected to very low temperatures, *e.g.* −210 °C when liquid nitrogen is used, during their preparation prior to the microscopic examination. Therefore, ESEM might allow for the microscopic examination using a high-resolution microscopy technique of food matrices with nearly no significant negative effects (artefacts created) related to the sample preparation, particularly artefacts related to freezing and the use of very low temperature during the microscopic examination.

ESEM has been recently used in a limited number of studies on the microscopic examination of cheese matrices.[5] The structures shown in the micrographs of natural cheese were compact, rough, and closed structures with small voids. In processed cheese, considerable differences were found between cryo-SEM and ESEM micrographs and this difference is due to the sample preparation necessary for cryo-SEM.[50,51] The ESEM results showed a porous protein matrix rather than a smooth-surfaced matrix in the cryo-SEM micrographs. However, ESEM was not useful for distinguishing starch and fat globules. This might be a result of a lower resolution of the micrographs obtained, compared to cryo-SEM, which is considered a disadvantage of using the ESEM technique for the microstructural analyses of cheese. The ESEM images of experimental cheeses at days 14, 28, 42, and 56 of storage are shown in Figure 8.7.[53] The micrographs show compact and closed structures with small voids in all experimental cheeses. This suggests that the microstructures of Halloumi cheeses kept in various NaCl and KCl mixtures did not differ compared with that of the control. However, voids disappeared slightly with prolonged storage, and this was more obvious in cheeses stored for 14 and 56 days as compared with those stored between periods close together, such as days 42 and 56. This may be attributed to a decrease in moisture content and an increase in proteolysis.[54] It has been reported that proteolysis in cheese produces small peptides that are soluble in the serum phase of cheese. Therefore, the increase in small peptides in Halloumi cheeses and the decrease in moisture content increased the compactness of Halloumi cheese texture.

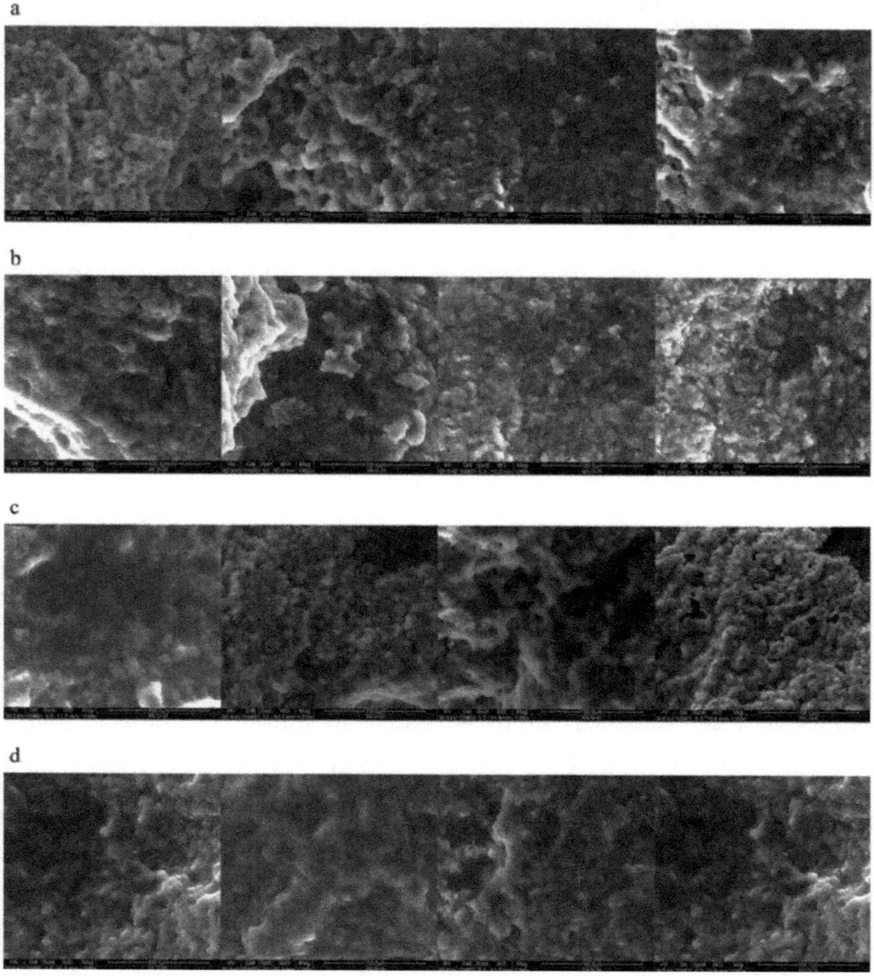

Figure 8.7 Environmental scanning electron microscopy images of salted Halloumi cheese. Reproduced from ref. 53 with permission from Elsevier, Copyright 2011.

8.3.1.3 Transmission Electron Microscopy

The TEM approach uses electrons that pass through the specimen, and the image formation is based on scattered electrons. Accordingly, specimens used for the examination by TEM need to be very thin (0.1–0.2 μm) to obtain a good EM resolution. TEM techniques practiced most frequently in cheese are the thin sectioning and replica methods. To examine cheeses using the thin sectioning technique, small samples (approximately $1\times1\times1$ mm^3) are chemically fixed, embedded, and sectioned, which is followed by the microscopic examination at similar magnifications, but at higher acceleration voltages (50–70 kV), to those used in SEM techniques. The fixation of the cheese sample is performed by using aqueous solutions of glutaraldehyde and osmium tetroxide for the sequential fixation of protein and fat components. This is followed by a dehydration using ethanol mixed with acetone and embedding of the sample by a low-viscosity epoxy resin such as Epon, Araldite, or Spurr's resin. The acetone increases the miscibility with the embedding agent used. Thin sections of ~0.1–0.2 μm are cut using an ultra-microtome with a glass knife. The sections are shed in a copper grid and stained with 2% aqueous uranyl acetate and lead citrate, followed by the microscopic examination. For the microstructural analysis of high fat cheeses, the replica technique is most often used as an alternative to thin sectioning. The samples of cheese ($1\times1\times1$ mm^3) are soaked in glycerol overnight before being plunged into liquid nitrogen, followed by freeze-fracturing, sublimation, and coating with platinum and carbon to create the replica. A carbon–platinum replica of the surface of the cheese is prepared by evaporating platinum at an angle of 20° to the surface, followed by the evaporation of a carbon layer from directly above the surface. This replica is cleaned by placing it into an acid solution, such as chromic acid, for a period of 2–3 days. Finally, the replica is washed in water, then in acetone, and mounted on copper/rhodium grids to be immediately examined by TEM, and the specimen of cheese can be dissolved away from the replica.

The main advantage of TEM in the microstructural analysis of cheese samples of fixed thickness prepared using an ultra-microtome is that it enables the highest resolution of all EM techniques. This can provide a basis for imaging and the use of software for image analyses. TEM also allows for the scrutiny of the ultrastructure and a higher consistency and accuracy of the results obtained in comparison to other EM techniques. However, there are numerous disadvantages that limit the use of TEM in cheese microstructure analysis due to the excessive sample preparation, time consumed, and high costs involved. Another disadvantage is related to the artefacts resulting from sample preparation. In addition to artefacts of fat solvation and shrinkage due to the steps of dehydration and freezing as explained in the sections of bright filed LM and cryo-SEM, other artefacts may occur due to the use of osmium tetroxide. The use of osmium tetroxide can lead to inadequate fixation of the fat and disruption of the protein, and therefore the structures may be misinterpreted. It can also happen that the replicas are hard to clean, difficult to handle and may not precisely reflect the surface topography of the cheese.

8.3.1.4 Fluorescence Microscopy

Fluorescence microscopy employs fluorescent light, which is defined as the emission of photons by atoms or molecules whose electrons are transiently stimulated to a higher excitation state by radiant energy from an outside source. In addition, there is a need for proper fluorochromes, *i.e.*, specific fluorescent dyes for staining the components of interest in the specimen.

Therefore, the micrographs obtained emit light from the excited molecules of the components of interest based on the dyes used. There are qualitative as well as quantitative methods for the analysis of the fluorescence emitted.

In microstructural analyses of cheese using fluorescence microscopy, a sample preparation is also required, which is the main drawback of the technique. Fluorescence provides a sensitivity that is not available in other forms of LM, allowing the detection of fluorescing compounds present in amounts as little as 10–18 mol. Fluorescence may be induced by a wide range of compounds such as fluorescent dyes that fluoresce only in specific chemical environments. For example, Nile red (a lipid-soluble dye) and Nile blue (a water-soluble dye) become intensely fluorescent in the hydrophobic medium of fat globules and protein within the cheese matrix, respectively. Nowadays, fluorescence microscopy is not commonly used in the microstructural analysis of cheese since such staining techniques are already applied in the analyses by confocal microscopy, in particular CLSM.

8.3.1.5 Confocal Laser Scanning Microscopy

CLSM is one of the most useful microscopy techniques for studying the microstructure of a wide variety of foods, in particular dairy products such as cheese. CLSM is considered to be a powerful tool since the laser scanning penetrates the cheese surface to visualize thin optical sections to obtain three-dimensional analysis of the cheese microstructure without disturbing the internal structure. The specimen is observed within a plane both transverse to and along with the optical axis. In this microscopy technique, it is not necessary to section the sample as in other LM techniques. Therefore, the artefacts associated with LM micrographs due to sample preparation are not present in CLSM micrographs. However, the drawback of using CLSM is related to the high costs of the equipment used and their maintenance.

CLSM standard equipment includes an argon laser with or without a helium–neon laser. This technique can provide focused images to a depth of up to several hundred microns. Therefore, the sequential sections obtained can be used for three-dimensional reconstruction of micrographs. In addition, several cheese components, such as protein and fat, and the location of microorganisms can be identified simultaneously using specific fluorescent labels or stains. The most frequently used fluorescent dyes in the cheese microstructure are fast green, rhodamine, and fluorescein isothiocyanate for the protein component and oil red O and Nile red for the fat. Recently, the development of more CLSM staining techniques has been proposed such as the provisional development of the pH sensitive fluorescent dye Oregon Green 514 carboxylic acid to determine localised variations in pH levels in the microstructure of the cheese matrix.

8.3.2 Dynamic Rheology

Dynamic oscillatory rheometry offers a method to probe the structure of cheese over a range of experimental time scales, extents of deformation, and temperatures, and provide indirect evidence for structural interactions. Small amplitude oscillations are imposed upon the cheese sample, such that the strains are within the linear viscoelastic region where any structural breakdown is largely reversible within the time scale of the experiment. Exceeding this limit causes the cheese structure to change and reform into a different conformation upon cessation of the oscillatory strain. The experimental time scale is defined as the reciprocal of the oscillation frequency. Large-time-scale experiments (at low frequencies) provide sufficient time for flow units within a sample to move and rearrange during the experiment, resulting in a liquid-like behaviour. Short-time-scale experiments do not provide sufficient time for the flow units to move, and thus the sample is more solid-like. The Deborah number is a measure of flowability and is equal to the ratio of the inherent relaxation time of a material to the experimental time scale.[63] One consequence of the Deborah number is that the elasticity will depend upon how fast the experiment is performed, *i.e.*, the higher the frequency of oscillation is, the more likely the material will be solid-like. The degree of solidness or fluidity is quantified by tan δ.[4] While δ approaches zero, the material is more solid- or gel-like, whereas higher values toward 90° indicate a more fluid-like substance. The transition between solid and liquid phases occurs at $\delta = 45$ degrees, and this criterion should be used to measure the point of gelation of a fluid product, rather than some small and arbitrary value of G' or G''.

8.3.3 Differential Scanning Calorimetry

Differential scanning calorimetry (DSC) is a thermoanalytical technique for monitoring changes in the physical or chemical properties of materials as a function of temperature by detecting the heat changes associated with such processes. In DSC, the measurement principle is the comparison of the rate of heat flow to the sample with that of an inert material which are heated or cooled at the same rate. Changes in the sample that are associated with absorption or evolution of heat cause a change in the differential heat flow which is then recorded as a peak. The area under the peak is directly proportional to the enthalpic change and its direction indicates whether the thermal event is endothermic or exothermic.

DSC is now generally accepted as the technique of choice to determine the energetics of protein folding/unfolding transitions and the thermodynamic mechanisms underlying those reactions. The melting properties of a fat can be obtained by DSC. Cow milk and water buffalo milk fats were separated into low-, medium-, and high-molecular weight (MW) fractions and DSC curves were obtained for each.[64] They attributed almost all the melting below 30 °C to low-MW and unsaturated high-MW triglycerides, which melt at lower temperatures than high-MW saturated triglycerides. The low-MW triglycerides contained butyric (and some caproic) acid esterified at position 3 on the glycerol molecule, and the high-MW unsaturated triglycerides contained oleic acid at position 3. DSC was used to measure the thermal properties of milk fat after the main stages involved in the manufacture of Emmental cheese. By heating the samples to 60 °C to eliminate their thermal history and cooling them at 2 °C min^{-1}, the liquid-to-solid phase transition of fat was investigated.[65] DSC was used to determine the ratio of solid to liquid fat, that is, the amount of fat that is crystallized, by dividing the partial enthalpy of melting of the fat for ripening temperature by the total enthalpy of melting of the same fat extracted from cheese. This study showed for the first time that milk fat is partially crystallized in Emmental cheese: about 55.7 ± 3.5% of fat is solid at 4 °C at the end of ripening. Polymorphic phase transitions of milk fat are also suggested during ripening of Emmental cheese.

8.3.4 Fluorescence Spectroscopy

Fluorescence spectroscopy has been efficaciously employed to assess molecular-level interactions between fat and proteins in numerous food-based emulsions. It has also been applied to observe structural changes in cheese and non-enzymatic browning in milk and other dairy products.[66] Tryptophan in the casein is a naturally occurring fluorescent substance. The fluorescent properties vary between hydrophobic and hydrophilic environments.[67] Fluorescence spectroscopy is also employed by researchers to observe the spectra of tryptophan. This enables prediction of the microstructure of any type of cheese. Simultaneously, it can also be used to determine Maillard browning and oxidative stability of processed cheese under storage.[68]

8.3.5 Fourier Transform Infrared Spectroscopy

Fourier transform infrared (FTIR) spectroscopy is a well-established technique that is widely used in the dairy industry to measure gross composition including lipid, protein, and moisture. This technique when combined with microscopy, which is known as FTIR microspectroscopy, provides insights into the distribution of major chemical components across the sample. To date, FTIR microspectroscopy has only had limited application in cheese products, where it has been used to map the spatial distribution of protein, lipid, and starch in imitation cheese. A limitation of this technique is the lower lateral resolution that can be achieved when compared to other commonly applied advanced microscopy techniques, including CLSM, with a diffraction-limited spatial resolution between 3 and 5 μm in the mid-infrared spectral region typically used for chemical analysis. Steps that can be taken to improve the signal-to-noise ratio of the data obtained at this resolution limit include the use of a synchrotron source, which provides a highly focused bright infrared light source, preventing energy loss when small apertures are used for high resolution mapping. Recent dairy studies have employed a related vibrational technique, confocal Raman microscopy, to examine the spatial distribution of key components in cheese. FTIR can provide information on the protein secondary structure. For this reason, Raman and FTIR microscopy techniques are often employed together to take advantage of the complementary information provided by each technique. FTIR microspectroscopy can also be applied in different modes of analysis, including transmission, attenuated total reflectance (ATR) and reflectance, whereas most of the analysis using Raman microscopy is conducted in the reflectance mode, providing information on only the surface components of cheese.

Synchrotron Fourier transform infrared (S-FTIR) microspectroscopy allows the label-free examination of the material microstructure but has not been widely applied to dairy products. S-FTIR microspectroscopy was applied to observe the microstructure of Mozzarella cheese and assess the protein and lipid distribution within individual cheese blocks.[69] The S-FTIR images in Figure 8.8 show the distribution of lipids and proteins in Mozzarella cheese samples, demonstrating the potential of the S-FTIR technique to characterise the spatial distribution of these components. Such information cannot be obtained with conventional single-point FTIR spectroscopy, where the sample area and spot size are limited to a single area, typically using a much larger spot size (~20–100 μm). Two high intensity peaks that were well separated from other spectral features were selected to create the S-FTIR maps; these were at 1745 cm^{-1} C=O ester carbonyl for lipids and 1645 cm^{-1} amide I for proteins.

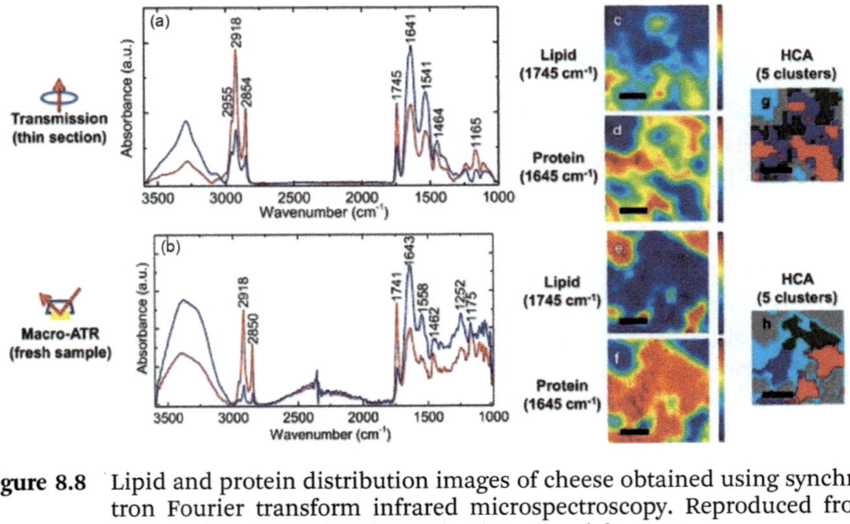

Figure 8.8 Lipid and protein distribution images of cheese obtained using synchrotron Fourier transform infrared microspectroscopy. Reproduced from ref. 69 with permission from Elsevier, Copyright 2019.

8.3.6 Magnetic Resonance Imaging and Nuclear Magnetic Resonance

Magnetic resonance imaging (MRI) has proved useful for examining the gross microstructure of cheese, including the location of holes and slits within the bulk of the cheese block, thus providing a non-invasive method for cheese quality control. NMR has been used to measure water mobility in cheese, which provides evidence for at least two phases of water, the more mobile bulk water trapped within the casein matrix and a less mobile water phase bound directly to the casein. Another NMR technique, which shows potential for application in food analysis, is fast field cycling (FFC) NMR relaxometry. It is based on the analysis of the behaviour of complex mixtures when the strength of the applied magnetic field is quickly changed, thereby allowing us to understand the molecular dynamics of chemical systems either as a whole or by the identification of the dynamics of each individual molecular component. NMR relaxometry has also been applied to several dairy products. For instance, it has been used to study curd syneresis, cheese ripening, water mobility in cheese, and cheese maturation.

Swiss-type semi-hard cheeses can be characterized by their bubbles, or "eyes", whose growth mechanisms have generally been explained empirically. In one study, the objective was to design a relevant multi-scale investigation strategy to determine if there is a relationship between eye growth kinetics and the surrounding cheese microstructure.[70] The magnetic resonance images taken during ripening showed different eye growth behaviours in cheese. First, small eyes situated under the rind were generally mechanical openings. These bubbles corresponded to the incorporation of small quantities of air that might occur during cheesemaking and that were trapped in the cheese after pressing. They grew very little during ripening and presented a non-spherical shape and cluster organization. Second, deeper within the body of the cheese, eyes remained small but well-formed and spherical, and contributed to cheese porosity uptake. Finally, the biggest eyes were detected in the core zone. At the end of ripening, the core eyes were mostly elliptical, oriented along a vertical axis. The body of the cheese was also vertically deformed during ripening, especially after the first week in the hot room (day 17). At the end of ripening, the top of the three cheeses had expanded. The bottom remained flat, except at the sides that were slightly elevated by the plastic film because of the cheese volume uptake, forming a dome shape. Therefore, it can be concluded that MRI provides the spatial distribution and growth kinetics of individual eyes during ripening. The eyes were smaller, less numerous, and growing more slowly under the rind than in the centre.

The potential of NMR to study cheese has been intensively reported on some specific topics, such as determining its geographic origin, molecular mobility of water, and water holding capacity. Italian Parmigiano Reggiano cheese was analysed for the identification of ripening and geographical discrimination using NMR.[6,71,72] During the ripening process, the chemical, physical, and enzymatic properties of cheese components undergo modifications. Researchers realized that the amino acid content may be influenced by the ripening and production location, with high contents of leucine and isoleucine in the early ripening stages and a high threonine content in the later ripening stages. When Parmigiano Reggiano cheese was compared with Grana-type cheese from Eastern Europe, the latter was rich in lactate, butanoate, acetate, leucine, and isoleucine, whereas the former contained large amounts of other compounds such as threonine. These free amino acids and fatty acids also contributed to the distinctive taste features of these cheeses. With NMR technology, other researchers were able to differentiate Asiago cheese produced in alpine farms from the same cheese produced commercially in larger operations. Results indicated that the alpine farm cheeses had a higher unsaturated fatty acid content (oleic, linoleic, linolenic, and conjugated linoleic acids) compared to the industrially produced cheese. Such differences in fatty acid composition can be attributed to the cattle being on high-concentrate diets, compared to cattle on Alpine farms where the production consists of milk from cattle grazing on natural pasture grasses at higher elevations. This further demonstrates the importance of the cattle's diet which has the ability to directly affect the quality of milk and milk products such as cheese.

8.3.7 X-ray Analysis

8.3.7.1 X-ray Diffraction

X-ray diffraction (XRD) has proven useful to identify mineral crystals in cheese. Two XRD peaks are present in Cheddar cheese, corresponding to D- and DL-enantiomers of calcium lactate pentahydrate. Different crystalline minerals have been identified in cheeses, brought about, in part, by the presence of microflora and pH gradients that develop during ripening. Powder XRD has identified tyrosine and leucine crystals in Parmigiano Reggiano and aged Gouda, calcium lactate pentahydrate in Cheddar cheese, and calcium carbonate hexahydrate crystals on the surface of a soft-washed rind cheese. Some of these crystals appear as a white haze, but individual cheese pearls can be as large as 6 mm in size. Other crystals identified include cysteine in Cheddar and calcium phosphate in Italian grana cheese. It is a personal preference of consumers whether these crystals are a defect in the cheese.

8.3.7.2 X-ray Microtomography

X-ray microtomography can be used to take sections of the internal structure of a food product and assemble into a three-dimensional image on a length scale of hundreds of nanometres. This technique is used to monitor eye growth in semi-hard cheese, fat in cream cheese, and ice crystal formation in frozen cheese. Information about the three-dimensional microstructure and ingredients of semi-solid foods such as cream cheeses can be obtained using X-ray microtomography (μCT).[73] Although a wide variety of imaging techniques exist, they are mostly invasive as they require sample preparation; hence, they cause the formation of artifacts. μCT, on the other hand, is a non-invasive and non-destructive technique that has several advantages over other methods, including the ability to image low-moisture materials. μCT uses the differences in X-ray attenuation, arising principally from differences in density within the specimen. A series of two-dimensional X-ray images are obtained as the sample is rotated. The series of slices, covering the entire sample, can be rendered into a three-dimensional image that can be presented either as a whole or as virtual slices of the sample at different depths and in different directions. Manipulation and analyses of μCT data using special software also allow reconstruction of cross-sections at depth increments as low as 15 μm and along any desired orientation of the plane of cut. A series of non-invasive μCT slices of the same sample in any direction can provide much more information than just one scanning electron microscopy or optical imaging picture.

8.4 Software for Cheese Microstructure Analysis

Image software used in microstructural analyses is dependent on the desired endpoint, *i.e.*, to capture, archive and/or process images and perform sophisticated image analyses. There are various commercially available software programs that can perform these functions for application in cheese microstructural analyses. Examples of these software programs are Adobe Photoshop, Image Pro®, MIP® (Microstructure Image Processing), and Imaris®. In addition, plug-in software for imaging functions is also available and depends on the main software used. Free image analysis software programs such as NIH Image or ImageJ for Macintosh or Windows and Linux users, respectively, are widely used. In addition, statistical software programs have also been used for the calculation of the data from the micrographs, *e.g.*, R statistical program. To date most of the research on the microstructural analyses of cheese has been performed on natural cheese products with limited research conducted on processed cheeses.

When comparing these programs for image analyses that are commercially available, Adobe Photoshop® is the most sophisticated, difficult to use, and accordingly the most expensive software for image analysis. This software has several updates, plug-in applications and advanced additional features for fully analysing images and graphics, mostly designed for advanced users such as software programmers. Image Pro®, for example, is a stand-alone software package that is easy to use, but has quite limited applications, *i.e.*, not suitable for advanced software programmers. The free image software programs (NIH® and ImageJ®) are based on Java, a programming language, and have their own features which make them suitable for advanced users. The features of these sophisticated programs, Adobe

Photoshop, and the free software (NIH and ImageJ) usually include the ability to read many image formats such as TIFF, PNG, GIF, JPEG, and BMP, calculate area and pixel value statistics, measure distances and angles, and create density histograms. In addition, these software programs support standard image processing functions such as logical and arithmetical operations between images, contrast manipulation, convolution, Fourier analysis, sharpening, smoothing, edge detection, and median filtering. These programs can support any number of images simultaneously, which is limited only by the available memory and specifications of the hardware and operating system of the computer used.

There are some common steps in the use of software for microstructural image analyses of food stuffs, including various cheese products. These steps are as follows.[5]

8.4.1 Image Acquisition

This refers to functions resulting in storage of a digital file representing an image in the hard disk of a computer. The acquired image, which could be in colour or grey scale, must be a 'true' representation of the subject of interest. The acquisition could be from an existing image or an image digitised directly from a light or electron microscope.

8.4.2 Calibration

Calibration refers to the process of defining a size or colour scale that can be correlated back to the image. The purpose of this step is to provide a quantitative measure (not only qualitative relationships) of distance, area, or colour value.

8.4.3 Image Enhancement

Image enhancement involves modifying the image with the aim of defining and increasing the visibility of the regions of interest. This enhancement typically includes several adjustments, such as contrast, sharpness, background correction, and edge enhancement. In addition, segmentation and counting is also performed. Segmentation or thresholding is mainly to separate the objects of interest from the other regions of the image, which can be considered as background. The segmentation results in its ultimate form as black and white, which is called the binary image. The binary image can be further prepared through erosion, dilation, or skeletonization for the next step of counting or measurement. Counting or measurement is the process of acquiring quantitative data based on the desired measurements. In most imaging programs, there is a choice of up to 20–30 measurement parameters. Examples of common measurements include parameters such as diameter, shape and size distribution related to the fat globules or pores.

8.4.4 Data Analyses

The data obtained are exported, using image analysis software, directly to other programs such as Excel or in a format that is recognisable by other programs *via* dynamic data exchange. These data are further processed to summarise or visualize the relevant aspects.

8.5 Conclusion

Much of the work on cheese structures has been carried out using various microscopy techniques. There remains a need to quantify this information such that it can be part of a model that combines microscopic observations, sensory properties, rheological measurements, and functional properties. The increase in the complexity of microscopes (*e.g.*, TEM and SEM in the early years to cryo-SEM and environmental SEM today) gives us more details about the structure under conditions that more closely approximate those of food products. Sensory and functional properties are important parts of a future model, particularly as these properties generate income for the dairy industry. More sophisticated instrumentation does not come cheaply, but it provides an opportunity to understand the structure at the basic molecular level and allows us to make predictions about structural development over the shelf-life of a food product. More importantly, it will provide a basis to develop new food ingredients to add value to milk products.

For future research on the cheese microstructure, the subsequent process can be recommended. First, there is a requirement for complete systematic studies on staining techniques by utilizing LM and CLSM, and elemental mapping by EM of minerals present inside the cheese matrix. Earlier research studies have been conducted with the aim of creating new staining protocols for minerals and pH mapping, and the application of X-ray microanalysis for elemental mapping. Nevertheless, these efforts were individual and not fully and/or systematically executed on various types of cheeses. Second, modern research is concentrating on the localisation of bacterial cells inside the cheese matrix. Therefore, additional methods for sample preparation ought to be developed for enhancing the study of localisation and the effect of the bacterial cells within the matrix. Lastly, the above-mentioned recommendations may possibly be further used in the microstructural analyses of cheeses during the stages of ripening. The findings might provide an understanding of different phenomena occurring during ripening and that correlate with the functional and biochemical changes.

Abbreviations

CLSM	Confocal laser scanning microscopy
DSC	Differential scanning calorimetry
EDSA	Energy dispersive X-ray spectrometry analysis

EM	Electron microscopy
ESEM	Environmental scanning electron microscopy
FTIR	Fourier transform infrared
G'	Elastic modulus
G''	Viscous modulus
HPP	High pressure processing
LAB	Lactic acid bacteria
LM	Light microscopy
MRI	Magnetic resonance imaging
MW	Molecular weight
NMR	Nuclear magnetic resonance
SEM	Scanning electron microscopy
S-FTIR	Synchrotron Fourier transform infrared
TEM	Transmission electron microscopy
UF	Ultrafiltration
XRD	X-ray diffraction
μCT	X-ray microtomography

References

1. P. F. Fox, T. P. Guinee, T. M. Cogan and P. L. H. McSweeney, *Fundamentals of Cheese Science*, Springer, New York, 2017.
2. C. I. Pereira, A. M. Gomes and F. X. Malcata, *Trends Food Sci. Technol.*, 2009, **20**, 213.
3. N. Joshi, K. Muthukumarappan and R. Dave, *J. Dairy Sci.*, 2004, **87**, 1975.
4. D. W. Everett and M. A. Auty, in *Cheese, Chemistry, Physics and Microbiology*, ed. P. L. H. McSweeney, P. F. Fox, P. D. Cotter and D. W. Everett, Elsevier, New York, 4th edn, 2017, p. 547.
5. M. El-Bakry and J. Sheehan, *J. Food Eng.*, 2014, **125**, 84.
6. L. Ong, X. Li, A. Ong and S. L. Gras, *Annu. Rev. Food Sci. Technol.*, 2022, **13**, 89.
7. M. Rowney, M. Hickey, P. Roupas and D. Everett, *J. Dairy Sci.*, 2003, **86**, 712.
8. D. W. Everett and M. A. Auty, *Int. Dairy J.*, 2008, **18**, 759.
9. M.-C. Michalski, B. Camier, V. Briard, N. Leconte, J.-Y. Gassi, H. Goudédranche, F. Michel and J. Fauquant, *Lait*, 2004, **84**, 343.
10. T. Truong, M. Palmer, N. Bansal and B. Bhandari, *Effect of Milk Fat Globule Size on the Physical Functionality of Dairy Products*, Springer, New York, 2016.
11. D. Waldron, W. Hoffmann, W. Buchheim, D. J. McMahon, H. D. Goff, S. V. Crowley, C. Moloney, J. O'Regan, F. Giuffrida, I. Celigueta Torres and P. Siong, in *Advanced Dairy Chemistry*, Springer, New Tork, 2020, vol. 2, p. 245.
12. A. Pastorino, R. Dave, C. Oberg and D. J. McMahon, *J. Dairy Sci.*, 2002, **85**, 2106.

13. C. J. Gamlath, T. S. Leong, M. Ashokkumar and G. Martin, *Food Hydrocolloids*, 2020, **109**, 106103.
14. B. Pendyala, A. Patras, V. V. S. Gopisetty, P. Vashisht and R. Ravi, *bioRxiv*, 2022, DOI: 10.1101/2022.02.22.481446v1.
15. A. Heck, S. Nöbel, B. Hitzmann and J. Hinrichs, *J. Food Biophys.*, 2021, **16**, 237.
16. S. V. Crowley, A. L. Kelly, J. A. O'Mahony and J. A. Lucey, *Colloids Surf., B*, 2019, **174**, 343.
17. J. Chandrapala and B. Zisu, in *Ultrasound Technology in Dairy Processing*, Springer, New York, 2018, p. 1.
18. B. K. S. Khanal, C. Budiman, M. P. Hodson, M. R. Plan, S. Prakash, B. Bhandari and N. Bansal, *J. Food Eng.*, 2019, **242**, 94.
19. H. A. Pushpadass, F. M. E. Emerald, B. Balasubramanyam and S. S. Patel, in *Milk-Based Beverages*, ed. A. M. Grumezescu and A. M. Holban, Woodhead Publ., Cambridge, MA, 2019, p. 373.
20. L. Nunes and G. M. Tavares, *Trends Food Sci. Technol.*, 2019, **90**, 88.
21. D. Carullo, F. Donsì and G. Ferrari, *LWT*, 2020, **130**, 109657.
22. Q. Li and Z. Zhao, *Food Chem.*, 2019, **291**, 231.
23. Q. A. Syed, A. Hassan, S. Sharif, A. Ishaq, F. Saeed, M. Afzaal, M. Hussain and F. M. Anjum, *Int. J. Food Prop.*, 2021, **24**, 871.
24. M. Kelly-Hayes, A. Beiser, C. S. Kase, A. Scaramucci, R. B. D'Agostino and P. A. Wolf, *J. Stroke Cerebrovasc. Dis.*, 2003, **12**, 119.
25. T. Huppertz, K. Hinz, M. R. Zobrist, T. Uniacke, A. L. Kelly and P. F. Fox, *Innovative Food Sci. Emerging Technol.*, 2005, **6**, 279.
26. J. Mayta-Hancco, A.-J. Trujillo, A. Zamora and B. Juan, *LWT*, 2019, **110**, 292.
27. W. Messens, D. Van de Walle, J. Arevalo, K. Dewettinck and A. Huyghebaert, *Int. Dairy J.*, 2000, **10**, 359.
28. M. Munir, M. Nadeem, T. M. Qureshi, C. J. Gamlath, G. J. Martin, Y. Hemar and M. Ashokkumar, *Ultrason. Sonochem.*, 2020, **67**, 105140.
29. D. Yadav, P. Sarkar, P. Patil and N. Mayur, in *Technological Interventions in Dairy Science: Innovative Approaches in Processing, Preservation, Analysis of Milk Products*, ed. R. S. Chavan and M. R. Goyal, CRC Press, Boca Raton, FL, 2018, p. 31.
30. A. Bezie, *J. Dairy Vet. Sci.*, 2019, **11**, 555822.
31. A. Enab, F. A. Hassan and M. Abd, *Int. J. Acad. Res.*, 2012, **4**, 79.
32. L. R. Ribeiro, B. R. D. L. Júnior and M. Cristianini, *Innovative Food Sci. Emerging Technol.*, 2018, **50**, 57.
33. M. Alinovi, M. Corredig, G. Mucchetti and E. Carini, *Food Res. Int.*, 2020, **137**, 109415.
34. M. Alinovi, G. Mucchetti, L. Wiking and M. Corredig, *Crit. Rev. Food Sci. Nutr.*, 2021, **61**, 3340.
35. D. Cais-Sokolińska, P. Bierzuńska, Ł. Kaczyński, H. Baranowska and J. Tomaszewska-Gras, *J. Food Eng.*, 2018, **222**, 226.
36. Q. Sun, B. Kong, S. Liu, O. Zheng and C. Zhang, *Foods*, 2021, **10**, 629.

37. J. Biegalski, D. Cais-Sokolińska, J. Tomaszewska-Gras and H. M. Baranowska, *Animals*, 2021, **11**, 2740.
38. P. Lamichhane, *Influence of Milk Pre-Treatment, and Changes in Cheese Structure and Carbon Dioxide Solubility on the Development of Split Defects in Cheese*, PhD thesis, University College Cork, Ireland, 2019.
39. M. C. Gonçalves and H. R. Cardarelli, *LWT*, 2019, **104**, 16.
40. B. M. Mehta, *Microstructure of Dairy Products*, ed. M. El-Bakry, A. Sanchez and B. M. Mehta, Wiley-Blackwell, Hoboken, NJ, 2018, p. 145.
41. S. Wu, J. Fitzpatrick, K. Cronin and S. Miao, *Food Chem.*, 2020, **332**, 127440.
42. A. Siddique and Y. W. Park, *J. Dairy Sci.*, 2019, **102**, 2890.
43. H. Li, A. Qin, H. Yu, Y. Han, S. Zheng, H. Li and J. Yu, *LWT*, 2020, **124**, 109185.
44. Z. Burdikova, Z. Svindrych, C. Hickey, M. G. Wilkinson, M. A. Auty, O. Samek, S. Bernatova, V. Krzyzanek, A. Periasamy and J. J. Sheehan, *Dairy Sci. Technol.*, 2015, **95**, 687.
45. C. Melilli, D. Carcò, D. Barbano, G. Tumino, S. Carpino and G. Licitra, *J. Dairy Sci.*, 2005, **88**, 2329.
46. I. Hernando, E. Llorca and A. Quiles, in *Handbook of Dairy Foods Analysis*, ed. F. Toldrá and L. M. L. Nollet, CRC Press, Boca Raton, FL, 2021, p. 323.
47. A. Aldalur, L. Ong, M. Á. Bustamante, S. L. Gras and L. J. R. Barron, *Food Bioprod. Process.*, 2019, **116**, 160.
48. A. S. Thomareis and G. Dimitreli, in *Processed Cheese Science and Technology*, ed. M. El-Bakry and B. M. Mehta, Elsevier, New York, 2022, p. 295.
49. S. H. Majid and G. M. Saleh, *Tikrit J. Agric. Sci.*, 2022, **22**, 35.
50. N. Noronha, E. Duggan, G. Ziegler, J. Stapleton, E. O'Riordan and M. O'Sullivan, *Food Res. Int.*, 2008, **41**, 472.
51. N. Noronha, E. D. O'Riordan and M. O'Sullivan, *Eur. Food Res. Technol.*, 2008, **226**, 385.
52. A. M. Donald, *Nat. Mater.*, 2003, **2**, 511.
53. M. M. Ayyash, F. Sherkat, P. Francis, R. Williams and N. P. Shah, *J. Dairy Sci.*, 2011, **94**, 37.
54. M. M. Ayyash, F. Sherkat and N. P. Shah, *J. Dairy Res.*, 2013, **80**, 7.
55. D. Huc, G. Moulin, F. Mariette and C. Michon, *Int. Dairy J.*, 2013, **33**, 10.
56. K. Soodam, L. Ong, I. B. Powell, S. E. Kentish and S. L. Gras, *Food Bioprocess Technol.*, 2014, **7**, 2912.
57. R. Awad, L. Abdel-Hamid, S. El-Shabrawy and R. Singh, *LWT–Food Sci. Technol.*, 2002, **35**, 54.
58. P. E. Macdougall, L. Ong, M. V. Palmer and S. L. Gras, *Int. Dairy J.*, 2019, **99**, 104548.
59. L. Ong, R. R. Dagastine, S. E. Kentish and S. L. Gras, *LWT–Food Sci. Technol.*, 2011, **44**, 1291.
60. B. Sameer, S. Ganguly, Y. Khetra and L. Sabikhi, *LWT*, 2020, **121**, 108944.

61. S. Karaman, M. T. Yilmaz, O. S. Toker and M. Dogan, *Int. J. Dairy Technol.*, 2016, **69**, 254.
62. A. H. Vollmer, I. Kieferle, N. N. Youssef and U. Kulozik, *J. Dairy Sci.*, 2021, **104**, 9505.
63. B. Bähler, M. Nägele, J. Weiss and J. Hinrichs, *J. Texture Stud.*, 2016, **47**, 58.
64. M. H. Tunick and E. L. Malin, *J. Am. Oil Chem. Soc.*, 1997, **74**, 1565.
65. D. J. McMahon, R. L. Fife and C. J. Oberg, *J. Dairy Sci.*, 1999, **82**, 1361.
66. S. Herbert, N. M. Riou, M. F. Devaux, A. Riaublanc, B. Bouchet, D. J. Gallant and É. Dufour, *Lait*, 2000, **80**, 621.
67. R. Karoui and É. Dufour, *Int. Dairy J.*, 2003, **13**, 973.
68. E. Dufour, M. Devaux, P. Fortier and S. Herbert, *Int. Dairy J.*, 2001, **11**, 465.
69. A. P. Pax, L. Ong, J. Vongsvivut, M. J. Tobin, S. E. Kentish and S. L. Gras, *Food Chem.*, 2019, **291**, 214.
70. D. Huc, F. Mariette, S. Challois, J. Barreau, G. Moulin and C. Michon, *Innovative Food Sci. Emerging Technol.*, 2014, **24**, 106.
71. M. F. Marcone, S. Wang, W. Albabish, S. Nie, D. Somnarain and A. Hill, *Food Res. Int.*, 2013, **51**, 729.
72. I. Hernando, E. Llorca and A. Quiles, in *Microstructure of Dairy Products*, ed. M. El-Bakry, A. Sanchez and B. M. Mehta, Wiley-Blackwell, Hoboken, NJ, 2018, p. 67.
73. J. Laverse, M. Mastromatteo, P. Frisullo and M. A. Del Nobile, *J. Dairy Sci.*, 2011, **94**, 43.
74. M. Alinovi, G. Mucchetti, U. Andersen, T. A. M. Rovers, B. Mikkelsen, L. Wiking and M. Corredig, *Foods*, 2020, **9**(5), 679.

CHAPTER 9

Cheese Rheology and Texture

MICHAEL H. TUNICK

Department of Food & Hospitality Management, College of Nursing & Health Professions, Drexel University, Philadelphia, PA 19104, USA
Email: mht39@drexel.edu

9.1 Definitions

Rheology and texture deal with the properties that a material exhibits upon application of a force. The term rheology, coined in the late 1920s,[1] describes the study of the flow and deformation of matter. Chemists and physicists often use rheology when investigating gases and liquids, but they can apply it to solids such as cheese. Rheological measurements come from instruments and not individuals.

Texture of food encompasses the appearance of food and its feel in the fingers and the mouth. The International Organization for Standardization (ISO) defines texture as "All the mechanical, geometrical, surface, and body attributes of a product perceptible by means of kinaesthesis and somesthesis receptors and (where appropriate) visual and auditory receptors from the first bite to final swallowing"[2] (see Table 9.1). Confusingly, the dairy industry uses "body" to denote the consistency of the product (including firmness, softness, brittleness, cohesiveness, crumbliness, rubberiness, elasticity, plasticity, pastiness, and curdiness) and describes the relative number, type, and size of openings observed visually or by the sense of touch as "texture".[3] This chapter will use the ISO definition.

Rheology and texture become especially important when cheese is an ingredient in another food. Table 9.2 shows some examples of the properties that cheese must have in certain applications.

Food Chemistry, Function and Analysis No. 40
Handbook of Cheese Chemistry
Edited by Michael H. Tunick

Published by the Royal Society of Chemistry, www.rsc.org

Table 9.1 Components of food texture.[2]

Term	Definition	Examples
Mechanical attributes	Attributes related to the reaction of the product to stress	Hardness, cohesiveness, viscosity, elasticity and adhesiveness
Geometrical attributes	Attributes related to the size, shape and arrangement of particles within a product	Denseness, granularity and conformation
Surface attributes	Attributes related to the sensations produced in the mouth by moisture and/or fat in and near the surface of the product	
Body attributes	Attributes related to the sensations produced in the mouth by moisture and/or fat in the substance of the product and the way in which these constituents are released	
Kinaesthesis	The sensation of the position, movement and tension of parts of the body perceived through nerves and organs in the muscles, tendons and joints	
Somesthesis	Sensations of pressure (touch), temperature, and pain perceived by the receptors located in the skin and lips, including oral mucosa, tongue and the periodontal membrane	

Table 9.2 Textural requirements of cheese when used as an ingredient.[4,5]

Attribute	Application	Example
Melts and flows with heating	Cooked food	Mozzarella, process cheese
Stretches with heating	Pizza	Mozzarella
Limited browning with heating	Pasta dishes (macaroni and cheese, lasagne)	Cheddar, Parmesan
Crumbles when scraped or rubbed	Soups and salads	Queso Fresco, Feta, Stilton
Slices easily	Sandwiches, slices on crackers	Cheddar, Swiss
Shreds easily	Pizza, pasta dishes	Cheddar, Swiss, Mozzarella
Not sticky when shaken	Sprinkling on food, coating snack food	Grated Parmesan, cheese powder
Viscous	Soups, sauces	Process cheese, cheese powder

The field of rheology and texture uses physical chemistry to describe the interactions leading to the observed mechanical behaviour, biochemistry to identify the enzymatic reactions that lead to these interactions, and analytical chemistry to evaluate them. Choosing the proper techniques, carrying out the appropriate procedure, and interpreting the results all lead to reliable data. Breakdown of the carbohydrates, lipids, and proteins in cheese is the primary force behind structural development. Results vary between and within cheese varieties owing to manufacturing procedures (such as cooking temperature and pressing conditions), packaging, and storage. To show how intricate and detailed this subject can be, Gunasekaran and Ak wrote an entire 430-page book titled Cheese Rheology and Texture.[5] This chapter will describe the most common rheological and textural methods and provide examples of their applications to cheese.

9.2 Empirical Tests

Scientists may attempt to determine texture by using three categories of instrumental techniques: empirical, imitative, and fundamental.[6] Empirical tests measure a particular characteristic and lack scientific rigour. Examples include assessing the hardness of cheese by pressing on it with a small metal sphere (ball compressor), measuring force by pushing a needle into the specimen (penetrometer), and determining resistance to a wire being passed through a block (curd tension meter).[7] Common empirical tests for heated cheese include measuring meltability by observing the spread or height decrease of a sample[8] and pulling on melting Mozzarella to calculate stretchability.[9] The procedures in empirical tests are relatively arbitrary, limiting their use, though they do provide information rapidly and inexpensively.

9.3 Imitative Tests: Texture Profile Analysis

Imitative tests mimic chewing and provide comparisons with similar samples, leading to standard values of food texture. The most popular of these is texture profile analysis (TPA), developed for food in the 1960s[10] and first applied to cheese in the following decade.[11] Scientists use universal testing machines or purpose-built texture analysers to conduct TPA. With both, the instrument lowers a flat plate that deforms a cylindrical or cubical specimen placed on a base. The plate returns at the same rate and repeats the procedure, roughly mimicking the action of biting twice on a piece of cheese. A typical test squeezes the cheese by 75%, breaking it and providing values for hardness and fracturability. Any cheese sticking to the plate as it returns upward registers as adhesiveness. When the plate comes back down, contacts the specimen again, and returns, values for springiness and cohesiveness result. Figure 9.1 displays a typical TPA curve for cheese and the textural parameters shown in Table 9.3 arise from the curve.

Although useful for comparisons of cheeses within a laboratory when experimental conditions are constant, TPA does not provide the fundamental information needed for interpreting cheese structure. Moreover, the nature of the test makes it difficult to distinguish whether the sample

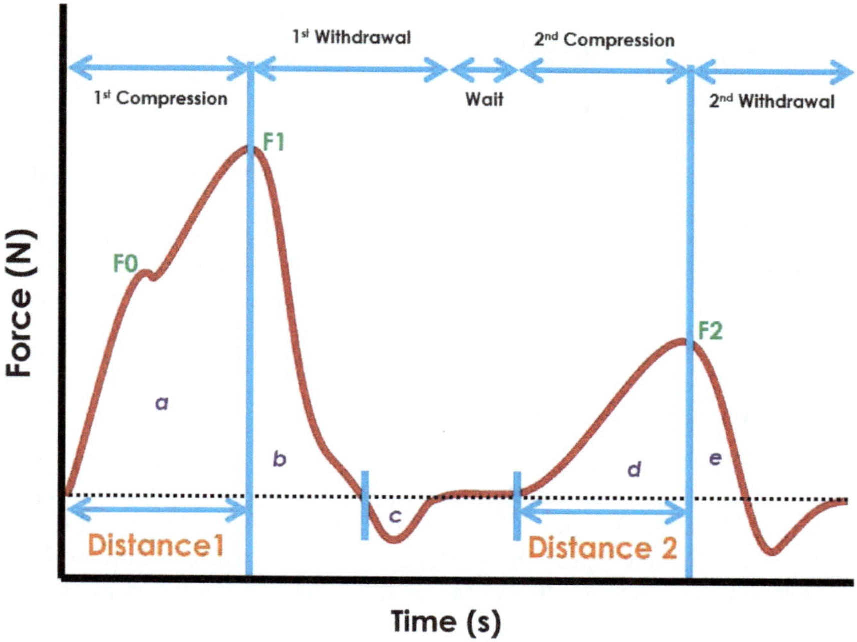

Figure 9.1 Texture profile analysis curve. Terms (with units): fracturability (N) = F_0, hardness (N) = F_1, adhesiveness (N s) = c, cohesiveness (unitless) = $(d+e)/(a+b)$, springiness (mm) = distance 2, chewiness (J) = hardness × cohesiveness × springiness. Reproduced with permission from the Centre for Industrial Rheology.

Table 9.3 Terms used in texture profile analysis of foods.[12,13]

Term	Definition	Source
Fracturability	Force required to fracture specimen	Force at the first significant break in the force–time curve
Hardness	Force required to attain required compression	Maximum force during the first compression (first peak)
Adhesiveness	Force required to overcome the attractive force between the specimen and plate	Force area of the negative peak following the first peak
Cohesiveness	Degree to which the specimen can be deformed before it breaks	Ratio of the positive force area of the second peak to that of the first peak
Springiness	Rapidity and degree of recovery from deformation	Height of specimen recovery from the end of the first peak to the start of the second
Chewiness	Work required to masticate a solid specimen to a state ready for swallowing	Product of hardness, cohesiveness, and springiness

fractured or merely deformed.[14] Despite these shortcomings, TPA is a common technique for assessing cheese.

9.4 Fundamental Analyses: Rheology

Fundamental tests are scientifically rigorous, measure intrinsic physical properties, and produce data in well-defined units that allow predictions of behaviour.[15] Rheological analyses are mathematically based, conducted under specific conditions, and describe and predict events occurring within the specimen. These are the fundamental tests used to analyse cheese.

9.4.1 Compression Tests

The simplest test is uniaxial compression, in which a plunger (such as a polymer or stainless steel plate) comes down vertically on a specimen at a given speed (load rate), squeezes it until it breaks (fractures), and lifts up. The procedure leads to values for stress, defined as the ratio of applied force to the original cross-sectional area of the specimen, and engineering strain, defined as the ratio of the height change to original height. Young's modulus is the ratio of stress to strain during compression and indicates the stiffness of the sample at fracture. A curve of force *versus* strain may show an inflection point where the structure begins to break down. Eventually the structure collapses faster than the stress build-up; the stress at this fracture point is the yield stress.[7] The work performed up to fracture equals the area under the curve.

Many publications featuring uniaxial compression of cheese are available in the scientific literature, partially owing to the simplicity and versatility of the technique. Table 9.4 shows a selection of these papers published since 2001, which demonstrates the utility of uniaxial compression in describing the effects of the type of milk, composition, manufacturing parameters, storage time, and other variables.

Tensile tests, which pull samples apart, are the opposite of compression tests. Aside from a lack of suitable grips for holding the specimen in place, the chewing of cheese does not involve pulling, so scientists rarely apply tensile testing to cheese, except during heating experiments.

9.4.2 Examination of Viscoelasticity

More complicated tests than those described earlier provide more in-depth results and involve viscosity and elasticity. Viscosity is the relationship between the stress applied to a material and the rate at which flow results. Samples of true liquids stop flowing upon removal of the stress. Elasticity is the ability of a material to recover its original shape spontaneously upon removal of the stress. Truly elastic samples completely recover. Cheeses exhibit both viscous and elastic properties, meaning they are viscoelastic,

Table 9.4 A selection of 21st century publications on uniaxial compression of cheese.

Cheese variety	Purpose of the study	Authors, year
Gruyère, Mozzarella	Examine lubricants and friction	Charalambides *et al.*, 2001[16]
29 varieties	Compare uniaxial compression, penetration, and puncture tests with sensory results	Breuil and Meullenet, 2001[17]
Processed	Observe the effects of fat composition	Jaros *et al.*, 2001[18]
Cheddar	Compare loading rates with sensory results	Moiny *et al.*, 2002[19]
26 varieties	Correlate compression with sensory measurements of hardness	Xiong *et al.*, 2002[20]
Sheep milk	Correlate fat content with instrumental and sensory results	Irigoyen *et al.*, 2002[21]
7 varieties	Correlate compression and sensory results	Goh *et al.*, 2003[22]
Tybo	Assess the effects of starter culture on hardness	Moreira *et al.*, 2003[23]
Pressure-treated Cheddar	Determine if pressure affects fracture	Wick *et al.*, 2004[24]
Swiss type	Determine if the starter culture composition affects fracture	Weinrichter *et al.*, 2004[25]
Iranian white	Note the effects of fat reduction	Madadlou *et al.*, 2005[26]
	Note the effects of coagulation temperature	Madadlou *et al.*, 2006[27]
	Note the effects of starter culture concentration	Khosrowshahi *et al.*, 2006[28]
	Note the effects of cream homogenization	Madadlou *et al.*, 2007[29]
	Note the effects of brine concentration	Madadlou *et al.*, 2007[30]
Canestrello Pugliese	Investigate changes during storage	Del Nobile *et al.*, 2007[31]
Teleme	Compare cow, goat, and sheep milk cheese	Pappa *et al.*, 2007[32]
Salami	Measure stress distribution	Balejko, 2007[33]
Pasteurized process	Analyse the effects of pH and emulsifying salts	Lu *et al.*, 2008[34]
Sheep	Observe changes during ripening	Juan *et al.*, 2008[35]
Cantal	Analyse the effects of cow breed and milking frequency	Martin *et al.*, 2009[36]
Ultrafiltered white	Correlate Ca, proteolysis, and hardness	Fathollahi *et al.*, 2010[37]
Murcia al Vino	Investigate the effects of various rennet pastes	Ferrandini *et al.*, 2011[38]
Feta	Examine the effect of adding whey protein concentrate and corn starch	Tashakor *et al.*, 2013[39]
Siahmazgi	Characterize cheese during ripening	Farahani *et al.*, 2014[40]

Table 9.4 *(Continued)*

Cheese variety	Purpose of the study	Authors, year
Minas Frescal	Assess variations in manufacture	Magenis *et al.*, 2014[41]
Gouda	Examine variability with position in specimen and ripening time	Vandenberghe *et al.*, 2014[42]
Pressure-treated starter-free fresh	Characterize product	Evert-Arriagada *et al.*, 2014[43]
Reduced- and full-fat Cheddar	Evaluate the effects of NaCl and fat reduction on cooking	McCarthy *et al.*, 2016[44]
Panela	Investigate the effects of goat/cow milk mixtures	Ramírez-López and Vélez-Ruiz, 2018[45]
Pecorino Romano	Assess the effects of the protein–fat ratio	Addis *et al.*, 2018[46]
Mozzarella	Examine the effects of stretching temperature	Gonçalves and Cardarelli, 2020[47]
Soft camel milk	Evaluate the effects of adding microbial transglutaminase to milk	Abou-Soliman *et al.*, 2020[48]
Cantal type	Investigate the effects of different feeds	Manzocchi *et al.*, 2021[49]

as are most foods. Cheese is normally a solid at room temperature, but soft cheeses such as Camembert can flow like a liquid and heated cheeses will flow as well, which demonstrates how viscosity and elasticity are both present.[50] Protein gels, including cheese, may have noncovalent linkages such as hydrogen bonds and hydrophobic interactions, covalent bonds such as disulphide bonds, and physical entanglements of the casein chains.[51] Dynamic viscoelastic measurements on casein gels such as cheese provide information on short-range interactions, in particular the conformation and structure of the casein particles within the sample.[52]

When applying a force to cheese, part of the mechanical energy supplied to it is stored in the material and part dissipates as heat. The stored energy is the elastic portion and the dissipated energy is the viscous portion. Scientists determine viscoelastic properties dynamically using a rheometer in the small amplitude oscillation mode, in which a sinusoidal oscillation (repeated slight back-and-forth wiggles) affects a cheese sample. The instrument varies stress or strain harmonically with time and then calculates the viscoelastic data from the determination of the complex modulus (G^*), a measure of the deformation of the sample, and the phase angle (δ), which is the angle that the strain and stress are out of phase. The elastic (or storage) modulus (G') is the energy stored per deformation and indicates solid-like or elastic behaviour, while the viscous (or loss) modulus (G'') is the energy dissipated per deformation cycle and indicates fluid-like or viscous behaviour. The equations relating these quantities are $|G^*|^2 = (G')^2 + (G'')^2$, $G' = G^*\ (\cos\delta)$, and $G'' = G^*\ (\sin\delta)$. The loss tangent ($\tan\delta$) is G''/G', and if the value is well below 1.0, the behaviour of the sample is more solid-like than liquid-like. The complex viscosity (η^*) is G^*/ω, where ω is the frequency of the oscillation.[51]

The rheological properties of cheese depend on the number, strength, and type of bonds between casein molecules as well as their spatial arrangement, although research on the arrangement aspect is lacking.[3] The response of the sample to the applied deformation depends on the timescale, the amount of time that a stress of a particular direction and magnitude acts upon a material.[53] A sample that recovers during a given time, usually a fraction of a second, exhibits a smaller rheological response than one that cannot recover. The timescale has applications in cheese manufacturing. As an example, Mozzarella is quite soft when first made, and the optimal speed of chopping it into pieces for packaging depends on time: cutting too quickly damages the curd and creates tiny pieces, but a slower speed results in a clean cut.[3] Camembert is sliceable with a knife over short timescales but flows over long ones.

Dairy food analysts typically use rheometers with cone-and-plate, parallel plate, and concentric cylinder measuring geometries.[54] Figure 9.2 shows an example of a rheometer and Figure 9.3 depicts the geometries and their applications.

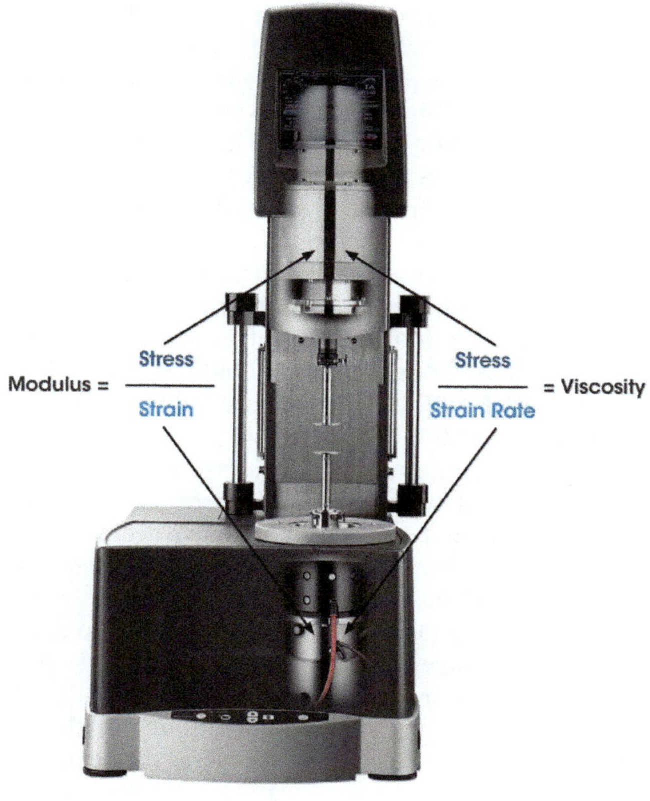

Figure 9.2 Cutaway view of a commercial rheometer. Courtesy of TA Instruments.

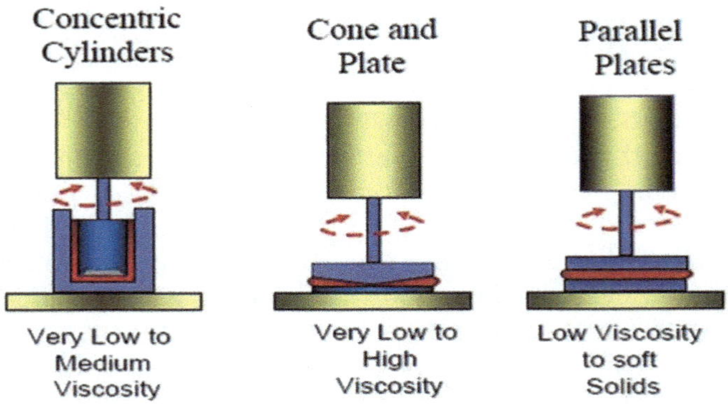

Figure 9.3 The three common geometries for investigating dairy food rheology. Low viscosity liquid samples such as milk typically go between concentric cylinders, higher viscosity samples such as melted cheese use the cone and plate configuration, and solid samples require parallel plates. Gluing cheese disks to the parallel plates prevents slippage.

Rheological analyses usually begin with locating the linear viscoelastic region (LVR), in which the cheese structure remains intact. Within the LVR, the moduli are a function of time alone and not the magnitude of stress or strain. Most often, scientists will vary strain levels at a constant frequency (a strain sweep) and locate the plateau in which the log of G' does not deviate by a significant amount from a constant value. The structure begins to change beyond a certain strain level and the $\log G'$ drops off, signifying the limit of the LVR. Subsequent analyses employ the same strain in that region (0.02%, for instance). When performing controlled stress tests, scientists use stress sweeps to locate the LVR. Operating in the LVR does not damage the specimen, making the procedure a non-destructive test.

Other types of sweeps include frequency, time, and temperature.[51] Frequency sweeps describe the time-dependent behaviour of a sample in the LVR and are used the most. For accuracy, ω should be limited to three orders of magnitude, such as 1–100 rad s^{-1}.[12] Figure 9.4 shows an example of a frequency sweep. When adding rennet to milk, a time sweep will show the point at which the curd forms, indicated by jumps in G' and G'' (Figure 9.5).

Temperature sweeps describe rheological behaviour upon heating. Cheeses soften when heated, resulting from a sharp decrease in the number and strength of bonds in the casein matrix. Flow may occur when G'' exceeds G', reflecting the loss of elasticity and a conversion to a more viscous structure. Cheeses "melt" when they flow and spread, though only the fat liquefies.[3] Temperature sweeps can produce a plot of $\log \eta^*$ *versus* the reciprocal of absolute temperature, and this straight line follows the Arrhenius equation. Multiplying the slope of the line by the gas constant yields the activation energy.[55] The method measures the degree of melting of cheese, with a high activation energy indicating rapid liquefaction of fat and collapse of the protein matrix.[56]

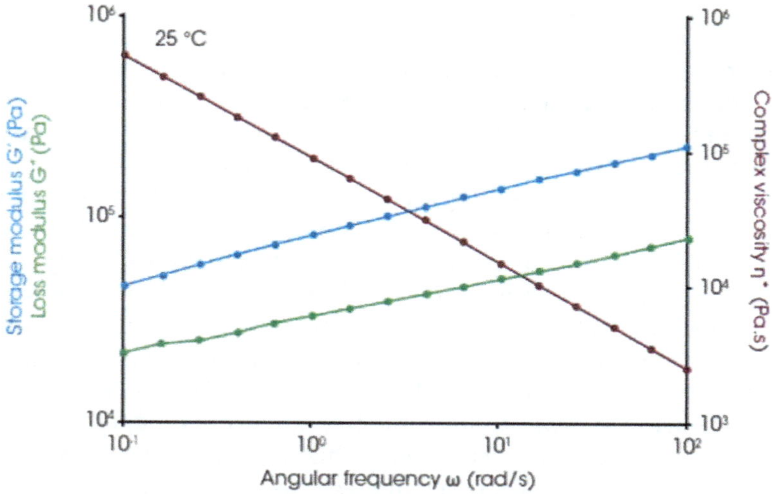

Figure 9.4 Frequency sweep of Cheddar cheese at 25 °C. Courtesy of TA Instruments.

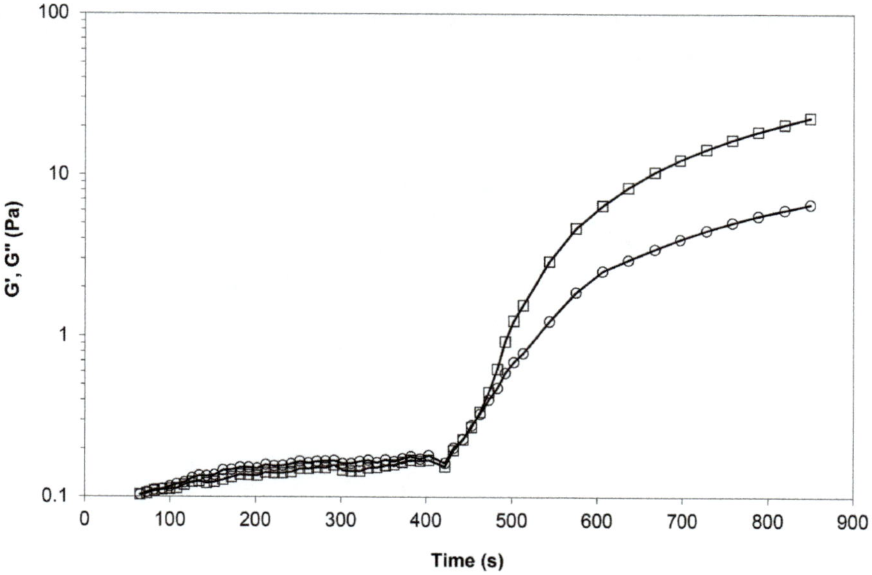

Figure 9.5 Time sweep of coagulating milk, with squares and circles representing the elastic modulus (G') and viscous modulus (G''), respectively. Rennet addition occurred at 0 s and analysis started at 60 s. Note that G'' values were greater than G' values until coagulation at 420 s, indicating a change from liquid-like to solid-like behaviour.

9.4.3 Transient Tests

Viscoelastic materials such cheese undergo creep, meaning they continue to deform under constant load (force) or stress.[57] Stress relaxation and creep recovery come under the category of transient tests. During a stress relaxation test, a specimen receives a sudden and constant strain, such as 0.1%, and the instrument monitors the resulting stress with time. The relaxation modulus, the ratio of stress to strain, is solely a function of time within the LVR. In a creep test, the application of a sudden constant stress leads to a creep curve of strain *versus* time. The removal of the stress produces a creep recovery curve. The ratio of strain to stress during the creep phase is the compliance, and the same ratio during recovery is the creep compliance. Figure 9.6 shows a creep curve and a creep recovery curve for Mozzarella. A hard cheese will strain less than a soft cheese and will recover more because of its higher rigidity and elasticity. Transient tests in cheese employ higher strains than dynamic tests, causing the breakage of short-range interactions and providing information on the long-range properties of the casein network.[12]

Only fat melts in heated cheese, as stated above, but protein bonding changes with temperature. Hydrophobic interactions strengthen with heat, which may reduce the size of the contact area between casein particles and therefore the gel strength. Electrostatic interactions decrease and electrostatic repulsions increase with temperature, also resulting in weakening of the casein matrix.[3] Transient tests of Mozzarella conducted at 20, 30, and 40 °C revealed that creep compliance values increased with temperature (and proteolysis), indicating the weakening of the structure.[58]

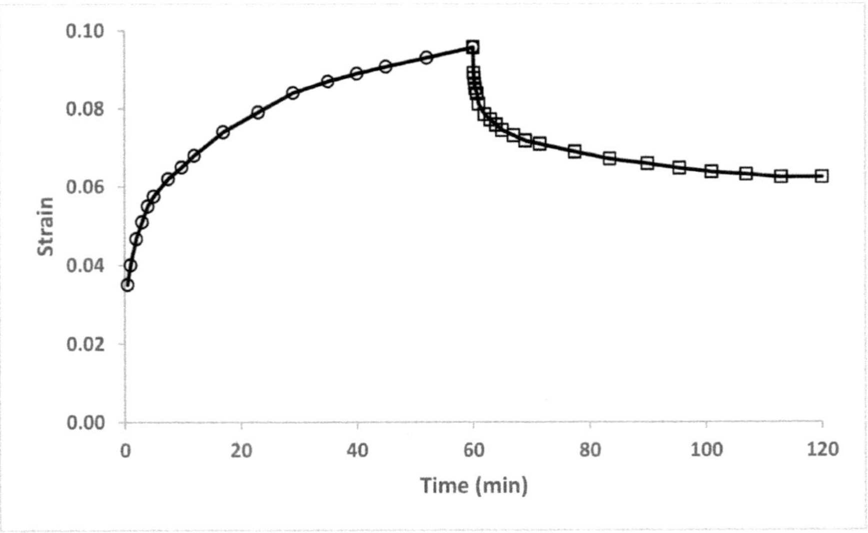

Figure 9.6 Creep recovery curve for Mozzarella, showing application of stress (creep, values indicated by circles), followed by removal of stress (relaxation, squares).

9.4.4 Torsion Tests

An analyst prepares samples for a torsion test by fashioning them into hourglass shapes. A torsion gelometer then twists a specimen until it breaks in the narrow centre, leading to values for the degree of rotation (shear strain) and magnitude of torque (shear stress) needed to fracture the sample. The shear rigidity is the ratio of stress to strain and indicates stiffness.[59] Torsion tests run on 17 fresh and aged cheeses showed significant correlations between shear stress and TPA hardness, and between shear strain and TPA cohesiveness.[59] Further tests showed significant correlations between shear stress and the amount of α_{s1}-casein.[60] A study of Cheddar cheeses at various fat levels related results from small amplitude oscillatory shear, transient tests, and torsion tests with sensory results. Increasing the fat content caused stiffening of the casein network while creating weak spots in the form of fat–protein interfaces. The weak spots increased strain weakening during deformation, resulting in lower fracture strain and smaller particles during fracture.[61]

9.5 Colour

Colour is our perception of light in the 400–800 nm range. Although not specifically cited as a component of texture in the ISO definition, colour is arguably a surface attribute and is definitely important for consumer acceptance. Consumers' initial perceptions of food occur within the first 90 s of observation, and they base some 60–90% of their assessment on colour.[62] Colour, flavour, and texture are the most important factors for consumer liking, so if the colour is not acceptable, people may not judge flavour and texture at all.[63] Some Cheddar manufacturers add annatto to their cheese to deepen its colour since many people associate orange colour with richer flavour (without realizing that the flavour is no different). Shininess, mottling, and chalkiness are other attributes of cheese appearance, which sensory panels usually define.[64]

The Commission Internationale de l'Éclairage defines the CIELAB three-dimensional organization of colours, where L^* values indicate luminosity (degree of lightness from black to white), a^* values green (−) and red (+), and b^* values blue (−) and yellow (+)[65] (CIE). Other colour models are RGB, an additive model in which red, green, and blue combine to represent colours, and CMYK, a subtractive model with partial masking of cyan, magenta, yellow, and key (black) on a white background. A CIELAB colorimeter is the instrument most often used for evaluating cheese and other foods. In a study of Emmental, the amount of colour increased continuously during 10 weeks of ripening and was constant during the final 6 weeks, and the orange hue of the cheese slightly increased in the last few weeks.[66] A study of low-fat Cheddar showed that colour correlated with consumer liking, with negative reactions when the product appeared too translucent or too white.[62]

9.6 Geometrical and Surface Attributes

Curd particles fuse and press together during cheese manufacture, forming curd granule junctions. These junctions, which contain more protein and less fat than the interior of the granules, are partially responsible for cheese texture.[67] Curd granules are observable by light microscopy and digital analysis.[68] On smaller scales, scientists use electron microscopy, confocal microscopy, and related instruments to study the microstructure of cheese (Chapter 8).

Swiss-type cheeses develop eyes, the holes formed by gas bubbles, and cheesemakers strive for their even distribution, regular shape, and range of diameters. Several non-invasive and non-destructive techniques reveal the internal structure of cheeses with eyes, including magnetic resonance imaging,[69] ultrasound,[70] acoustic pulses,[71] and radiography and computed tomography.[72]

Crystals of calcium lactate may form on the surface of cheese, causing consumer rejection. When microflora converts some of the L(+) lactic acid to the less soluble D(−) lactic acid, crystal formation often results. Smooth surfaces and tight packaging help prevent this textural defect from occurring.[73]

9.7 Sensory Testing

The sensory analyst determines the objectives of the project and the test, screens the samples, designs and conducts the test, analyses the data, and interprets and reports the results.[74] The analyst is also mindful of the test room location, design, preparation and training areas, *etc.* The colour of the lights may influence responses, as may air temperature, humidity, and flow. The sample containers should have an appropriate size and shape and the code numbers and order of presentation should be random to avoid bias. The analyst must also consider issues such as panellist fatigue, time of day, preconceived notions, and many other factors.[75]

The second edition of Sensory Evaluation of Dairy Products devotes six of its 17 chapters to different categories of cheese.[76] Sensory evaluation encompasses observing, manipulating, smelling, and consuming cheese using all of the senses and then using descriptive terms to characterise the response in a scientific manner.[64] Food companies use sensory panels consisting of a wide range of people, from consumers "off the street" to highly trained professionals. Panellists may develop their own lexicon for describing texture or use one previously defined. Table 9.5 gives a summary of texture terms derived from testing of a number of cheese varieties. A term that may be unfamiliar is rugosity, the ratio of actual surface area to geometrical surface area, and a measure of roughness. Texture panels have used rugosity as an indicator of smoothness and roughness of the surface of sheep milk cheese.[78]

Table 9.5 Compilation of the common descriptors of cheese texture.[61,64,77–80]

Term	Definition
Adhesiveness[a]	Degree to which mass sticks to the mouth after five chews
Chewiness	Degree of mastication required to break up cheese
Cohesiveness[a]	Degree to which mass holds together after five chews
Creaminess	Extent of breakdown to semi-liquid texture, assessed between the tongue and palate during chewing
Crumbliness	Extent to which the structure breaks up in the mouth during the first 2–3 chews
Crustiness	Force required to break through cheese crust during the first bite, assessed using front teeth
Curdiness	Degree to which mealy texture is perceived in the mouth during chewing
Degree of breakdown[a]	Amount of sample breakdown after five chews
Dry	Degree of dryness or moistness sensed in the mouth during mastication
Firmness[bc]	Various definitions: extent of resistance offered by cheese assessed during the first five chews using front teeth; force required to squeeze a 1.5 cm^3 cube flat between the thumb and index finger (hand firmness); force required to take the first bite, assessed using front teeth; force required to completely bite through cheese, assessed using molars (bite firmness)
First-bite sticky	Sticky sensation during the first bite
Fracturability[c]	Degree to which the sample fractures when completely bitten through with molars
Friability	Ability of the sample to break into numerous pieces during chewing
Grainy	Extent to which granular structures are formed during the 2nd half of chewing
Hardness	Force required for the first bite
Mealy	Feeling when the sample breaks into small pieces that are difficult to gather for swallowing
Moist (or humidity)	Perceived moisture
Mouth coating	Degree of coating on the tongue and palate during chewing
Oily	Any oily, greasy, or fatty mouthfeel
Porosity	Air spaces or pores on the surface
Rate of recovery[b]	Rate at which the sample returns to its original shape after being depressed by 30% between the thumb and index finger
Residual mouthfeel	Degree of graininess in the mouth just before swallowing
Residual smoothness (or smoothness of mouth)	Degree of smoothness in the mouth after expectorating the sample
Rubberiness	Extent to which cheese returns to its initial form during the first 2–3 chews, or degree of springiness while biting the sample
Rugosity	Perception of a granular surface
Sandiness	Sensation associated with small sand particles
Slimy	Soft, moist, and sticky

Table 9.5 (*Continued*)

Term	Definition
Smoothness (or smoothness of mass)[a]	Smoothness against the palate after 5 chews
Springiness[b]	Amount of recovery after the sample is depressed by 30% between the thumb and index finger
Stickiness	Stickiness against the palate and teeth during chewing
Viscous	Mouthfeel associated with highly viscous fluids

[a] Term may be preceded by "chewdown".
[b] Term may be preceded by "hand".
[c] Term may be preceded by "bite".[61]

A relatively new descriptive method of gathering sensory information is Temporal Dominance of Sensations (TDS), which was developed in 1999. In TDS, panellists note a complete list of attributes on a monitor and assess which of these attributes they perceive as dominant at a given time. They give a score to this dominant sensation, but when the panellists decide that the dominant attribute has changed, they score the new dominant attribute, continuing until the perception ends.[81] Scientists have applied TDS to studies of cream cheese,[82] Minas,[83] Prato,[84] Gouda,[85] Mozzarella,[86] and Fourme d'Ambert.[87]

9.8 Basis of Food Texture Terms

When a person chews food, saliva and the proteins within it become the driving force in the agglomeration of food particles into a bolus.[88] This "oral processing" includes wetting and coating, hydration by water molecules diffusing into void spaces within the food, and granulation into small pieces.[89] The sensory perception and feedback during mastication regulates the chewing pattern, formation of the bolus, and swallowing, influencing texture as well as flavour.[90] The composition, structure, and physical properties of cheese change greatly during chewing and play a large role in swallowing time, which is the most important factor for explaining the properties of the bolus and overall texture of cheese.[91]

Various properties of cheese and their relationship with biting and chewing explain the texture terms noted in Table 9.5. The ratios of moisture, protein, and fat obviously affect texture.[92] In most cases, hardness is related to the number and type of interactions per unit of sample. Cheese becomes less firm and more deformable with aging owing to proteolysis of the casein network.[55] The pH of cheese affects hydration of casein, reaching a maximum at pH 5.35, where softness should be the highest.[93] Disulphide and hydrogen bonds tend to give elasticity to protein-based food such as cheese,[89] which manifests itself in springiness and rubberiness. Fat may create weak spots where fracture begins[61] and oiliness results from release of liquid fat and contributes to creaminess.[91] The mouth begins to feel dry if a relatively large amount of water is absorbed, although high fat levels

Table 9.6 A selection of 21st century publications wherein instrumental methods had significant correlations with sensory tests for describing texture.

Cheese variety	Technique	Sensory attributes	Authors, year
29 varieties	Penetrometer	Hardness, springiness, cohesiveness	Breuil and Meullenet, 2001[17]
Salers	Uniaxial compression (UA) (stress at 20 and 80% compression)	Elasticity, deformability, friability	A. Lebecque *et al.*, 2001[97]
26 varieties	UA (amount and rate of deformation)	Hardness	Xiong *et al.*, 2002[20]
Roncal	UA (maximum strain and stress)	Firmness	Irigoyen *et al.*, 2002[21]
Cheddar, Swiss, Gouda	Texture profile analysis (TPA)	11 texture terms	Adhikari *et al.*, 2003[98]
Mozzarella, Monterey Jack, process cheese	TPA, small amplitude oscillatory shear (SAOS), torsion	Hand firmness and springiness, first bite firmness, chewdown adhesiveness	Brown *et al.*, 2003[77]
Paneer	TPA	Firmness, chewiness, gumminess	Khamrui *et al.*, 2004[99]
Cheese analogue	SAOS, creep test	Firmness, rubberiness, stickiness, curdiness	Pereira *et al.*, 2005[100]
Cheddar	TPA	9 texture terms	Everard *et al.*, 2006[101]
Manchego	UA (hardness, strain and force at fracture, stiffness, Young's modulus)	12 texture terms	Viñas *et al.*, 2007[102]
Cream cheese	SAOS, penetration	Firmness, stickiness, gumminess, cohesiveness, difficulty to dissolve, difficulty to spread	Brighenti *et al.*, 2008[103]
Ezine	TPA	10 texture terms	Aday *et al.*, 2010[104]
9 Polish cheeses	TPA, shear, penetration	Hardness, elasticity	Półtorak *et al.*, 2015[105]
Cheese spread	Penetration	Hardness, adhesiveness, spreadability	Ghorbel *et al.*, 2016[106]
Cream cheese	TPA, SAOS	Viscosity, firmness, thickness, cohesiveness, mouth coating, creaminess	Ningtyas *et al.*, 2019[82]
Spreadable whey cheese	Stress relaxation	Firmness, smoothness, spreadability	Chatziantoniou *et al.*, 2019[107]
White brined cow, goat, and sheep cheeses	TPA	Fracturability, hardness, adhesiveness, chewiness	Baysal and Ozcan, 2020[108]
Palermo	TPA	Roughness, elasticity, superficial and mouth moisture	Álvarez and Fresno, 2021[109]

decrease this effect.[94] Interactions between the bolus and the inside of the mouth generate adhesion and cohesion, from which the sensation of stickiness arises.[89] In contrast, mouthcoating is the sensation created if the cheese forms a film that lines the mouth around swallowing, most commonly seen in soft products such as cream cheese.[89] Perceived grittiness or sandiness increases with the number and size of particles present.[95]

9.9 Relating Instrumental Data and Sensory Evaluations

Ideally, instrumental tests will be closely and significantly related to sensory tests and eliminate the time and cost of maintaining sensory panels. However, in the real world, food has a complex composition and structure and mixes with saliva while being manipulated inside the mouth, making it nearly impossible for instruments to duplicate oral processing.[96] Model foods have shown significant correlations between oral processing and Young's modulus, for instance, but real non-homogeneous foods such as cheese do not show similar relationships. Foods with similar Young's moduli such as young Gouda and beef have completely different fracture mechanics resulting in contrasting chewing durations and numbers of chews.[89]

Despite the obstacles, food scientists have attempted to match instrumental results with some aspects of cheese texture. The statistical tools commonly used to determine rheology–sensory relationships include regression, correlation, and principal component analyses, but no universal method for developing these relationships exists yet. Predicting all of the important textural attributes of cheese clearly requires several instrumental tests.[17] Table 9.6 shows an assortment of 21st century papers relating instrumental techniques with sensory evaluation of texture. Half of these studies found correlations between sensory and TPA results, an indication of the popularity of this technique.

9.10 Summary

The determination of the rheological and textural properties of the many varieties of cheese has been ongoing for many years and will continue to be a subject of research for many more. Relating sensory and instrumental data continues to be a key area of interest as scientists try to find rheological techniques that will accurately predict textural attributes.

References

1. M. Reiner, *Phys. Today*, 1964, **17**, 62.
2. ISO 5492, *Sensory Analysis – Vocabulary*, 2008, https://www.iso.org/obp/ui/#iso:std:iso:5492:ed-2:v1:en.
3. J. A. Lucey, M. E. Johnson and D. S. Horne, *J. Dairy Sci.*, 2003, **86**, 2725.

4. P. F. Fox, P. L. H. McSweeney, T. M. Cogan and T. P. Guinee, *Fundamentals of Cheese Science*, Springer, New York, NY, 2000, p. 310.
5. S. Gunasekaran and M. M. Ak, *Cheese Rheology and Texture*, CRC Press, Boca Raton, FL, 2002.
6. G. S. Scott-Blair, *Adv. Food Res.*, 1958, **8**, 1.
7. M. H. Tunick and E. J. Nolan, in *Physical Chemistry of Food Processes*, ed. I. C. Baianu, Van Nostrand Reinhold, New York, NY, 1992, vol. 1, p. 273.
8. J. Park, J. R. Rosenau and M. Peleg, *J. Food Sci.*, 1984, **49**, 1158.
9. C. Apostolopoulos, *J. Dairy Res.*, 1994, **61**, 405.
10. H. H. Friedman, J. E. Whitney and A. S. Szczesniak, *J. Food Sci.*, 1963, **28**, 390.
11. F. Shama and P. Sherman, *J. Texture Stud.*, 1973, **4**, 344.
12. M. H. Tunick, *J. Dairy Sci.*, 2000, **83**, 1892.
13. R. Di Monaco, S. Cavella and P. Masi, *J. Texture Stud.*, 2008, **39**, 129.
14. E. A. Foegeding and M. A. Drake, *J. Dairy Sci.*, 2007, **90**, 1611.
15. A. J. Rosenthal, in *Food Texture: Measurement and Perception*, Aspen, Gaithersburg, MD, 1999, p. 1.
16. M. N. Charalambides, S. M. Goh, S. L. Lim and J. G. Williams, *J. Mater. Sci.*, 2001, **36**, 2313.
17. P. Breuil and J. F. Meullenet, *J. Texture Stud.*, 2001, **32**, 41.
18. D. Jaros, J. Petrag, H. Rohm and F. Ulberth, *Appl. Rheol.*, 2001, **11**, 19.
19. V. Moiny, J. F. Meullenet and R. Xiong, *J. Texture Stud.*, 2002, **33**, 237.
20. R. Xiong, J. F. Meullenet, J. A. Hankins and W. K. Chung, *J. Food Sci.*, 2002, **67**, 877.
21. A. Irigoyen, M. Castiella, A. I. Ordonez, P. Torre and F. C. Ibáñez, *J. Sens. Stud.*, 2002, **17**, 145.
22. S. M. Goh, M. N. Charalambides, S. Chakrabarti and J. G. Williams, *Int. Symp. Food Rheo. Struct.*, 2002, **3**, 267.
23. M. Moreira, E. Storani, A. Bevilacqua and G. De Antoni, *Ital. J. Food Saf.*, 2003, **15**, 287.
24. C. Wick, U. Nienaber, O. Anggraeni, T. H. Shellhammer and P. D. Courtney, *J. Dairy Res.*, 2004, **71**, 107.
25. B. Weinrichter, H. Sollberger, W. Ginzinger, D. Jaros and H. Rohm, *Food/Nahrung*, 2004, **48**, 73.
26. A. Madadlou, A. Khosroshahi and M. E. Mousavi, *J. Dairy Sci.*, 2005, **88**, 3052.
27. A. Madadlou, A. Khosroshahi, S. M. Mousavi and Z. E. Djome, *J. Dairy Sci.*, 2006, **89**, 2359.
28. A. Khosrowshahi, A. Madadlou, M. E. Mousavi and Z. Emam-Djomeh, *J. Dairy Sci.*, 2006, **89**, 3318.
29. A. Madadlou, M. E. Mousavi, Z. Emam-Djome and M. Zargaran, *Int. Dairy J.*, 2007, **17**, 547.
30. A. Madadlou, M. E. Mousavi and J. Farmani, *J. Food Eng.*, 2007, **81**, 330.
31. M. A. Del Nobile, S. Chillo, P. M. Falcone, J. Laverse, S. Pati and A. Baiano, *J. Food Eng.*, 2007, **83**, 621.
32. E. C. Pappa, I. Kandarakis and H. Mallatou, *J. Food Eng.*, 2007, **79**, 143.

33. J. A. Balejko, *Pol. J. Food Nutr. Sci.*, 2007, **57**, 275.
34. Y. Lu, N. Shirashoji and J. A. Lucey, *J. Food Sci.*, 2008, **73**, E363.
35. B. Juan, V. Ferragut, B. Guamis and A. J. Trujillo, *Int. Dairy J.*, 2008, **18**, 129.
36. B. Martin, D. Pomies, P. Pradel, I. Verdier-Metz and B. Rémond, *J. Dairy Sci.*, 2009, **92**, 4730.
37. I. Fathollahi, J. Hesari, S. Azadmard and S. Oustan, *Int. J. Nutr. Food Eng.*, 2010, **4**, 844.
38. E. Ferrandini, M. B. López, M. Castillo and J. Laencina, *Food Chem.*, 2011, **124**, 583.
39. A. Tashakori, S. Ali, Y. Ardakani and M. Daneshi, *Am. J. Food Sci. Technol.*, 2013, **1**, 25.
40. G. Farahani, H. Ezzatpanah and S. Abbasi, *LWT–Food Sci. Technol.*, 2014, **58**, 335.
41. R. B. Magenis, E. S. Prudêncio, C. B. Fritzen-Freire, M. P. Stephan, A. S. do Egito and H. Daguer, *Food Control*, 2014, **45**, 22.
42. E. Vandenberghe, S. Choucharina, S. Luca, B. De Ketelaere, J. De Baerdemaeker and J. Claes, *J. Food Eng.*, 2014, **142**, 31.
43. K. Evert-Arriagada, M. M. Hernández-Herrero, B. Guamis and A. J. Trujillo, *LWT–Food Sci. Technol.*, 2014, **55**, 498.
44. C. M. McCarthy, M. G. Wilkinson, P. M. Kelly and T. P. Guinee, *Int. Dairy J.*, 2016, **56**, 74.
45. C. Ramírez-López and J. F. Vélez-Ruiz, *J. Food Sci.*, 2018, **83**, 1862.
46. M. Addis, M. Pes, M. Fiori, G. Nieddu, S. Furesi and A. Pirisi, *Small Rumin. Res.*, 2018, **162**, 1.
47. M. C. Gonçalves and H. R. Cardarelli, *J. Food Process. Preserv.*, 2020, **44**, e14703.
48. N. H. I. Abou-Soliman, S. Awad and M. I. El-Sayed, *Food Nutr. Sci.*, 2020, **11**, 153.
49. E. Manzocchi, *et al.*, *J. Dairy Sci.*, 2021, **104**, 5285.
50. J. H. Prentice, K. R. Langley and R. J. Marshall, in *Cheese: Chemistry, Physics and Microbiology*, ed. P. F. Fox, Springer, New York, NY, 1993, vol. 1, p. 303.
51. M. H. Tunick, *J. Agric. Food Chem.*, 2011, **59**, 1411.
52. S. P. F. M. Roefs, T. van Vliet, H. J. C. M. van den Bijgaart, A. E. A. de Groot-Mostert and P. Walstra, *Neth. Milk Dairy J.*, 1990, **44**, 159.
53. T. van Vliet, in *Food Emulsions and Foams; Interfaces, Interactions and Stability*, ed. E. Dickinson and J. M. Rodriguez Patino, Royal Society of Chemistry, Cambridge, UK, 1999, p. 307.
54. L. Ma and G. V. Barbosa-Cánovas, *Food Sci. Technol. Int.*, 1995, **1**, 3.
55. M. H. Tunick, E. J. Nolan, J. J. Shieh, J. J. Basch, M. P. Thompson, B. E. Maleeff and V. H. Holsinger, *J. Dairy Sci.*, 1990, **73**, 1671.
56. M. H. Tunick, *Int. Dairy J.*, 2010, **20**, 680.
57. S. Purkayastha, M. Peleg, E. A. Johnson and M. D. Normand, *J. Food Sci.*, 1985, **50**, 45.

58. M. L. Olivares, S. E. Zorrilla and A. C. Rubiolo, *J. Texture Stud.*, 2009, **40**, 300.
59. M. H. Tunick and D. L. Van Hekken, *J. Dairy Sci.*, 2002, **85**, 2743.
60. M. H. Tunick and D. L. Van Hekken, *J. Texture Stud.*, 2003, **34**, 219.
61. N. R. Rogers, M. A. Drake, C. R. Daubert, D. J. McMahon, T. K. Bletsch and E. A. Foegeding, *J. Dairy Sci.*, 2009, **92**, 4756.
62. R. Wadhwani and D. J. McMahon, *J. Dairy Sci.*, 2012, **95**, 2336.
63. F. J. Francis, *Food Qual. Prefer.*, 1995, **6**, 149.
64. C. M. Delahunty and M. A. Drake, in *Cheese: Chemistry, Physics and Microbiology*, ed. P. F. Fox, P. L. H. McSweeney, T. M. Cogan and T. P. Guinee, Elsevier, Oxford, UK, 3rd edn, 2004, vol. 1, p. 455.
65. Commission Internationale de l'Éclairage, *Color Res. Appl.*, 1978, **2**, 5.
66. H. Rohm and D. Jaros, *Z. Lebensm.-Unters. Forsch.*, 1996, **203**, 241.
67. M. Kalab, R. J. Lowrie and D. Nichols, *J. Dairy Sci.*, 1982, **65**, 1117.
68. M. Rüegg and U. Moor, *Food Microstruct.*, 1987, **6**, 35.
69. M. Rosenberg, M. J. McCarthy and R. Kauten, *Food Struct.*, 1991, **10**, 185.
70. J. J. Eskelinen, A. P. Alavuotunki, E. Hæggström and T. Alatossava, *J. Dairy Sci.*, 2007, **90**, 4071.
71. P. Schuetz, *et al.*, *Int. Dairy J.*, 2013, **31**, 150.
72. M. Gonzalez, N. Pérez, E. Budelli and P. Lema, *Proc. Meet. Acoust.*, 2019, **38**, 032002.
73. P. Rajbhandari and P. S. Kindstedt, *J. Dairy Sci.*, 2014, **97**, 1885.
74. J. P. Erhardt, *Food Technol.*, 1978, **32**(11), 57.
75. M. C. Meilgaard, G. V. Civille and B. T. Carr, *Sensory Evaluation Techniques*, Taylor and Francis, Boca Raton, FL, 5th edn, 2016.
76. S. Clark, M. Costello, M. A. Drake and F. Bodyfelt, *The Sensory Evaluation of Dairy Products*, Springer, New York, NY, 2nd edn, 2009.
77. J. A. Brown, E. A. Foegeding, C. R. Daubert, M. A. Drake and M. Gumpertz, *J. Dairy Sci.*, 2003, **86**, 3054.
78. P. Bárcenas, *et al.*, *Int. Dairy J.*, 2007, **17**, 1139.
79. C. Hernández-Morales, A. Hernández-Morales, E. Aguirre-Mandujano and A. Villegas de Gante, *Int. J. Dairy Technol.*, 2010, **63**, 552.
80. L. M. Barden, M. A. Drake and E. A. Foegeding, *J. Sens. Stud.*, 2012, **27**, 286.
81. N. Pineau, *et al.*, *Food Qual. Prefer.*, 2009, **20**, 450.
82. D. W. Ningtyas, B. Bhandari, N. Bansal and S. Prakash, *Food Res. Int.*, 2019, **120**, 586.
83. R. M. Bemfeito, J. F. Rodrigues, J. G. e Silva and L. R. Abreu, *J. Dairy Sci.*, 2016, **99**, 7886.
84. H. L. A. Silva, *et al.*, *J. Dairy Sci.*, 2018, **101**, 8837.
85. A. Thomas, *et al.*, *Food Res. Int.*, 2017, **99**, 426.
86. J. F. Rodrigues, C. S. Gonçalves, R. C. Pereira, J. D. S. Carneiro and A. C. M. Pinheiro, *J. Dairy Sci.*, 2014, **97**, 4733.
87. C. Bord, D. Guerinon and A. Lebecque, *J. Sens. Stud.*, 2019, **34**, e12509.
88. B. J. Teubl, *et al.*, *Clin. Oral Invest.*, 2018, **22**, 929.

89. Q. Guo, *Compr. Rev. Food Sci.*, 2021, **20**, 2941.
90. E. A. Foegeding, C. J. Vinyard, G. Essick, S. Guest and C. Campbell, *J. Texture Stud.*, 2015, **46**, 152.
91. A. Saint-Eve, M. Panouillé, C. Capitaine, I. Déléris and I. Souchon, *Food Hydrocolloids*, 2015, **46**, 144.
92. M. E. Carunchia Whetstine, P. J. Luck, M. A. Drake, E. A. Foegeding and P. D. Gerard, *J. Dairy Sci.*, 2007, **90**, 3091.
93. R. C. Lawrence *et al.*, in *Cheese: Chemistry, Physics and Microbiology*, ed. P. F. Fox, P. L. H. McSweeney, T. M. Cogan and T. P. Guinee, Elsevier, Oxford, UK, 3rd edn, 2004, vol. 1, p. 71.
94. L. Laguna, P. Varela, A. Salvador and S. Fiszman, *Food Res. Int.*, 2013, **51**, 544.
95. M. R. Sainani, H. K. Vyas and P. S. Tong, *J. Dairy Sci.*, 2004, **87**, 2854.
96. H. S. Joyner, *Curr. Opin. Food Sci.*, 2018, **21**, 7.
97. A. Lebecque, A. Laguet, M. F. Devaux and É. Dufour, *Lait*, 2001, **81**, 609.
98. K. Adhikari, H. Heymann and H. E. Huff, *Food Qual. Prefer.*, 2003, **14**, 211.
99. K. Khamrui, S. Dutta and K. Dave, *Milchwissenschaft*, 2004, **59**, 640.
100. R. B. Pereira, R. J. Bennett and M. S. Luckman, *J. Sens. Stud.*, 2005, **20**, 434.
101. C. D. Everard, O. J. O'Callaghan, T. V. Howard, C. P. O'Donnell, E. M. Sheehan and C. M. Delahunty, *J. Texture Stud.*, 2006, **37**, 361.
102. M. A. G. Viñas, C. Ballesteros, P. J. Martín-Alvarez and L. Cabezas, *J. Sens. Stud.*, 2007, **22**, 462.
103. M. Brighenti, S. Govindasamy-Lucey, K. Lim, K. Nelson and J. A. Lucey, *J. Dairy Sci.*, 2008, **91**, 4501.
104. M. S. Aday, C. Caner and Y. K. Yuceer, *Akad. Gıda*, 2010, **8**, 6.
105. A. Półtorak, *et al.*, *Int. J. Food Sci. Technol.*, 2015, **50**, 999.
106. D. Ghorbel, N. Ben Bettaïeb, F. Ghrib, M. Ben Slema and H. Attia, *Int. J. Food Prop.*, 2016, **19**, 1513.
107. S. E. Chatziantoniou, A. S. Thomareis and M. G. Kontominas, *Eur. Food Res. Technol.*, 2019, **245**, 2401.
108. S. Baysal and T. Ozcan, *Int. Food Res. J*, 2020, **27**, 1132.
109. S. Álvarez and M. Fresno, *Animals*, 2021, **11**, 58.

CHAPTER 10

Artisanal Cheese Chemistry

B. VALLEJO-CORDOBA,* R. REYES-DÍAZ,
L. M. BELTRÁN-BARRIENTOS, A. F. GONZÁLEZ-CÓRDOVA AND
A. HERNÁNDEZ-MENDOZA

Centro de Investigación en Alimentación y Desarrollo, A.C., Carretera Gustavo Enrique Astiazarán Rosas # 46, La Victoria CP 83304, Hermosillo, Sonora, Mexico
*Email: vallejo@ciad.mx

10.1 Introduction

Cheese is one of the most appreciated foods worldwide since it represents more than just food. Actually, in many countries, cheese is consumed as an essential ingredient to elaborate different types of dishes; therefore, these products are part of local gastronomy that may cross cultural borders. Currently, cheese making is one of the most important industries worldwide. According to FAO statistics, global milk production reached nearly 906 million tonnes in 2020 and international trade in cheese products reached 2.8 million tonnes.[1] According to the FAO, global cheese production in 2018 reached values above 23 million tonnes, which represented approximately 30% of the total milk produced in the world in 2018.[2] Special attention is often given to artisanal cheeses which are characterized by their uniqueness and typicity.[3] In fact, artisanal cheeses achieve social prestige with enough influence to be consumed not only by the local population, but also throughout the urban areas,[4] since they are appreciated to be part of several traditional dishes.[5] Nevertheless, to the best of our knowledge, there are no

Food Chemistry, Function and Analysis No. 40
Handbook of Cheese Chemistry
Edited by Michael H. Tunick

Published by the Royal Society of Chemistry, www.rsc.org

reports on the estimation of milk used for the production of artisanal cheeses, since these data are not easily quantified.

For better understanding, the term artisan or artisanal cheeses refers to those produced primarily by hand, in small batches, with particular attention paid to the tradition of the cheese maker's art with as little mechanization as possible during production.[4,6] In fact, artisanal cheeses have been made for several decades in small family farms and in small quantities. However, due to globalization, artisanal products have competed and coexisted with industrial products. Thus, artisanal producers have a big challenge competing against multinationals, with high capacity for production and an incomparable financial power. Fortunately, although artisanal cheeses are produced by small family farmers living in less favored areas, they have gained popularity in urban areas, not only due to their quality and flavour, but also due to the fact that they represent experiences, ancient cultures, and a strong and direct relationship with nature that leads back to conscious consumers.[6] Artisanal cheeses are distinctive products usually made from raw milk as a result of traditional techniques applied by dairy farmers all over the world. Although cheese has been made in this manner for thousands of years, this style of cheese making has recently gained popularity for their added value claimed by the growing demand of consumers.[4]

Additionally, it has been extensively reported that raw-milk artisanal cheeses are sources of a great variety of microorganisms, such as molds, yeasts, and bacteria that confer special characteristics in flavour, aroma, and texture. Furthermore, several scientific studies have widely reported that diverse lactic acid bacteria (LAB) isolated from some Latin American artisanal cheeses may show important beneficial effects related to major chronic diseases, such as hypertension, cancer, diabetes, and obesity, among others.[5] Hence, consuming artisanal cheeses may provide an added benefit to human health that goes beyond nutrition as long as its safety is assured. Although artisanal cheeses are popular worldwide and are extensively reported in the literature, less attention was given to Hispanic cheeses, particularly Mexican cheeses. Since artisanal Mexican cheese is an important ingredient of traditional Mexican gastronomy that was recognized by UNESCO in 2010 as a cultural treasure and an Intangible Cultural Heritage of Humanity,[7] and a public policy was enacted aimed at promoting national gastronomy,[8] it became most relevant to focus attention in the most representative and unique artisanal Mexican cheeses.

Thus, in this chapter we will describe how the manufacturing process may affect the chemical, physical, and microbiological characteristics of selected Mexican artisanal cheeses, including their potential use as functional foods due to the rich variety of bioactive compounds generated through artisanal fermentation.

10.2 Artisanal Cheese Chemistry

Of particular attention in artisanal cheeses are their flavour, aroma, and texture influencing the purchase decisions of consumers. These characteristics are the result of the chemical composition of the milk used as the raw material and also the specific making procedure of the cheese type, including ripening.[9,10] Additionally, in the case of raw-milk cheeses, microorganisms as well as natural enzymes present in raw milk are considered to be responsible for enhancing desirable flavour characteristics;[11] therefore, these cheeses tend to ripen faster and develop a more intense flavour than those made from pasteurized milk.[12,13]

In general, freshly made cheese may have weak flavour and unpleasant texture; however, after curd or cheese ripening, other flavours and different textures may develop. Regardless of the cheese variety, artisanal raw milk cheeses develop a more diversified and intense flavour than pasteurized milk cheeses, due to more development of raw milk microflora, differences in proteolysis patterns, higher lipolysis, and a larger concentration of volatiles.[14] Proteolytic and lipolytic changes may involve non-microbial and microbial enzymes, and an earlier production and higher concentration of volatiles are present in raw milk cheeses. Compounds such as volatile fatty acids and esters are always higher in raw milk cheeses, whereas ketones and aldehydes are intermediary metabolites that may be produced and degraded.[14]

During cheese ripening, different chemical and microbiological changes may occur. Chemical changes are grouped into primary (lipolysis, proteolysis, and metabolism of residual lactose and of lactate and citrate) or secondary (metabolism of fatty acids and of amino acids [AA]).[15] The metabolism of lactose is the first glycolytic process during the early stages of cheese ripening. In fact, although approximately 96% of lactose in milk is lost in the whey, a substantial amount of lactose may remain in the curd. On the other hand, it has been reported that flavour in young cheese is directly associated with lactic acid produced during cheese ripening.[16] Lastly, citrate is also lost in the whey, since it is mostly present in the soluble phase. However, it has been reported that even at low concentrations in cheese, citrate is utilized by bacteria and specific aroma compounds may be produced.[15]

On the other hand, lipids may be degraded to free fatty acids by the action of endogenous and/or exogenous lipases, and these free fatty acids are important flavour compounds or precursors for volatile flavour compounds.[17] However, in cheeses made with pasteurized milk, endogenous lipases are inactivated by heat treatment, thus affecting cheese flavour. In contrast, in raw milk cheeses, endogenous lipases participate in lipolysis.[18] Finally, proteolysis is the most complex biochemical change that occurs during cheese making, and most importantly during cheese ripening. In general, during this particular chemical process, cheese texture is directly affected by the hydrolysis of the para-casein matrix, the curd increases the water-binding capacity, and finally the pH is increased by the production of NH_3 due to the catabolism of the released AA.[19]

In artisanal cheeses, similar basic chemical and microbiological changes may occur; nevertheless, these are determined by the particular manufacturing process and the diversity and composition of the microbial flora. Interestingly, some unique manufacturing practices for artisanal cheeses made with raw milk from the tropics, Chiapas or Bola de Ocosingo cheeses from southern Mexico, involve curd ripening for several days, before salting, molding, and pressing, thus generating a highly acidic curd that results in a soft crumbly cheese with a very distinctive rich acid flavour due to a long acid-enzymatic coagulation period.[20] In contrast, Cocido cheese, from northern Mexico, is a pasta-filata cheese made with raw milk that is inoculated with natural whey as a starter culture during renneting to induce acidification and assist coagulation and curd formation, in order to confer a mild flavour and reduce manufacturing time.[5] On the other hand, Fresco cheese is a soft high moisture cheese manufactured throughout Mexico, rich in LAB native to the raw milk being used[21] that confers its characteristic mild milky flavour when consumed before ten days after manufacturing.[22] At the other end of the spectrum, Cotija cheese is a low moisture salty cheese produced from raw milk from crosses of zebu cattle (*Bos taurus indicus*) in central Mexico.[23]

Because of these peculiar manufacturing practices for Mexican artisanal cheeses, classification based on texture and moisture content is more suitable than cheese classification based on maturity.[24] Thus, Mexican artisanal cheeses may be classified according to their moisture content resulting in different textures and chemical and microbiological compositions. Soft cheeses with a moisture content of more than 67% are the most consumed, particularly, Fresco cheese; nevertheless, another less known soft cheese is Bola de Ocosingo cheese, Chiapas. Semi-soft and semi-hard cheeses, with a moisture content from 54 to 63%, include Chihuahua, Crema de Chiapas, and Poro de Tabasco cheeses. Finally, hard cheeses with a moisture content from 49 to 56% are the least consumed such as Cotija or Añejo cheeses.[24]

10.3 Fresco Cheese

Fresco (fresh) cheese is one of the most popular cheeses in Latin American countries that may be immediately consumed after production since it is unripened.[25] When this cheese is made using artisanal procedures, raw cow's milk is traditionally used and the product may present different variants in size, shape, and composition due to small variations in cheese making and the used ingredients according to each producing region. Therefore, in some regions it is also known as Aro, Molido, or Ranchero cheese.[3] Additionally, other cheeses named as Panela, Canasto, or Blanco are made using relatively the same procedures resulting in similar characteristics to those of Fresco cheese; thus, it would not be too adventurous to say that we are referring to the same cheese product. On the other hand, although this cheese is one of the most popular artisanal Latin American cheeses,[3,25] it is scarcely characterized considering that its composition may vary among regions due to the lack of standard cheese making procedures. In this sense, different factors that may determine its chemical and microbiological composition may be considered such as the quality of raw milk, cheese making procedures, the lack of GMP, and shelf conditions during commercialization.

In particular, Fresco cheese from Sonora, Mexico, has been widely characterized by its chemical and microbial composition,[21,26] as well as its aroma profile associated with volatiles.[22,27] This cheese is usually made with raw whole bovine milk using traditional methods in a small scale, often with commercial rennet as the curdling agent (some cheese makers still use natural rennet). Milk coagulation is carried out under uncontrolled temperature and environmental conditions at each farm, in which acidification may occur by the presence of indigenous LAB present in the raw milk. After a firm coagulum is formed, the curd is cut into small squares and set to rest to let the curd precipitate. Then, whey is drained off by any separation procedure such as decanting, filtering and/or hand pressing. When a drained curd is obtained, salt (approximately 30 g/100 L) is added and manually homogenized. Afterwards, the salted curd is molded by stuffing into molds of different shapes and sizes and slightly pressed, and then left to rest overnight for the whey to be expelled. When more whey is lost and a harder product is obtained, it is usually known as Oreado cheese. Finally, the cheeses are taken from the molds and are ready to be consumed.[3,22,26,28]

Furthermore, Fresco cheese is often packaged by putting them in plastic bags and commercialized in the local markets with a very short shelf-life. Also, it is sold in local stores under refrigerated shelves with a shelf-life of approximately 10 days under such conditions.[3,22,26,28] The final product is soft and creamy with a bright-white color, crumbles easily, and its flavour is slightly reminiscent of fresh, salty milk (see Figure 10.1). Additionally, some vegetables may be added as extra ingredients, according to the region where it is prepared.[3] Since Fresco cheese does not melt and has a mild flavour, it is mostly consumed crumbled in small pieces in many traditional Mexican foods such as soups and tortilla-based foods, such as tacos, tostadas, and enchiladas, among other dishes.

Fresco cheese from Sonora is a soft, full fat cheese that presents the following chemical composition: 51 to 63% moisture, 13.6 to 25.2% protein, 13 to 26.5% fat, 0.01 to 2.5% carbohydrates, 2.2 to 3.4% ash, 0.8 to 1.8% NaCl, and 1 to 3.2% acidity, with pH 5–5.6.[21] As a result of the raw milk used, the microbial composition presents high counts of aerobic mesophilic bacteria, molds, and yeasts. Coliforms and pathogens may also be present surpassing in some cases the permissible limits, revealing the lack of GMP. Fortunately, LAB such as *Lactococcus* spp., *Streptococcus* spp., and *Lactobacillus* spp. are also present in high counts.[21,26] In fact, these LAB may play an important role against undesirable microorganisms and may improve the typical flavour characteristics of this cheese. In this sense, Fresco cheese has been reported to present yeasty, milk fat, yogurt, dirty socks, and barny flavour notes. Also, a wide diversity of volatiles were identified such as esters, acids, alcohols, ketones, and aldehydes, from which methyl esters (C4, C6, C8, and C10), acids (C4, C6–C10, and C12), 3-methyl-1-butanol, phenylethyl alcohol, and acetoin were found as key aroma compounds.[22]

Figure 10.1 Artisanal Fresco cheese.

Although the volatile concentrations of five commercial raw milk Fresco cheeses were reported for these different groups of compounds, the concentrations of 3-methyl-1-butanol (3.48 ± 0.1 μg g^{-1}) and isovaleric acid (3-methyl-butanoic acid) (0.69 ± 0.02 μg g^{-1}) were highlighted as important aroma contributors. Particularly, the alcohol 3-methyl-1-butanol, associated with Fresco cheese aroma, may be one of the key compounds that may serve for screening LAB strains for their suitability as starter cultures. In this regard, specific *Lactococcus lactis* inoculated in pasteurized milk cheese presented from 3.4 to 13.13 $\mu g\,g^{-1}$ of 3-methyl-1-butanol, thus showing potential as starter cultures in Mexican Fresco cheese manufacture; however, the effect of other volatiles in cheese aroma could not be disregarded.[22] Indeed, it has been reported that this alcohol derived from the catabolism of leucine by *Lactococcus lactis* imparts a strong cheesy aroma.[29] Also, isovaleric acid (3-methyl-butanoic acid), which was found in raw milk Fresco cheese but was absent in pasteurized milk cheese, was reported to be associated with descriptors such as dirty socks, sweaty, fecal, and barny. Moreover, hexanoic acid was found at the highest concentration of all acids and was related to goaty and cheesy sharp notes.[22] Since it has been reported that the catabolism of AA is the most important pathway for volatile production and is strain specific,[29] the selection of specific LAB strains may be based on volatile production and its association to AA profiles.

In a previous study, during 8 weeks of refrigerated storage of Fresco cheese, the activity of spoilage bacteria contributed to proteolysis, appearance of volatile compounds presumably generated from lipolysis and decreases in lactose and pH levels.[27] However, when Fresco cheese is still within no more than two weeks after production, that is when it is acceptable for consumption, volatile production may be the result of the metabolic activity of native LAB.[22] Also, protein degradation in cheeses may take place not only by the proteolytic activity of microorganisms, but also by the activity of the rennet enzymes. Thus, protein degradation by the activities of rennet enzymes in addition to cell-envelope proteinases and peptidases from LAB result in small peptides and free AA that may also be available as flavour precursors.[22]

Interestingly, small dairy peptides not only serve as flavour precursors, but may also possess bioactivity. Indeed, Fresco cheese may contain numerous peptides released by proteolysis either by native LAB or by added starter cultures. In this regard, Fresco cheese made with specific LAB presented ACE inhibitory peptides.[30] Also, radical scavenging peptides have been identified in water-soluble extracts derived from Fresco cheese,[31] showing its potential as an antioxidant food.

10.4 Chihuahua Cheese

Chihuahua is a semi-hard cheese mostly consumed in northern Mexico and in the United States since this type of cheese is an essential ingredient used by the Hispanic population in their cuisine[32] (see Figure 10.2). In fact, since Chihuahua cheese melts easily, it is the most popular cheese used to make quesadillas, stuffed poblano chili peppers, as well as a melted topping in green sauce enchilada, among other traditional Mexican foods.

This cheese has been traditionally made by the Mennonites settled in Cuauhtémoc, Chihuahua, Mexico in the 1920s.[33] It has been reported that a worker of a German pharmacist from the Mormon colonies in Chihuahua learned the technique and started to make this cheese in this region.

Figure 10.2 Artisanal Chihuahua cheese.

Mennonites have had a great influence in cheese manufacturing improving all agro-industrial sectors related to this product including raising dairy cattle, making milking more hygienic, and perfecting the cheese making procedures. Moreover, empirical knowledge built up through trial and error has allowed Mennonites to manufacture a product that met their own standards.[33,34] Therefore, this cheese is also known as Menonita cheese; interestingly, in Mennonite communities it is called Chester cheese or

Cheddar cheese, but out of the state of Chihuahua, this cheese is recognized as Chihuahua cheese and in this way the product appropriates its name.[34] At present, this cheese is now made, at small and industrial scales, by both Mennonites and non-Mennonites throughout Mexico.[35] Also, this type of cheese made with pasteurized milk is made in the USA under the trade name Chihuahua™.[33]

The Chihuahua cheese making procedure resembles that of Cheddar, but the product is consumed no longer than two months after manufacture.[33] Cheese making is not quite standardized; it is made in artisanal cheese factories, since it depends on the availability of tools or special infrastructure. Nevertheless, all the procedures are carried out by following four essential processes: coagulation or curdling of the milk, dehydration of the clot or curd (dewatering), cheddaring, and ripening.[33]

This cheese is made from raw or pasteurized (standardized or whole) bovine milk using artisanal procedures or at an industrial scale. Calcium chloride, potassium (or sodium) nitrate, annatto tincture and a starter culture (for pasteurized milk) are added, and milk is set to rest for 30–35 min at ~32 °C. Rennet (15 mL/100 L) is used to form a firm coagulum that is cut into squares using 6.4–25.4 mm curd knives; after reposing for 5 min, the curd is worked at 30–45 °C for 30–90 min and whey is drained. The resulting curd grain is left to rest for 30–40 min while whey draining occurs to achieve dehydration of the clot that is further cut into blocks of ~10–15 cm. After this, cheddaring is carried out, which is one of the most crucial steps for the development of the unique characteristics of this cheese. In this regard, cheddaring consists in resting the curd slabs for ~15–90 min while stacking and flipping until the curd acidification reaches 30–35 °D (pH 5.2–5.5) in the leached whey. Then the curd slabs are milled, salted (~2–2.5%, referring to the drained mass), and molded in 4–10 kg cylindrical or rectangular prism molds which are then pressed (120–550 kPa) for ~16–20 h. After pressing, the pieces are demolded and maintained in a fresh room (15–20 °C) for ~3–6 days and finally packaged and stored at 0–20 °C for a short ripening period of ~15 to 30 days.[33,36]

Chihuahua cheese is minimally ripened and has the following chemical composition: 32 to 44% moisture, 23.2 to 27.6% protein, 32.3 to 34.3% fat, 2.8 to 3.6% ash, and 1 to 1.5% NaCl, with a pH of 4.8 to 5.2. It has a crumbly texture and is an uncooked paste with good melting properties due to its pH and possesses typical flavour notes such as salty, sour, diacetyl, cooked, whey, bitter, and milk fat.[32,33,36,37] Many of these characteristics are strongly influenced by its microbial composition which has been characterized and identified by DNA sequencing following denaturing gradient gel electrophoresis (DGGE) of PCR products. In fact, LAB were one of the most abundant groups, mainly those of the genera *Lactococcus*, *Lactobacillus*, *Streptococcus*, *Leuconostoc*, and *Enterococcus*.[3] Additionally, sequence analysis allowed the identification of potential pathogens such as *Escherichia coli* and *Staphylococcus aureus* in Chihuahua cheeses made from raw and pasteurized milk.[38]

In an attempt to relate proteolysis to cheese flavour, descriptive flavour analysis was performed and young flavours like cooked, whey, milk fat, and diacetyl were identified. The basic flavours, *i.e.*, sour, bitter, salty, and astringent, were also identified in all cheeses.[39] Gel electrophoresis and reverse phase high performance liquid chromatography (RP-HPLC) showed that while cheese samples from one manufacturer showed little protein degradation over time, samples from other manufacturers degraded extensively when stored at 4 °C for 8 weeks. Water-soluble cheese proteins analyzed by mass spectrometry revealed the presence of short, hydrophobic peptides in quantities correlating with bitterness.[39] This study was carried out aiming towards the development of a pasteurized milk cheese using a starter culture that mimics the taste and functionality of Chihuahua cheese, thus allowing it to be sold in the US.

Furthermore, it is important to disclose that this cheese has been a good source of different LAB strains capable of producing compounds with different bioactivities such as antithrombotic, hypocholesterolemic, and antihypertensive peptides,[40,41] and gamma-aminobutyric acid.[42]

10.5 Bola de Ocosingo and Crema de Chiapas Cheeses

Bola de Ocosingo is an emblematic soft cheese produced in Ocosingo, Chiapas, southern Mexico, and was the first cheese to achieve a collective mark distinction in Mexico, and a wider protection, such as a PDO recognition, has also been sought.[3] Nevertheless, despite being a good candidate to achieve this recognition, it is a big challenge for the producers since it is also reported that the rules established for the collective mark are scarcely fulfilled, obtaining products of different qualities.[43] This cheese is unusual and atypical because of its peculiar characteristics of shape and texture: it is a spherical ball of bright-yellow creamy-texture cheese with milky, salty, and fresh taste covered with two successive hard layers of stringy-paste curd made from skim milk, which functions as a protective rind (see Figure 10.3). Thus, a final product that weighs 700 g is composed of 450 g of the creamy cheese inside and 250 g of the stringy-paste cheese used as a rind.[3,44,45]

The historical origin of Bola de Ocosingo cheese is uncertain; however, it has a peculiar cheese making process that it is believed it was developed following a recipe for Edam cheese; interestingly, the spherical form may have had some inspiration from this Dutch cheese.[46] Traditionally, Bola de Ocosingo cheese is made using non-standardized artisanal procedures that vary among cheese makers and begin with the use of raw bovine milk, to which cream (obtained from a separate batch of milk) is added in order to make the internal creamy-texture paste.

The process begins when rennet and acidic whey (obtained from the previous day) are added to curdle the milk that is cut and left to repose up to 20 h at room temperature (~30 °C). Then, the curd is placed in a muslin

Figure 10.3 Artisanal Bola de Ocosingo cheese.

cloth sack and hung to let the whey drain for ~5 days. Afterwards, salt (~3–4%) is added and the curd is kneaded to be placed in a new sack and left to rest for another 21 days (this period could be less in drought or up to 45 days in the rainy season) at room temperature, although some producers may carry out this step at 5 °C. During this ripening period, the paste is kneaded periodically, approximately every 3 days, and the sacks are changed to avoid contamination. After this, the resulting creamy cheese is compacted by hands to form a ball that is completely lined with one layer of a stringy-paste cheese. This stringy-paste is made from milk that was previously skimmed, fermented (or acidified with vinegar), curdled for 3–5 h, cut, drained, and left to rest for ~12 h until the acidity becomes sufficient to achieve a spun paste that is put in boiling water for ~5 min and kneaded to get stringy. After the first stringy layer becomes hard, a second layer is applied to form a hard rind of cheese whose function is to protect the filling, favoring its commercialization at room temperature.[3,44,47–49] Finally, the cheese is ready to be marketed without other packaging than the hardened paste rind that is not traditionally consumed. One label sticker is usually attached on the surface of the hardened paste rind; however, some producers may additionally use a labeled-plastic wrap.

Since Bola de Ocosingo cheese is formed from two cheeses, the internal creamy-texture cheese and the stringy-paste rind, it is convenient to describe both pieces. In this sense, López *et al.*[44] characterized the physicochemical and microbial composition of this cheese. The authors reported that the external rind cheese has the following composition: 45.9% moisture, 45.3% protein, 1.5% fat, 5.3% ash, and 2.6% NaCl, with pH 5.5; the internal cheese contains 35.7% moisture, 22% protein, 39.3% fat, 2.8% ash and 2.3% NaCl, with pH 4.5. The external rind showed the following microbiological composition: 5.17 $\log_{10}$ CFU per g aerobic mesophilic bacteria, 0.76 $\log_{10}$ CFU per g total coliforms and <3 MPN per g fecal coliforms. Similarly, the internal cheese showed 3.8 $\log_{10}$ CFU per g aerobic mesophilic bacteria, 0.34 $\log_{10}$ CFU per g total coliforms and <3 MPN per g fecal coliforms. Besides, another study applied high-throughput sequencing to evidence the presence of a low proportion of *Escherichia coli*/*Shigella flexneri* species detected in almost all cheeses sampled. However, several LAB species such as *Streptococcus* spp., *Lactococcus* spp., and *Lactobacillus* spp. were identified and predominated during processing, with their prevalence varying from one producer to another.[47] It is well known that LAB species impact positively the typical sensory characteristics; however, to the best of our knowledge, there is no scientific report describing its sensory attributes by descriptive analysis. Nevertheless, it was found that LAB counts decreased in Bola de Ocosingo cheese according to its ripening time and resulted in higher concentrations of low molecular weight peptides with angiotensin-converting enzyme inhibitory activity.[45]

On the other hand, Crema de Chiapas cheese is also a traditional product from the state of Chiapas mostly consumed in southern Mexico. It is characterized for being an unripened, soft and pressed paste with a strongly acidic taste and rich aroma and a crumbly texture (see Figure 10.4). It has some similarities to Bola de Ocosingo since both are manufactured from raw milk from Zebu-Brown Swiss cows from the tropics and are the result of a prolonged acid-enzymatic coagulation that occurs under similar environmental conditions. Nevertheless, commercialization of Bola de Ocosingo is rather regional, whereas Crema de Chiapas has reached other markets beyond its region of production. Due to the crumbly texture of Crema

Figure 10.4 Artisanal Crema de Chiapas cheese.

de Chiapas cheese, it makes it a versatile ingredient that may be used as topping on typical Mexican dishes such as enchiladas, chilaquiles, flautas (hard shell tacos), and empanadas, among other tortilla-based dishes.

Crema de Chiapas cheese manufacturing starts with raw milk acidification for 3 to 5 h at ambient temperature (≥25 °C); then, coagulation takes place by adding commercial rennet. After 2 to 8 h, a highly acidic curd is produced, cut and left overnight. Then, the curds are placed in cotton bags which are hanged for whey draining. When the curds reach a pH of 4.5, they are kneaded, salted, molded, and pressed in a rustic press.[3] The most peculiar distinctive feature of this cheese is that it is marketed in rectangular pieces covered with a 3-layer colorful package consisting of a plastic film, an aluminum foil, and yellow or red cellophane. As a result of this traditional process, the product obtained contains 52.5 to 59.4% moisture, 40.5 to 47.4% total solids, 33.8 to 38.7% protein, 20.8 to 33.4% fat, 4.9 to 8.1% ash, and 2.0 to 6.6% salt and has a pH of 4.3 to 4.7.[50] These wide ranges in chemical composition reveal the lack of standardization of the artisanal Crema de Chiapas cheese process.

Crema de Chiapas cheese also has a collective mark[51] and has potential for obtaining a PDO since it is an important economic activity of the region. Due to the rich LAB consortia, the dynamics were characterized by DGGE of PCR products derived from DNA extracted from cheese after 15 and 30 days of ripening. The dominant genera were *Lactobacillus*, *Streptococcus*, and *Lactococcus*, as identified by DNA sequencing of the 16S rRNA gene and were different in structure according to the season and region.[52,53] Also, halotolerant or halophilic LAB were isolated, identified, and characterized using phenotypic and genotypic techniques in Crema de Chiapas and Cotija cheeses. Moreover, technological traits such as acidifying, proteolytic, and lipolytic activities of these LAB were determined to assess their potential use as indigenous starter cultures.[54]

Moreover, although consumption of Crema de Chiapas cheese is greatly appreciated in southern Mexico, its flavour and aroma profile and volatiles have not been reported. Nevertheless, Crema de Chiapas cheese showed high proteolytic activity and a high concentration of free AA associated with specific LAB that were isolated and characterized.[20] Interestingly, this cheese presented a remarkably high concentration of gamma-aminobutyric acid concentration (0.78 $mg\,kg^{-1}$), which is a bioactive compound with several beneficial effects. In addition, this cheese presented angiotensin-I converting enzyme inhibition (ACEI). Furthermore, when specific highly proteolytic LAB strains isolated from this cheese were inoculated in milk, fermented milk also presented ACEI and antioxidant activity.[20]

10.6 Cotija Cheese

Cotija cheese is hard, crumbly, friable, and ripened for at least 3 months. Besides the uncooked paste, it has a salty-acid taste, and when it is extensively ripened, it is in white-yellow color with an intense cheesy aroma (see Figure 10.5). It is found in a wide variety of forms, but it is usually pressed into a cylindrical shape that can weigh up to 20–30 kg. This cheese is traditionally made with artisanal procedures using raw bovine milk from Holstein, creole Holstein, Brown Swiss-Zebu, or Holstein-Zebu crossbred cattle in different states of Mexico, such as Chiapas, Tabasco, Jalisco, and Michoacán.[36] The salt added during the manufacturing process contributes

Figure 10.5 Artisanal Cotija cheese.

to its flavour and firm and crumbly texture, thus making grated Cotija cheese suitable for sprinkling over foods.

In 2005, Cotija cheese was granted a collective mark (CM): "Cotija Cheese, Region of Origin". The CM distinction has also been assigned to Crema de

Chiapas and Bola de Ocosingo cheeses from Chiapas and Poro de Balancán cheese from Tabasco.[51] Although the name of Cotija cheese is linked to the town of Cotija, Michoacán, the CM "Cotija Cheese, Region of Origin" is assigned to cheese makers from Sierra de Jalmich, a region located along the adjacent area within the states of Jalisco and Michoacán. It includes the municipalities of Santa María del Oro, Jilotlán de los Dolores and Quitupan in Jalisco and Cotija, Tocumbo and Buenavista Tomatlán in Michoacán, where it has been produced for more than 400 years, mainly during the months of July to October.[55]

The CM distinction was sought in order to distinguish the authentic cheese from its imitations which can serve for the commercial protection of products whose quality is linked to origin. This distinction constitutes a point of reference for producers of different artisanal cheeses, allowing them to value the traditional products emphasizing their history, region of origin and the distinctive intrinsic attributes that make them genuine.[56] Besides the CM distinction, the most symbolically important achievement for the recognition and fame of Cotija cheese was when a two year-old piece won the award for the foreign product category in a world contest celebrated in Cremona, Italy in 2006, emphasizing its regional and national cultural heritage.[36,55]

Behind the typical characteristics of Cotija cheese, there are some key production decisive stages that stand out, such as the work of the curd, which requires great care and patience, the amount of salt added and the ripening conditions. Nevertheless, it is important to disclose that Cotija cheese production is non standardized and may vary among regions, dairies, or cheese makers. Overall, during its processing, raw milk is used with no addition of a starter culture; thus, acidification usually occurs along the process by the presence of indigenous bacteria from raw milk. Moreover, milk is curdled usually using commercial rennet; this procedure is carried out at ~30 °C until a firm coagulum is attained (~2 h, which may vary since the rennet dose is often uncontrolled). The curd is cut and crushed by hand or by a large spoon or knife and left to rest for a better separation of whey that is carefully drained on fiber mats. The drained clot is cut in small pieces with a knife, salt (~4–6%) is added and the curd is kneaded by hand until a homogeneous mixture is achieved; then, it is placed in wooden, cylindrical molds fastened by belts or by twine from maguey fibers that are tied around the mold, maintaining pressure for ~24 h. Afterwards, the molds are placed under pressure for 8–12 days more at room temperature, until the whey is completely drained. Finally, it is demolded and left for a ripening process for at least 3 months in fresh rooms.[3,36,57,58]

Furthermore, some semi-industrial dairies apply a ripening period of 1–3 months in refrigerated storage, and it is worth mentioning that pasteurized milk may often be used. Since ripening is usually carried out under uncontrolled environmental conditions, it is important to highlight that the Jalmich region is located at an altitude of 700–1700 m with an average annual temperature of 20 to 25 °C, an average annual precipitation of

900 to 1200 mm per year, and a relative humidity of 50 to 95%. These parameters are of utmost importance for ripening since they provide intrinsic typical characteristics that make Cotija cheese a unique mature Mexican cheese.[3,36,57,58]

Artisanal Cotija cheese is characterized as containing 30 to 38.2% moisture, 61.7 to 70% total solids, 22.8 to 34% fat, 28.2 to 30.1% protein, 6.45 to 8.26% ash, and 4.11 to 4.63% salt.[36,57] These wide compositions reveal the lack of standardized cheese making procedures. Additionally, low moisture and high salt content may probably give this product a low water activity, which does not promote optimum growth of microorganisms such as indicator microorganisms, aerobic mesophilic bacteria, LAB, or pathogens.[57]

In this sense, different yeast species such as *Kluyveromyces lactis*, *Kluyveromyces marxianus*, *Pichia guilliermondii*, *Rhodotorula mucilaginosa*, *Galactomyces reessii*, and *Galactomyces geotrichum*, and members of the Lactobacillaceae family such as *Pediococcus pentosaceus*, *Lactobacillus brevis*, *Lactobacillus plantarum*, and *Lactobacillus casei* have been identified in this cheese.[57] Additionally, it was reported that the metagenome of Cotija cheese is composed mainly of *Lactobacillus* spp., *Leuconostoc* spp., and *Weissella* spp., and more than 500 non-dominant genera grouped in 31 phyla including halophilic bacteria and archaea; interestingly, pathogenic bacteria such as *Salmonella*, *Listeria monocytogenes*, *Brucella*, and *Mycobacterium* were not found.[58] On the other hand, Morales *et al.*[54] found halophilic or halotolerant LAB which may play a role in the ripening of Cotija cheese; thus, *Lactobacillus acidipiscis*, *Tetragenococcus halophilus*, *Weissella thailandensis*, and *Lactobacillus pentosus* were identified by phylogenetic reconstruction based on 16S rRNA gene sequences.

Most recently, the microbiological characteristics and behaviour of *Staphylococcus aureus*, *Salmonella* spp., *Listeria monocytogenes*, and staphylococcal toxins during elaboration and maturation of Cotija cheese were reported. Although *Salmonella* spp. and *Listeria monocytogenes* were not detected in any sample after 60 days of ripening, *Staphylococcus aureus* was detected at the end of the study, as well as LAB (7.2 log CFU per g).[23] The metabolic activity of the microbiota present during ripening generates compounds such as peptides, bacteriocin, peroxide, and organic acids that determine the predominance of LAB that are also responsible for their production, as well as the formation of compounds responsible for the organoleptic characteristics of Cotija cheese.[23] These findings reveal the complexity in the ripening process of artisanal Cotija cheese that makes it unique and suggest that if good quality raw materials and GMP are used, along with a three-month ripening process, it may be possible to obtain a safe, good quality cheese.

Interestingly, although the microbial and chemical composition of Cotija cheese has been documented, its sensory characteristics and volatiles have not been reported. For this reason, flavour studies relating sensory descriptors to the volatiles present in Cotija cheese are needed in order to evaluate the role of LAB in the ripening process and their impact on the sensory quality of the product. Additionally, it is well known that, during ripening, bioactive compounds may be released due to proteolysis displayed by the microbial populations, resulting in unique peptide profiles that exert biological activities. In this regard, antioxidant and angiotensin-converting enzyme inhibiting peptides were released and increased through the ripening of Cotija cheese.[59]

10.7 Oaxaca and Cocido Cheeses

Oaxaca cheese is one of the most popular artisanal Mexican cheeses. Its name is linked to the state of Oaxaca, where it likely originated, but it is traditionally made using artisanal procedures in various regions of Mexico. However, Oaxaca cheese is also produced industrially (by small, medium, or large companies) using standardized and automated processes since it is highly demanded. It is an unripened cheese of a soft, cooked, stretched, spun pasta-filata type that melts with heating. This cheese is also known as de Hebra, de Bola, or Quesillo (see Figure 10.6). Other pasta-filata cheeses from Mexico are Asadero (made in different states such as Durango, Coahuila, Chihuahua, Aguascalientes, Jalisco, and Guanajuato), Guaje (San Luis Potosí), Cocido (Sonora), and Quesadilla (Sonora).[3,5,60] Due to their typical sensory attributes such as flavour, aroma, and texture, but mostly due to their melting properties, all these pasta-filata cheeses are appreciated as essential ingredients in several traditional Mexican dishes, but mainly in "quesadillas" that are a traditional Mexican dish highly consumed all over the country.

At the industrial level, Oaxaca cheese is usually made from pasteurized milk; however, artisanal cheeses are traditionally made from raw bovine milk from Zebu-Brown Swiss or Zebu-Holstein cattle livestock.[61] Its preparation requires high technical expertise and knowledge since it presents several critical points that are essential, such as reaching a suitable acidity of the milk and curd, determining an adequate thread point of the curd, and an appropriate heating/kneading/stretching of the paste. The manufacturing process may vary among regions, dairies, and cheese makers, and depending on whether the factories are artisanal or industrial.[36,62,63]

The cheese making begins when the milk is acidified by different methods, such as fermenting by native microorganisms in raw milk, by adding specific starter cultures, or the use of organic acids (acetic, lactic, or propionic acid) or fermented whey to accelerate the time of curd acidification. In this last aspect, it requires strict pH control in the curd (5.2 to 5.4 or 37 °D) to reach an optimum condition for melting and stretching. After acidification, the milk is curdled by adding the coagulant that is thoroughly mixed; for this purpose, commercial rennet is usually applied (15 mL/100 L of milk or according to the supplier) and left to rest at 30–32 °C until a firm coagulum is obtained. Then, the coagulum is cut into ~1–2 cm^3 cubes with stainless-steel harps in a longitudinal and transversal manner or by using curd knives or large wooden spoons. Thereafter, the curd is worked by gentle agitation for

Figure 10.6 Artisanal Oaxaca cheese.

~10 min, followed by reposing for 5–25 min. Subsequently, most of the whey is drained and the clot is left to rest for half an hour to 4 h until reaching a pH of 5.2 to 5.4 (35–38 °D in leached whey). The optimum acidity degree is empirically established and handmade curd stretching tests are conducted by immersing the curd portions in hot water (~70 °C) until an optimum stretching point is achieved when stretching occurs without deformation or breaking. Afterwards, the curd is kneaded in hot water (72–75 °C) and strands that are handmade with a 3 to 6 cm width are formed and immediately submerged in cold water or left to rest on the table at room temperature for ~20 min for cooling. The resulting strips are then rubbed with salt and looped around to form tangled balls of different sizes or weights. Finally, the cheese is ready to be consumed and some producers cover the final product with a labeled-plastic wrap and store it under refrigeration for its commercialization.[36,62,63]

Since artisanal Oaxaca cheese is made in different regions with non-standardized processes, products with different chemical compositions are obtained; overall, this cheese is characterized as containing 49.1 to 51.3% moisture, 48.7 to 50.9% total solids, 19.9 to 29.8% fat, 18.5 to 25.4% protein, 3.5 to 3.7% ash, and 1.6 to 4.3% salt.[36,64,65] Oaxaca cheese made from raw milk is characterized by its flavour notes like milky, buttery, acidic, and fermented.[3] On the other hand, a sensory descriptive analysis was carried out and the most representative attributes such as visual (waxy), tactile (moist and oily), orthonasal aroma (empyreumatic and cow), basic taste (salty), mouth texture (chewable, gummy, adherent, fibrous, moist, and fatty), and auditory (squeaky) were included.[66] In fact, it was reported that, in contrast to pasteurized milk cheese, raw milk enhances cowy, buttery, smoked, and fermented flavours; regarding texture, it has low hardness and adhesiveness and high springiness, cohesiveness, and chewiness.[67] Nevertheless, some changes can occur during storage, such as decreases in pH, hardness, elasticity, and whiteness, and increased meltability.[68]

Aroma and flavour are strongly related to the volatile compounds present in the food matrix; therefore, the volatile profile of Oaxaca cheese made from raw milk has been previously studied. Results showed that this cheese contains propanoic acid (propionic acid), 6-heptyloxan-2-one (δ-dodecalactone), 6-pentyloxan-2-one (δ-decalactone), ethanoic acid (acetic acid), 3-hydroxybutan-2-one (acetoin), 5-octyloxolan-2-one (γ-dodecalactone), ethyl dodecanoate (ethyl laurate), 2-hydroxypropanoic acid (lactic acid), decanoic acid (capric acid), octanoic acid (caprylic acid), 2-(4-methyl-1,3-thiazol-5-yl)ethanol (sulfurol), decanal, 2,3-pentanedione, and butanal (butyraldehyde).[69] Interestingly, most of these volatiles have also been reported in Fresco cheese.[22] Thus, it would be interesting to reveal which of these are the key compounds that give Oaxaca cheese its typical aroma and flavour by using multivariate analysis of sensory and chemical data, thus allowing the differences that make this product an exquisite and unique product among artisanal Mexican cheeses to be established.

Also, the microbial composition of this cheese has been characterized and, despite the heat treatment applied to the paste, high counts of aerobic mesophilic bacteria, LAB (*Lactococcus* spp., *Lactobacillus* spp., and *Enterococcus* spp.), molds and yeasts, coliforms, *Escherichia coli*, *Staphylococcus aureus*, and *Rhodotorula* spp. have been detected.[68,70] Fortunately, some of these counts decreased according to storage time when monitored for 24 days under vacuum at 8 °C.[68] Nevertheless, species identification using molecular techniques is needed throughout the production process, mainly because high microorganism counts have been detected.

On the other hand, Cocido cheese is particularly popular in Sonora, Mexico and along the Mexican border with the US (see Figure 10.7). This cheese is classified as a semi-hard, medium-fat, pasta-filata cheese and continues to be produced with traditional artisanal procedures, very similar to those of Oaxaca cheese. Particular steps are important during its production, such as the use of raw whole bovine milk, a natural whey starter culture (leached whey previously fermented for 24 h at room temperature)

Figure 10.7 Artisanal Cocido cheese.

that is usually added to the milk as an acidifying agent to reach a pH of 4.6, and the application of heat treatment (50–60 °C for 20 min) to the paste. Unlike Oaxaca cheese, Cocido is not a stretched paste and has a typical shape of the mold into which the cooked curds are poured to form circular cheeses of different sizes and weights. The resulting product has a non-stringy texture and good melting properties which give this cheese its unique characteristics that differ from those of other artisanal Mexican cheeses.[5]

Cocido cheese contains 53.7 to 57.7% moisture, 40.2 to 46.2% total solids, 16.7 to 20.6% protein, 15.7 to 19.6% fat, 2.2 to 2.7% ash, and 2.0 to 2.5% salt, with a pH of 5.5 to 5.9. These high variations in chemical composition indicate the non-standardized cheese making procedures. Its microbial composition showed the presence of microorganisms such as LAB, aerobic mesophilic bacteria, molds, yeasts, total coliform bacteria, fecal coliform bacteria, *E. coli*, and *S. aureus*. Fortunately, pathogens such as *Salmonella* spp. and *Listeria monocytogenes* were not found and staphylococcal enterotoxins were absent.[5] However, since it is a non-ripened and high-moisture product, it is highly perishable; thus, strategies such as starting with good-quality raw milk, an efficient curd heat treatment (>60 °C) and applying GMP throughout the process may result in a safe product.

To date, Cocido cheese has been scarcely studied and further studies are needed to fully characterize this type of pasta-filata cheese, in particular LAB species that may contribute to its characteristic flavour and aroma. In this sense, multivariate analysis of sensory descriptors derived from quantitative descriptive analysis and volatile content determination may reveal the key compounds involved. Although bioactive compounds have not been reported in Cocido cheese, it has been a good source of potential probiotic LAB strains that, when used to ferment milk, have presented immunomodulatory and anti-inflammatory activities[71–73] as well as the ability to produce bacteriocins.[74] Thus, these specific LAB strains may be used as potential starter cultures with different technological characteristics for designing functional dairy products.

10.8 Conclusions and Future Perspectives

Most popular artisanal Mexican cheeses have been characterized in terms of physico-chemical and microbiological composition; however, full chemical characterization of compounds responsible for their sensory characteristics is still lacking, especially regarding peptides and volatiles associated with taste and aroma. Since Fresco, Chihuahua, and Oaxaca cheeses not only are produced from raw milk, but have also been produced at an industrial scale from pasteurized milk, most research has been focused on these cheese varieties, perhaps due to their availability throughout the country and their popularity among the Hispanic population in the United States. Nevertheless, cheeses that are only produced from raw milk, such as Bola de Ocosingo and Crema de Chiapas, as well as Cotija and Cocido cheeses, remain to be fully characterized beyond their microbiological composition. Full chemical and sensory characterization of Mexican artisanal cheeses is essential for promoting CM and appellations of origin, so that the originality and genuineness of these products as unique ingredients in Mexican gastronomy are preserved. Moreover, the implementation of quality standards and distinctive quality seals may help consumers to differentiate authentic genuine Mexican artisanal cheeses from imitation products available in the market. Finally, it is well known that native lactic acid bacteria in artisanal cheeses can produce a great variety of bioactive compounds, particularly peptides with different health benefits. Thus, it may be feasible to add value to artisanal cheeses by potentiating their health benefits beyond basic nutrition.

Abbreviations

AA	Amino acids
CM	Collective mark
FAO	Food and Agriculture Organization
GMP	Good manufacturing practices
IDF	International Dairy Federation
LAB	Lactic acid bacteria
PDO	Protected designation of origin

References

1. FAO, 2020, http://www.fao.org/3/cb4230en/cb4230en.pdf (accessed November 2021).
2. FAO, 2021, http://www.fao.org/faostat/es/#data/QCL (accessed November 2021).
3. A. F. González-Córdova, C. Yescas, Á. M. Ortiz-Estrada, M. de los Á de la Rosa-Alcaraz, A. Hernández-Mendoza and B. Vallejo-Cordoba, *J. Dairy Sci.*, 2016, **99**, 3250.
4. M. N. Biango-Daniels and B. E. Wolfe, *J. Dairy Sci.*, 2021, **104**, 6283.
5. P. F. Cuevas-González, P. Y. Heredia-Castro, J. I. Méndez-Romero, A. Hernández-Mendoza, R. Reyes-Díaz and B. Vallejo-Cordoba, *J. Sci. Food Agric.*, 2017, **97**, 4459.
6. G. Licitra, M. Caccamo and S. Lortal, in *Raw Milk. Balance between Hazards and Benefits*, ed. L. A. Nero and A. F. De Carvalho, Elsevier Academic Press, London, 2019, p. 175.
7. UNESCO, 2010, http://www.unesco.org/culture/ich/es/RL/la-cocina-tradicional-mexicana-cultura-comunitaria-ancestral-y-viva-el-paradigma-de-michoacan-00400 (accessed November 2021).
8. SECTUR, 2015, https://patrimonioculturalyturismo.cultura.gob.mx/documentos/pdf/Politica_de_fomento_a_la_gastronomia_nacional.pdf (accessed November 2021).
9. P. L. H. McSweeney, in *Improving the Flavour of Cheese*, ed. B. C. Weimer, CRC Press, Woodhead Publishing Limited, Boca Raton, FL, Cambridge, England, 2007, p. 1.

10. V. V. Mistry, in *Improving the Flavour of Cheese*, ed. B. C. Weimer, CRC Press, Woodhead Publishing Limited, Boca Raton, FL, Cambridge, England, 2007, p. 239.
11. IDF, 2016, https://fil-idf.org/wp-content/uploads/2016/12/Raw-milk-cheeses.pdf (accessed November 2021).
12. A. L. Kelly, in *Cheese Problems Solved*, ed. P. L. H. McSweeney, Woodhead Publishing Limited and CRC Press LLC, Boca Raton, FL, 2007, p. 20.
13. M. C. Broome, in *Improving the Flavour of Cheese*, ed. B. C. Weimer, CRC Press, Woodhead Publishing Limited, Boca Raton, FL, Cambridge, England, 2007, p. 177.
14. H. P. Bachman, M. T. Frohlich-Wyder, E. Jakob, E. Roth, D. Wechsler, E. Beuvier and S. Buchin, in *Encyclopedia of Dairy Sciences*, ed. J. W. Fuquay, Elsevier Academic Press, Mississippi State, MS, 2011, p. 652.
15. P. L. H. McSweeney, A. A. Hayaloglu, J. A. O'Mahony and N. Bansal, *Aust. J. Dairy Technol.*, 2006, **61**, 69.
16. P. F. Fox, T. P. Guinee, T. M. Cogan and P. L. H. McSweeney, *Fundamentals of Cheese Science*, Aspen Publ., Gaithesburg, MD, 2000.
17. Y. F. Collins, P. L. H. McSweeney and M. G. Wilkinson, *Int. Dairy J.*, 2003, **13**, 841.
18. P. F. Fox and P. L. McSweeney, in *Cheese: Chemistry, Physics and Microbiology*, ed. P. L. H. McSweeney, P. F. Fox, P. Coter and D. Everett, Elsevier Academic Press, London, 2017, p. 5.
19. P. L. H. McSweeney, in *Cheese: Chemistry, Physics and Microbiology*, ed. P. F. Fox, P. L. H. McSweeney, T. M. Cogan and T. P. Guinee, Elsevier Applied Science, Amsterdam, 2004, p. 347.
20. C. R. Gonzalez-Gonzalez, J. Machado, S. Correia, A. L. McCartney, J. S. Elmore and P. Jauregi, *LWT*, 2019, **111**, 449.
21. J. I. Méndez-Romero, R. Reyes-Díaz, L. Santiago-López, A. Hernández-Mendoza, B. Vallejo-Cordoba, S. G. Sáyago-Ayerdi, B. Gómez-Gil and A. F. González-Córdova, *Int. J. Dairy Technol.*, 2021, **74**, 359–370.
22. R. Reyes-Díaz, A. F. González-Córdova, M. C. Estrada-Montoya, J. I. Méndez-Romero, M. A. Mazorra-Manzano, H. Soto-Valdez and B. Vallejo-Cordoba, *J. Dairy Sci.*, 2020, **103**, 242.
23. M. A. Olea-Rodríguez, P. Chombo-Morales, K. Nuño, O. Vázquez-Paulino, Z. Villagrán-de la Mora, L. E. Garay-Martínez, J. Castro-Rosas, A. Villarruel-López and M. R. Torres-Vitela, *Appl. Sci.*, 2021, **11**, 8154.
24. M. Centeno-Rodríguez, M. Gutiérrez-Cárdenas, M. Jaime-Patlán, E. Meza-Plaza, K. Montecillos-Ramírez, W. Rojas-Salinas and C. Ozuna, *Agro productividad*, 2020, **13**, 9.
25. D. L. Van Hekken, M. H. Tunick, L. N. Leggett and P. M. Tomasula, *J. Dairy Sci.*, 2012, **95**, 5527.
26. M. J. Torres-Llanez, B. Vallejo-Cordoba, M. E. Díaz-Cinco, M. A. Mazorra-Manzano and A. F. González-Córdova, *Food Control*, 2006, **17**, 683.
27. M. Tunick, D. L. Van Hekken, L. Guo and P. M. Tomasula, in *Hispanic Foods: Chemistry and Bioactive Compounds*, ed. M. Tunick and E. González de Mejía, ACS Books, Washington, DC, 2012, p. 10.

28. D. L. Van Hekken and N. Y. Farkye, *Food Technol.*, 2003, **57**, 32.
29. L. M. Beltrán-Barrientos, H. S. García, R. Reyes-Díaz, M. C. Estrada-Montoya, M. J. Torres-Llanez, A. Hernández-Mendoza, A. F. González-Córdova and B. Vallejo-Cordoba, *Foods*, 2019, **8**, 645.
30. M. J. Torres-Llanez, A. F. González-Córdova, A. Hernández-Mendoza, H. S. García and B. Vallejo-Cordoba, *J. Dairy Sci.*, 2011, **94**, 3794.
31. J. E. Aguilar-Toalá, M. J. Torres-Llanez, A. Hernández-Mendoza, R. Reyes-Díaz, B. Vallejo-Cordoba and A. F. González-Córdova, *J. Food Sci. Technol.*, 2022, **59**, 2705.
32. D. L. Van Hekken, M. A. Drake, F. J. Molina-Corral, V. M. Guerrero Prieto and A. A. Gardea, *J. Dairy Sci.*, 2006, **89**, 3729.
33. M. H. Tunick, D. L. Van Hekken, F. J. Molina-Corral, P. M Tomasula, J. Call, J. Luchansky and A. A. Gardea, *Int. J. Dairy Technol.*, 2008, **61**, 62.
34. J. A. López and C. Vargas, 2011, https://elibros.uacj.mx/omp/index.php/publicaciones/catalog/book/63 (online accessed November 2021).
35. La Crónica de Chihuahua, 2011, http://www.cronicadechihuahua.com/Alcanza-el-queso-menonita-Norma.html (accessed November 2021).
36. A. Villegas, *Tecnología quesera*, Trillas S.A. de C.V., México, 2012, vol. 2.
37. D. L. Van Hekken, M. H. Tunick, P. M. Tomasula, F. J. Molina and A. A. Gardea, *Int. J. Dairy Technol.*, 2007, **60**, 5–12.
38. J. A. Renye Jr., G. A. Somkuti, D. L. Van Hekken and V. M. Guerrero, *J. Dairy Sci.*, 2011, **94**, 3311.
39. M. Paul, A. Nuñez, D. L. VanHekken and J. A. Renye, *J. Food Sci. Technol.*, 2014, **51**, 3432.
40. L. M. Beltrán-Barrientos, A. F. González-Córdova, A. Hernández-Mendoza, E. H. Torres-Inguanzo, H. Astiazarán-García, J. Esparza-Romero and B. Vallejo-Cordoba, *J. Dairy Sci.*, 2018, **101**, 1.
41. M. A. Rendon-Rosales, M. J. Torres-Llanez, A. F. González-Córdova, A. Hernández-Mendoza, M. A. Mazorra-Manzano and B. Vallejo-Cordoba, *Nutrients*, 2019, **11**, 2150.
42. A. Santos-Espinosa, L. M. Beltrán-Barrientos, R. Reyes-Díaz, M. Á. Mazorra-Manzano, A. Hernández-Mendoza, G. A. González-Aguilar, S. G. Sáyago-Ayerdi, B. Vallejo-Cordoba and A. F. González-Córdova, *Ann. Microbiol.*, 2020, **70**, 12.
43. A. Espejel, *Revista CIFE: Lecturas de Economía Social*, 2019, **20**, 137.
44. R. López, A. Hernández-Montes, A. Villegas-De Gante and A. Santos-Moreno, *Rev. Cienc. Nat. Agropecu.*, 2015, **2**, 345.
45. M. López, Masters thesis, Departamento de Ingeniería Agroindustrial, Posgrado en Ciencia y Tecnología Agroalimentaria, Universidad Autónoma Chapingo, 2018.
46. C. Yescas, 2009, http://lactography.blogspot.mx/2009/11/cheese-of-week-queso-de-bola-de.html (accessed November 2021).
47. A. Aldrete-Tapia, C. M. Escobar-Ramírez, M. L. Tamplin and M. Hernández-Iturriaga, *Front. Microbiol.*, 2018, **9**, 2598.

48. R. Vázquez-Velázquez, M. Salvador-Figueroa, L. Adriano-Anaya, G. DeGyves-Córdova and A. Vázquez-Ovando, *CyTA-J. Food*, 2018, **16**, 460.
49. M. Agudelo-López, A. Cesín-Vargas, A. Espinoza-Ortega and B. Ramírez-Valverde, *Rev. Mex. Cienc. Pecu.*, 2019, **10**, 104.
50. P. A. Romero-Castillo, G. Leyva-Ruelas, J. G. Cruz-Castillo and A. Santos-Moreno, *Rev. Mex. Ing. Quím.*, 2009, **8**, 111.
51. C. Yescas, in *The Oxford Companion to Cheese*, ed. C. Donnelly, Oxford University Press, New York, NY, 2016, p. 467.
52. S. C. Rangel-Ortega, Masters thesis, Centro de Investigación en Alimentación y Desarrollo, A.C., 2011.
53. S. C. Rangel-Ortega, D. Ercolini, A. F. González-Córdova and B. Vallejo-Cordoba, Abst. 132-02, IFT Annual Meeting and Expo, New Orleans, LA, 2012.
54. F. Morales, J. I. Morales, C. H. Hernández and H. Hernández-Sánchez, *Appl. Biochem. Biotechnol.*, 2011, **164**, 889.
55. T. Poméon, El Queso Cotija, México, 2007, https://www.fao.org/fileadmin/templates/olq/documents/Santiago/Documentos/Estudios%20de%20caso/Cotija/Queso_COTIJA_Mexico.pdf (accessed November 2021).
56. F. Cervantes and A. Villegas, in *La leche y los quesos artesanales en México. Centro de Investigaciones Económicas, Sociales y Tecnológicas de la Agroindustria y la Agricultura Mundial (CIESTAAM)*, ed. M. Á. Porrúa, Universidad Autónoma Chapingo, Consejo Nacional de ciencia y Tecnología (CONACyT), México, 2012, vol. 1.
57. R. Flores-Magallón, A. A. Oliva-Hernández and J. A. Narváez-Zapata, *Food Sci. Biotechnol.*, 2011, **20**, 997.
58. A. Escobar-Zepeda, A. Sanchez-Flores and M. Quirasco, *Food Microbiol.*, 2016, **57**, 116.
59. L. Hernández-Galán, A. Cardador-Martínez, D. Picque, H. E. Spinnler, M. López-del-Castillo and S. T. Martín, *J. Food Res.*, 2016, **5**, 85.
60. A. Z. Villegas and D. Messner, *Ing. Agríc. Biosist.*, 2015, **7**, 35.
61. F. Cervantes-Escoto, A. Villegas de Gante, J. A. Cesín-Vargas and A. Espinoza-Ortega, *Los Quesos Mexicanos Genuinos, Patrimonio cultural que debe rescatarse*, Mundi Prensa México, México City, México, 2008.
62. B. R. Aguilar-Uscanga, M. Montero-Lagunes, J. De la Cruz, J. R. Solís-Pacheco and H. S. García, *Agrociencia*, 2006, **40**, 569.
63. E. M. De Oca-Flores, O. A. Castelán-Ortega, J. G. Estrada-Flores and A. Espinoza-Ortega, *Int. J. Dairy Technol.*, 2009, **62**, 535.
64. I. Caro, S. Soto, L. Fuentes, N. Gutiérrez-Méndez, B. García-Islas, K. E. Monroy-Gayosso and J. Mateo, *Food Nutr. Sci.*, 2014, **5**, 366.
65. E. M. De Oca-Flores, A. Espinoza-Ortega and C. M. Arriaga-Jordána, *Rev. Mex. Cienc. Pecu.*, 2019, **10**, 367.
66. J. R. Sandoval-Copado, J. Orozco-Villafuerte, D. Pedrero-Fuehrer and M. A. Colín-Cruz, *J. Dairy Sci.*, 2016, **99**, 7075.

67. A. Villanueva-Carvajal, M. Esteban-Chávez, A. Espinoza-Ortega, C. M. Arriaga-Jordán and A. Dominguez-Lopez, *CyTA-J. Food*, 2012, **10**, 63.
68. L. Fuentes, J. Mateo, E. J. Quinto and I. Caro, *J. Dairy Sci.*, 2015, **98**, 2833.
69. J. Sandoval-Copado, J. Orozco-Villafuerte, M. M. García-Fabila and M. de los Á Colín-Cruz, *J. Chromatogr. Sep. Technol.*, 2018, **9**, 6.
70. G. Castro-Castillo, F. E. Martínez-Castañeda, Á. R. Martínez-Campos and A. Espinoza-Ortega, *Rev. Soc. Venez. Microbiol.*, 2013, **33**, 105.
71. A. Reyes-Díaz, V. Mata-Haro, J. Hernández, A. F. González-Córdova, A. Hernández-Mendoza, R. Reyes-Díaz, M. J. Torres-Llanez, L. M. Beltrán-Barrientos and B. Vallejo-Cordoba, *Nutrients*, 2018, **10**, 691.
72. L. Santiago-López, A. Hernández-Mendoza, V. Mata-Haro, B. Vallejo-Cordoba and A. F. González-Córdova, *Food Agric. Immunol.*, 2018, **29**, 911.
73. L. Santiago-López, A. Hernández-Mendoza, V. Mata-Haro, B. Vallejo-Córdoba, A. Wall-Medrano, H. Astiazarán-García, M. del, C. Estrada-Montoya and A. F. González-Córdova, *Nutrients*, 2018, **10**, 1039.
74. P. Y. Heredia-Castro, R. Reyes-Díaz, M. Á. Rendón-Rosales, L. M. Beltrán-Barrientos, M. J. Torres-Llanez, M. C. Estrada-Montoya, A. Hernández-Mendoza, A. F. González-Córdova and B. Vallejo-Cordoba, *J. Dairy Sci.*, 2021, **104**, 4033.

CHAPTER 11

Production, Consumption, and Nutritive Value of Cheeses from Cows and Other Mammals

YOUNG W. PARK

Georgia Small Ruminant Research & Extension Center, Fort Valley State University, The University System of Georgia, Fort Valley, GA 31030, USA
Email: parky@fvsu.edu

11.1 Introduction

Cheese has wide ranges of flavours, moisture, and textures, and various shapes by coagulation of the milk protein casein.[1] Cheese is a manufactured dairy product made mostly from the milk of cows, but also from other mammals, including goats, sheep, buffalo, yaks, camels, reindeer, *etc.* Cheese was considered primarily as a concentrated form of milk with the benefit of a prolonged shelf life in the early centuries.[2] However, cheese is a highly nutritious and versatile food that can play an important role in a well-balanced diet. Cheese is a rich source of essential nutrients, particularly proteins, bioactive peptides, amino acids, fat, fatty acids, vitamins, and minerals. Ripened cheese is free of lactose and therefore suitable for the nutrition of lactose-intolerant individuals. Lactic acid, free fatty acids, free amino acids, carbonyls, and a variety of smaller molecular compounds are formed, and these influence the nutritive properties of this fermented food.[3]

Food Chemistry, Function and Analysis No. 40
Handbook of Cheese Chemistry
Edited by Michael H. Tunick

Published by the Royal Society of Chemistry, www.rsc.org

Compared to other dairy products, the *per capita* consumption of cheese has been increasing, and it has a healthy and positive image in many markets and consumers around the world.[1] The high levels of essential amino acids in cheese are known to contribute to growth and development of the human body, especially in growing children. No known relationship has been shown between the consumption of cheese and any incidence of diseases, though significant amounts of saturated and *trans* fatty acids are found in cheeses. Conjugated linoleic acid and sphingolipids present in cheese may have anti-carcinogenic properties. The high concentration of calcium in cheese is well known to contribute to the formation and maintenance of strong bones and teeth, and prevention of osteoporosis, and also shows a positive effect on blood pressure and losing weight in combination with low-energy diets.[1,2]

Although cow milk cheeses are produced and consumed in greatest volumes, large amounts and many different varieties of goat milk and sheep milk cheeses are produced worldwide, depending on the diversity of locality, milk composition and manufacturing techniques.[4] Goat milk cheeses especially vary in composition due to the high seasonal variations in milk composition, modifications of manufacturing procedures, and a multitude of aging times and conditions.[4,5] Much of the difference among cheeses is attributed to the nature of physical and chemical changes during ripening,[4,6] which are influenced by the cultures, chemicals, or flavour ingredients added to the curd during manufacturing.[1,5,7] The ratio between protein and water and the type of ripening are the main variables for cheese classification.[2] According to the General Standard for Cheese (Codex Standard 283-1978 of Codex Alimentarius),[8] a cheese variety can be designated by a combination of two descriptive terms: (i) the percentage of moisture on a fat-free basis that reflects firmness and (ii) the principal ripening (such as unripened, ripened, mould ripened, and brine).[1] Thus, the classification of goat milk cheeses is more difficult than its bovine counterparts due to their diversities in moisture, texture, composition, manufacturing procedures, and localities.[1]

Milk proteins are basically composed of 80% caseins and 20% whey proteins, and cheese curd is produced by coagulation of the casein fractions of milk proteins by acidification to around pH 4.6 at about 31 °C or by limited proteolysis using rennets.[4,9] Rennet-coagulated cheeses account for approximately 75% of total cheese production. Some minor varieties of cheeses such as ricotta are produced by coagulating the whey proteins by heating at 90 °C.[9] With regard to the nutritive values of cheeses, they are a great source of protein, Ca, and fat, and also contain high amounts of vitamin A, vitamin B_{12}, and riboflavin, along with P.[10,11] Cheese made from the milk of 100% grass-fed animals contains the highest amounts of nutrients and also omega-3 fatty acids, conjugated linoleic acid (CLA),[12,13] and vitamin K. Examples of some varieties of cow milk and goat milk cheeses are displayed in Figures 11.1 and 11.2.

Figure 11.1 Some varieties of cow milk cheeses.[11]

Figure 11.2 Some varieties of goat milk cheeses.[1,4]

The manufacture of all types of cheeses in general involves a similar protocol, with various modifying steps to make a product with the desired characteristics.[4,9] Besides bovine milk, cheeses are produced from different dairy species for human consumption, produced in different regions of the world, which significantly contribute to human wellbeing.[1,9] The purpose of this chapter is to review and delineate the nutritive values of different varieties of cheeses produced worldwide, the composition of the products, and their association with cheese quality.

11.2 World Cheese Production and Consumption

As shown in Table 11.1, global cheese production amounted to about 21.222 million tonnes (metric tons) in 2020.[14] The European Union was by far the top producer of cheese worldwide, with a production volume of around 10.35 million metric tons of cheese that year.[14,16] The total volume of cheese production in the world is dominated by bovine milk products, while significant amounts of goat and sheep milk cheeses are produced in Europe, especially in Southern Europe, and are regulated by the schemes of Protected Designation of Origin or Protected Geographical Indication (PDO or PGI), which link cheese varieties with specific territories.[1] Since a different variety of cheese has its specificity, the legal definition of a specific cheese is important for consumers, producers as well as regulatory agencies.[1,17]

Cheese from cow's milk delivered to dairies represents more than 80% of the global natural cheese production. The rest is made up of farm and homemade products, but also cheeses made from other milks (sheep, goat, buffalo, *etc.*). Europe and North America are responsible for more than 75% of the world cheese production.[18,19] The European Union (EU) is the largest producer of cheese in the world over 10 million tonnes annually, followed by the US with over 6 million tonnes.[20] The EU and US account for more than 75% of the world cheese production, which showed dynamic growth through 2021 where the production approached 22 million tonnes[14,20] (Table 11.1).

More than 1000 varieties of cheese with a great diversity of flavours, textures, and forms have been recorded or reported, where most of them are produced locally having minor importance, while the major internationally accepted varieties include Cheddar (about 2 million tonnes annually), Mozzarella (pizza cheese, about 1.5 million tonnes annually), Gouda, Emmental, Grana type, Camembert, and Quark cheeses.[9] Fluid milk and cheese are the most important and highly consumed dairy products, which account for about 40 and 30% of total milk production, respectively. On a worldwide basis, the principal cheese-producing regions are Europe, North and South America, Australia, and New Zealand, where the proportion of milk used for cheese is as high as 70–90% in certain European countries, such as Italy, France, Germany, and Denmark.[21] Cheese production has been increased annually by approximately 2% for several decades due to its positive nutritional image, convenience, functionality, and price,[9] while the

Table 11.1 Cheese production and consumption in selected countries since 2016, including estimates for 2021.[15]

Country	2016	2017	2018	2019	2020	2021
Production (in thousands of tonnes)						
Argentina	552	514	444	523	488	537
Australia	344	348	366	364	385	395
Belarus	275	260	275	300	346	360
Brazil	745	771	760	770	750	760
Canada	445	497	510	515	510	515
China	251	249	276	282	283	300
European Union	9810	10 050	10 160	10 210	10 340	10 450
Japan	47	46	45	44	45	47
Korea, South	25	35	37	40	43	44
Mexico	375	396	419	437	455	460
New Zealand	360	386	370	365	350	365
Philippines	2	2	2	2	2	2
Russia	865	951	970	983	1035	1060
Ukraine	186	190	192	187	180	175
Total non-US	14 282	14 695	14 826	15 022	15 222	15 470
US	5525	5733	5914	5959	6000	6218
Total	19 807	20 428	20 740	20 981	21 222	21 688
Domestic consumption (in thousands of tonnes)						
Argentina	508	485	380	430	441	473
Australia	275	291	293	297	305	315
Belarus	71	72	72	73	75	76
Brazil	785	799	785	795	776	796
Canada	458	504	537	539	539	550
China	348	357	384	397	409	430
European Union	9094	9297	9386	9394	9482	9560
Japan	304	324	329	346	344	346
Korea, South	136	159	155	166	170	180
Mexico	496	511	526	551	564	575
New Zealand	42	40	38	38	38	38
Philippines	32	40	40	42	46	50
Russia	1069	1141	1200	1231	1318	1340
Taiwan	32	32	32	31	32	34
Ukraine	187	188	198	205	221	220
Total non-US	13 837	14 240	14 355	14 535	14 760	14 983
US	5379	5494	5674	5751	5766	5950
Total	19 216	19 734	20 029	20 286	20 526	20 933

EU alone recorded a 5% increase in cheese production during 2015 to 2019[14] (Figure 11.3).

World cheese production is expected to maintain the positive growth trend from the last decade and the expansion from 2012 to 2020 was expected to be 24%, equalling 4.9 million tonnes of cheese (Table 11.1). The major uncertainties are the global economic development and the development in the global demand and supply situation for milk and dairy products in general.[16,19] Among the EU countries, Germany is the highest cheese exporter with an annual average of 1 million tonnes, which

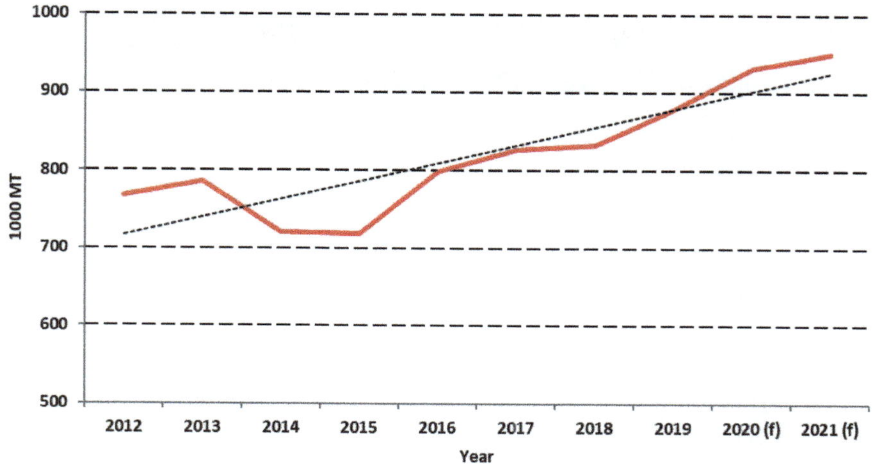

Figure 11.3 EU cheese exports, showing an annual growth of 5% (dashed line), in thousands of metric tons.[15]

Table 11.2 Cheese export summary for some major exporters (in thousands of tonnes).[15]

	2019	2020 Forecast	2021 Forecast	2020–2021 Change (%)
Australia	160	155	170	10
Belarus	244	270	290	7
European Union	879	931	950	2
New Zealand	335	316	345	9
United States	357	359	368	3
Total	1975	2031	2123	5

represents 18.6% of the total world export.[22] Other exporters are the Netherlands, France, New Zealand, and Italy.[22] These countries have modern production technology and tradition which has resulted in a large range of high-quality products.

The short shelf-life and perishability of milk make it necessary to preserve the fluid milk for extended periods of time for later consumption in cheeses, especially in tropical regions, and 50% of the 132.4 million tons of milk annually produced in India, mainly buffalo milk, is used for cheese.[23] Furthermore, cheeses can be made from a mixed or blended milk of goats with other species such as sheep, cow, and buffalo milks in some countries. Although mixing milks is prohibited in the US, it is fairly common in certain countries as a means of extending the supply of goat milk and of reducing the cost of the finished products.[1,4]

As shown in Table 11.1, in terms of the total volume of cheese production by individual countries, the US has produced by far the highest tonnage of cheeses, followed by EU, New Zealand, Belarus, and Australia (Tables 11.1, 11.2 and Figure 11.3). The IDF report on world cheese production[21] showed

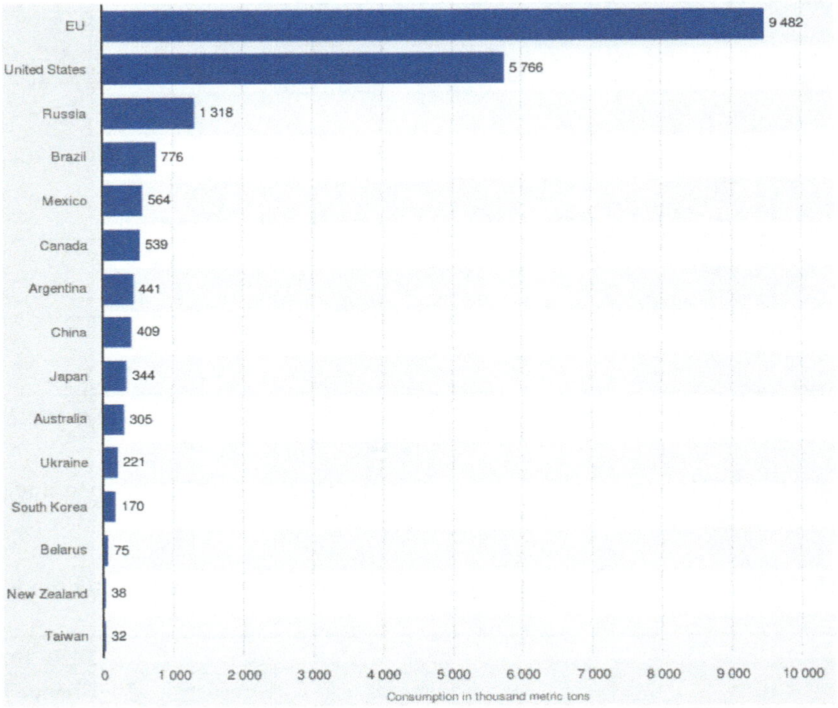

Figure 11.4 Annual consumption of cheese worldwide in 2020 by selected countries.[15]

approximately 17 million tonnes were produced per annum, where approximately 35% of total milk produced was used for cheese production, indicating that the increase was at a rate of 2–5% per annum. Europe is the principal cheese producing region with an annual production of approximately 8 million tonnes, followed by North America[9,14] (Table 11.1 and Figure 11.3).

Consumption of cheese in the world varies widely between countries (Table 11.1 and Figure 11.4). The EU shows the highest amount of consumption, followed by the US and Russia, which follows the similar trend of cheese production volumes of the world. In contrast, the Asian countries, especially the East Asian region such as China, Japan, South Korea, and Taiwan reveals significantly lower cheese consumption compared to those of the traditional Western dairy producing countries. Due to high consumption of milk and dairy products, especially in cheeses, the EU and US show far greater consumption of cheeses compared to the rest of the world. There have been consistent increases in cheese consumption in most countries over recent years.[9] Cheese is the major product growth segment within the dairy sector along with fermented milks. This elevation in cheese consumption is attributable to several factors, including a positive dietary image, convenience and flexibility in use, and a great diversity of flavours and textures. In addition, cheese can be used as a major component of a

Table 11.3 Cheese trade summary for selected countries (in thousands of tonnes).[15]

Country	2016	2017	2018	2019	2020	2021
Exports (in thousands of tonnes)						
Argentina	53	44	61	61	70	60
Australia	167	171	172	160	155	170
Belarus	205	189	211	244	270	290
European Union	799	828	833	879	931	950
New Zealand	355	343	322	335	316	345
Others	55	59	63	56	61	60
Total non-US	1634	1634	1662	1735	1803	1875
US	287	340	349	357	359	368
Total	1921	1974	2011	2092	2162	2243
Imports (in thousands of tonnes)						
China	97	108	108	115	126	130
Japan	258	273	286	303	300	300
Korea, South	110	125	124	131	131	132
Mexico	126	122	123	121	119	125
Russia	223	226	250	273	310	300
Others	328	337	324	340	373	388
Total non-US	1142	1191	1215	1283	1359	1375
US	165	138	138	139	126	126
Total	1307	1329	1353	1422	1485	1501

meal, a dessert, a component of other foods, or as a food ingredient. Cheese can be consumed as a food without preparation or used in various cooking processes.[1,9]

With respect to the cheese trade, some countries such as New Zealand, Australia, Belarus, Brazil, and Argentina are major exporters, and other Asian countries such as China, Japan, and South Korea only import their cheeses from foreign countries with no exports (Tables 11.2 and 11.3). The trade of annual cheese export by the EU has grown 5% annually during the period of 2015 through 2019 (Figure 11.4), suggesting that the demand for cheese products has been consistently growing and the trend appears to be continuing. World imports are increasing more than exports (at a rate of 4.9%), and the USDA Foreign Agricultural Service forecast that there will be increases in exports and imports in the world[15] (Tables 11.2 and 11.3). The major cheese exporting countries such as Australia, Belarus, EU-28, New Zealand, and the US have shown 10, 7, 2, 9 and 3 percent increases, respectively, during the 2020–2021 period (Table 11.2). With a production of 625 000 tonnes and a participation of 12%, Germany is also the world's main cheese importer, even though it produces the highest volume of cheese among the EU countries. Major importers are also Italy, Great Britain, Russia, and the Netherlands, which constitute more than 40% of world cheese imports.[18]

11.3 Milk Composition and Cheese Quality

The nutrient composition of a cheese is dependent on that of the original milk used for its manufacture. The composition of milk varies with diet, breed, animals within breed, parity, environmental conditions, feeding and management conditions, season, locality, and stage of lactation.[23–26] Significant variations occur in milk composition and yield during different seasons and stages of lactation of milking animals.[1,25,27] The total solids, fat, and protein contents of the milk are high in early lactation, fall rapidly and reach a minimum during the 2nd and 3rd months of lactation, and then increase towards the end of lactation (Figure 11.5).[26,28]

Milk drawn from the lacteal glands is highly perishable and easily changed negatively by various factors from improper handling, such as feeding of milking animals, handling of animals prior to and during milking, handling of the milk during and after milking, cooling, transportation, pasteurization, processing, packaging of milk, and processing utensils and equipment.[4,29] The quality of milk largely depends on the farm producers as well as workers at dairy processing plants.[4,29]

The quality of cheese is directly affected by the quality of the original milk used for production of the cheese, which must meet the criteria to have an acceptable quality of cheese.[1] To safeguard quality milk production, at least five major parameters are routinely monitored by various regulatory agencies in commercial milk production channels: (i) nutritional constituents in milk, (ii) somatic cell counts related to mastitis, (iii) bacterial counts related to sanitary practices, (iv) adulteration and pesticide residue contents, and (v) flavour, taste, appearance, and temperature of the raw milk.[30]

In non-bovine species milk products, researchers[25,26] reported that the nutrient compositions of different species' milk and the corresponding cheeses markedly vary from region to region. These variations are also mainly due to several factors, including the analysed samples of individuals

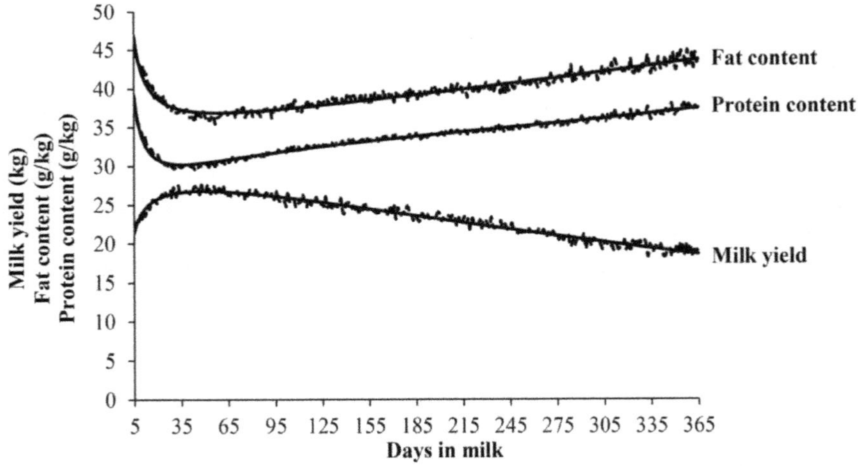

Figure 11.5 Changes in the composition and yield of Holstein milk with days of lactation. Reproduced from ref. 28 with permission from Elsevier, Copyright 2017.

or bulk milk, the analytical procedure used, lactation period, animal breeding, feeding regimen, and so on. In comparison to other species, camel milk has the lowest pH value and mare milk has the highest. The titratable acidity (TA, as percent lactic acid) of camel milk is lower than that of cow, buffalo, sheep, and goat, but higher than that of human, ass, and mare milks, suggesting the TA would significantly affect the composition and quality of cheeses.[25]

As shown in Table 11.4, the basic nutrient composition of cow milk is similar to that of goat milk. Buffalo milk, sheep milk, and Bactrian camel milk contain significantly higher total solid content, which also reveals higher fat and protein contents in the milks of these species than those in cow and goat milks. These compositional differences are directly related to the yield of cheese in buffalo, sheep, and camel milk cheeses. As in the case of cow milk, the composition of goat and other species' milks varies with diet, breed, animals within breed, parity, environmental conditions, feeding and management conditions, season, locality, and stage of lactation.[23,24,32] Caprine milk, on average, contains 12.2% total solids, consisting of 3.8% fat, 3.5% protein, 4.1% lactose and 0.8% ash (Table 11.4), indicating that it has more fat, protein, and ash, and less lactose than bovine milk. It is known that significant variations occur in milk composition and yield during different seasons and stages of lactation of milking cows (Figure 11.5), and the same phenomena would occur in goat and other species' milks. The fat, total solids, and protein contents of dairy species milk are high in early lactation, fall rapidly and reach a minimum during the 2nd to 3rd months of lactation, and then increase towards the end of lactation (Figure 11.5). These changes in milk composition result in an inverse relationship between the yield of milk and the percentage composition of these components.[10,32] Goat milk

Table 11.4 Physiochemical parameters of the milk of various species.[25]

	Species, mean values							
Constituents	Cow	Buffalo	Sheep	Goat	Camel	Ass	Mare	Human
Water, %	87.78	83.81	82.95	87.30	84.81	90.79	89.74	88.66
Total solids, %	12.25	16.19	17.05	12.12	15.19	9.16	10.16	11.34
Fat, %	3.60	6.75	5.95	4.15	5.32	0.95	1.01	2.80
Milk solids-not-fat, %	8.65	9.44	11.10	7.97	9.87	8.21	9.15	8.54
Protein, %	3.24	4.18	5.25	3.02	4.09	1.86	2.31	1.97
Fat:casein	1.43	2.09	1.47	1.79	1.77	1.51	1.04	3.94
Casein, %	2.51	3.22	4.06	2.32	3.01	0.63	0.97	0.71
Whey protein, %	0.73	0.96	1.19	0.70	1.02	1.23	1.34	1.26
Whey protein:casein	0.24	0.27	0.29	0.28	0.34	1.94	1.38	1.77
Lactose, %	4.65	4.45	4.91	4.21	4.95	5.95	6.40	6.30
Ash, %	0.76	0.81	0.94	0.74	0.81	0.40	0.44	0.27
pH	6.68	6.70	6.79	6.70	6.55	6.85	7.01	6.90
Titratable acidity, %	0.18	0.18	0.19	0.17	—	0.08	0.10	0.06
Specific gravity	1.03	1.04	1.04	1.03	1.03	1.03	1.03	1.03
Chloride, %	0.12	0.12	0.11	0.12	—	0.03	0.03	0.04
Energy, kcal L^{-1}	701	1035	1043	721	920	430	480	619

contains slightly less total casein, but higher non-protein nitrogen than the cow counterpart (Table 11.4). The most remarkable difference in basic composition between goat (or cow) milk and human milk is in protein and ash contents (Table 11.4). These compositional differences are also reflected in corresponding species' cheeses.

Figure 11.6 shows that fat and crude protein (CP) contents were similar throughout the year, except during two periods: from late May (week 8) to late September (week 21) when the fat content was lower (3.0%) than the CP content (3.2%), and from late November (week 34) to late March (week 52) when the fat content was higher (around 4.2%) than the CP content (about 3.6%). Total solids were significantly ($P<0.01$) correlated to fat ($r=0.96$), casein ($r=0.87$), and time of year ($r=0.72$). These data in Figure 11.6 reveal that the trend of basic nutrient data is similar to that in Figure 11.5, where the concentrations of basic nutrients were high in early lactation, decreased in mid-lactation, and increased toward the end of lactation, which is reflected in cheese composition.[28]

Camel milk has the lowest pH value and mares' milk has the highest. The TA of camel milk is lower than that of cow, buffalo, sheep, and goat milks, but higher than that of human, ass, and mare milks (Table 11.4). The specific gravity of camel milk is near that of cow and goat milks. The total solid content of Bactrian camel milk is higher than those of most dairy species milk, except buffalo and sheep milks. Bactrian milk has also higher contents of casein and whey protein compared to the milks of other species, except sheep milk. The ratio of whey protein to casein in camel milk is higher than

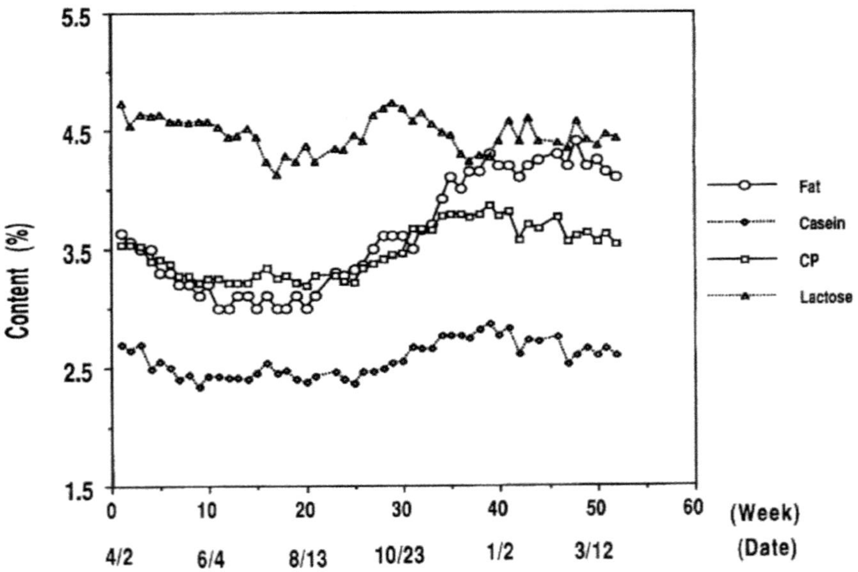

Figure 11.6 Changes in the contents of fat, casein, crude protein (CP), and lactose of commingled goat milk during 52 weeks of study. Reproduced from ref. 31 with permission from Elsevier, Copyright 2001.

that in cow, buffalo, sheep, and goat, but lower than that in human, ass, and mare milk proteins. This may explain why the coagulum of camel milk is softer than that of cow and buffalo milks.[25] The fat globule diameter and the ratio of fat to casein can affect the smoothness of the formed curd, where the fat-to-casein ratio is higher in camel milk than that in cow milk. The lactose contents in human, mare, and ass milks are higher than those in cow, goat, and other dairy species' milks (Table 11.4). These differences in the composition and physicochemical parameters of milk can significantly influence the quality of cheese products.

In a comparative study of the effect of the feed type on the composition of a cow cheese, two types of feeds, control diet (total mixed rations, TMR) and TMR plus native pasture, were investigated to compare the composition and aroma compounds of Ragusano cheese.[33] Significant impacts of the day of cheese making and pasture feeding on cheese composition were observed (Table 11.5). The presence of various plant species and the mix of maturity levels among plant species were changing and could have influenced milk and cheese composition. In addition, the stage of lactation had impacts on milk composition during the 45 day period of cheese making by milk composition in both groups of cows. Mean milk casein and fat from pasture-fed and control diet cows were 2.55 and 2.62% casein and 3.44 and 3.30% fat, respectively. The mean milk casein content was relatively constant during the cheese-making period for both TMR and pasture-fed cow groups, but the fat content of the milk increased during the period of the study (7 months). The milk casein-to-fat (C/F) ratio was 0.74 for pasture and 0.79 for TMR, with the ratio staying relatively constant during the 45 day cheese-making period for TMR, while the C/F ratio for the pasture-fed group decreased from about 0.80 to 0.68. In general, the differences in moisture, fat, FDB, and PDB for pasture and TMR cheeses were due to the differences in the milk C/F ratio (Table 11.5).

The nutrient composition of a cheese depends on the cheese type, the type of milk used such as whole milk, skim milk or whey, and on the type of precipitation and coagulation. The milk components in cheesemaking are usually concentrated through the removal of whey.[1,3] All water-insoluble components are retained in the curd, so that solid components in most ripened cheeses are increased approximately 10 times compared to those in the original milk. This indicates that 3.2% fat in milk is increased to 32% fat and 2.3% protein in milk to 23% protein in cheese for fresh Cheddar cheese.[3] Water-soluble compounds are lost in whey, resulting in lower lactose, soluble protein and salts in cheese curd than those in whey. A whey cheese or powder naturally possesses these soluble nutrients in a concentrated form.

Table 11.6 displays the composition of selected cheese and fermented milk foods. Not all cultured milk products have similar nutritional qualities. In other fermented milk products such as yogurt, extra milk solids are added, while others vary in processing temperatures, starter cultures and lactic acid concentration.[3] Buttermilk and yogurt are supplemented with iron, vitamins, and other nutrients. As far as the nutrient compositions of

Table 11.5 Influence of feed type on the composition of Ragusano cheeses ($n = 4$) at 4 and 7 months of age made from milk produced by cows fed pasture and total mixed ration (TMR).[a] [33]

Component	Pasture	TMR	Feed	Day	Day×day	Day×feed	R^2
4 months							
Moisture, %	38.29	39.57	NS[b]	***	NS	NS	0.76
Fat, %	28.20	26.94	NS	**	NS	NS	0.70
Protein, %	26.57	26.98	NS	NS	NS	*	0.55
Salt, %	3.41	3.23	NS	NS	NS	NS	0.33
pH 4.6/TN,[c] %	14.18	15.48	NS	NS	NS	NS	0.49
12% TCA/TN,[d] %	12.72	14.15	NS	NS	NS	NS	0.41
NaCl/M,[e] %	8.95	8.18	NS	NS	NS	NS	0.46
FDB,[f] %	45.68	44.58	NS	NS	NS	NS	0.46
PDB,[g] %	43.07	44.64	NS	***	NS	***	0.90
pH	5.42	5.51	NS	NS	NS	NS	0.34
7 months							
Moisture, %	34.49	35.48	NS	***	**	NS	0.87
Fat, %	30.03	28.28	NS	***	NS	*	0.88
Protein, %	28.14	28.82	NS	***	*	NS	0.87
Salt, %	4.10	4.30	NS	NS	*	NS	0.52
pH 4.6/TN, %	MD[h]	MD	—	—	—	—	—
12% TCA/TN, %	MD	MD	—	—	—	—	—
NaCl/M, %	12.08	12.23	NS	**	**	NS	0.74
FDB, %	45.78	43.82	NS	***	NS	*	0.79
PDB, %	43.00	44.66	NS	*	*	*	0.85
pH	5.36	5.42	NS	NS	NS	NS	0.52

[a] $*P < 0.05$; $**P < 0.01$; $***P < 0.001$.
[b] NS = not significant, $P > 0.05$.
[c] pH 4.6/TN = pH 4.6 acetate buffer soluble nitrogen as a percentage of total nitrogen.
[d] 12% TCA/TN = 12% trichloroacetic acid soluble nitrogen as a percentage of total nitrogen.
[e] NaCl/M = [(sodium chloride)/moisture)] × 100.
[f] FDB = fat on dry basis.
[g] PDB = protein on dry basis.
[h] MD = missing data.

cheeses are concerned, hard, semi-hard and soft cheeses are determined by the moisture content of the cheeses. Hard cheeses such as Parmesan contain less water (30%) and higher fat and protein than other semi-hard and soft cheeses. Cheddar and Swiss cheeses contain about 40% moisture, while fat and protein and carbohydrate levels are closer to each other, and the calorie content of Cheddar is higher than that of Swiss due to the difference in fat content between the two cheeses. Process cheeses, process cheese foods, and cheese spreads contain less fat and protein compared to the original cheeses due to the dilution of other ingredients and addition of water in the processed products.

11.4 Cheese as a Nutrient Source in the Human Diet

Cheese is a major dairy food providing humans with general nutrients and some specific beneficial effects on human nutrition, especially in Western

Table 11.6 Comparison of selected cheese and fermented milk foods (per 100 g portions).[3]

Cheese type	Calories (kcal)	Water (g)	Fat (g)	Protein (g)	Carbohydrates (g)
Blue	368	40	30.5	21.5	2.0
Brick	370	41	30.5	22.2	1.9
Camembert	299	52	24.7	17.5	1.8
Cheddar	398	37	32.2	25.0	2.1
Creamed cottage	106	78	4.2	13.6	2.9
Uncreamed cottage	86	79	0.3	17.0	2.7
Cream	374	51	37.7	8.0	2.1
Limburger	345	45	28.0	21.2	2.2
Parmesan	393	30	26.0	36.0	2.9
Swiss	370	39	28.0	27.5	1.7
Pasteurized process					
Cheddar	370	40	30.0	23.2	1.9
Swiss	355	40	26.9	26.4	1.6
Cheese food	323	43	24.0	19.8	7.1
Cheese spread	288	48	21.4	16.0	8.2
Cultured spread					
Yogurt	50	89	1.7	3.4	5.2
Buttermilk	36	91	0.1	3.6	5.1
Sour cream	211	73	18.0	3.0	4.3

countries. Although the nutritional values of different varieties of cheese vary considerably, it is generally a rich source of fat, protein, fat-soluble vitamins, and minerals, particularly Ca and P. Most varieties of cheese have a higher protein : fat ratio compared to other major foods, where cheese is especially significant in the human diet in the Western world, as well as Australia and New Zealand. As an example, France's annual *per capita* cheese consumption was 24.3 kg per person in 2007.[9] Hence, cheese contributed approximately 13% of the recommended energy (2000 kcal per day), which is about 30% of the daily protein requirement for men (56 g per day) and 50% of the recommended Ca intake (1000 mg per day) in the human diet.[9]

In a comparative study of the nutritive values of more than 30 varieties of commercially manufactured goat cheeses from 11 states in the US, Park[7] found that most varieties of goat cheese contained higher moisture contents than those of corresponding or similar types of cow milk cheese. The same study showed that goat Cheddar cheese had higher moisture and protein and lower fat content than the cow cheese counterpart listed in the USDA's Agricultural Handbook.[34] Other varieties such as Blue, Camembert, and Feta cheeses had comparable levels of basic nutrients in caprine cheeses compared to the corresponding bovine cheeses with some variations. The mean percentages of moisture, fat, protein, and ash for plain soft, semi-soft, hard, pepper, garlic, and herb cheeses were also reported in the same study. Garlic cheeses contained significantly greater moisture than the other varieties of cheeses. The percentages of ash, protein and fat were higher ($P < 0.01$) in semi-soft and hard cheeses than those in soft plain and spice-added goat cheeses.

The inclusion of cheese in the diet increases diversity and variety with a nutrient-rich diet in affluent countries. Most people prefer to consume cheese because of its organoleptic and functional attributes or its convenience and flexibility of use.[9] Cheese is also used as an ingredient in both the food service (catering) and industrial cheese sectors.[35] In food service, cheese is extensively used in the preparation of ready-to-eat meals or snacks such as pizza, cheeseburgers, tortillas, macaroni cheese, tacos with cheese, quiche, lasagne, toasted sandwiches, sauces, salads, and desserts which are sold mainly through fast-food outlets, restaurants and delicatessen counters.[9] Secondly, cheeses are used in the industrial sector as the commercial manufacture of assembled foods, including frozen and chilled pizza, prepared sandwiches, cheese salads, formulated foods (*e.g.*, gratins, prepared meals, processed cheese products, co-extruded products, cheesecake, dairy desserts) and also as food ingredients such as cheese sauces, cheese powders, enzyme modified cheeses, ready-to-use grated/shredded cheeses, dried grated cheeses, and freeze-dried cheese pieces.[9] Cheese as a food ingredient is used mainly by the manufacturers of formulated foods including soups, dried cheese sauces, dehydrated potato mixes, infant meals, pasta dishes, snack coatings, and bakery products, whereas formulated and assembled cheese-containing foods are sold mainly through retail and food service outlets.[9]

Selected nutrient (moisture, fat, protein, Na, and Ca) compositions of some major varieties of cheeses are shown in Table 11.7. Cheese contains high-quality proteins in the range of 4–40%. Most cheese varieties are

Table 11.7 Nutrient contents of selected cheese varieties.[9,34,36]

Variety	Moisture (%, w/w)	Protein (%, w/w)	Fat (%, w/w)	Na (mg/100 g)	Ca (mg/100 g)
Brie	48.6	19.2	26.9	703	184
Camembert	50.7	20.9	23.7	842	388
Cheddar	37.2	25.4	33.1	710	730
Reduced-fat Cheddar	43.0	17.0	33.4	745	934
Cheshire	38.5	24.2	31.9	710	473
Cottage cheese (creamed)	79.1	13.8	3.9	393	94
Danish Blue	43.0	18.4	37.3	1297	465
Edam	43.8	26.0	25.4	785	731
Emmental	35.7	28.7	29.7	256	936
Feta	56.5	15.6	20.2	1116	492
Gouda	40.1	24.0	31.0	819	721
Gruyére	33.6	27.3	34.5	467	666
Mozzarella	49.8	25.1	21.0	589	710
Parmesan	30.6	34.9	26.0	1108	1097
Processed cheese (block)	49.1	18.3	23.3	1351	531
Ricotta	73.2	11.3	10.3	141	272
Roquefort	41.3	19.7	32.9	1500	560
Stilton	40.5	21.6	33.7	1360	420

produced using rennet- or acid-coagulated procedures, where the proteins in curd are coagulated.[37] In non-rennet coagulated cheese, the incorporation of whey proteins in heat-acid coagulated varieties and some cheeses made using UF (ultrafiltration) retentate, the nutritional value of its proteins may be enhanced.

Soft cheeses such as Cottage and Ricotta contain very high levels of moisture (>73% of water) and very low levels of fat and protein, as well as low levels of sodium and calcium (Table 11.7). Most semi-soft and hard cheeses contain high levels of fat and protein, and high calcium and sodium contents. Due to the implication of fats, especially saturated fat, in health problems, some reduced fat cheeses have been studied and produced for marketing. However, the reduction of fat in cheese products has certain problems of consumer acceptability owing to the deficiency in flavour, texture, appearance, *etc.*[38] In terms of the Na concentration of cheeses, some varieties such as Danish Blue, Feta, Parmesan, Roquefort, and Stilton cheeses have extremely high Na contents, while other major popular cheese varieties contain medium levels of Na (500–800 mg/100 g cheese).

11.5 Fat, Cholesterol, and Volatiles

Cheese is frequently regarded as a high-fat food, but its fat content varies from nearly 0 (Cottage) to 38% (cream). A typical Western diet provides 2000 kcal per day, 40% of which is derived from 88 g fat.[9] A serving of 50 g of Cheddar cheese would contain approximately 16 g fat, indicating a substantial proportion of the fat intake is from cheese in affluent Western societies.[36] Cheese fat generally contains approximately 65% saturated fatty acids (FA), 30% monounsaturated FA, and 5% polyunsaturated FA (Table 11.8). Production of reduced fat cheeses has been attempted to reduce the content and proportion of saturated FA in cheese, while only limited success has been achieved because of reduced organoleptic and functional properties of the reduced cheese products.[9,38,39]

Goat milk has a unique characteristic in that its ratio of lauric acid (12:0) to capric acid (10:0) is significantly lower than that of cow milk (0.46 *vs.* 1.16).[40] The ratio becomes proportionally larger with increased substitution of cow milk in lieu of goat milk. The detection of the extent of adulteration of goat or sheep milk or cheese with cow milk or cheese has been used in the dairy industry.[40,41] The remarkably high concentrations of palmitic (16:0) and oleic (18:1) acids in goat and cow milk fats are not species specific, but rather are common to most mammals (Table 11.8).

The cholesterol content of cheese is a function of its fat content and ranges from approximately 10 to 100 mg per 100 g, which is low compared with many other animal-derived foods. From a nutritional viewpoint, the cholesterol in cheese is less important than the level of saturated FA.[9,10] Milk lipids containing cholesterol esters of 14:0, 16:0 and 18:1 acids are significantly higher in goat milk compared to cow milk, which is reflected in the respective cheeses (Table 11.8). Human milk contains these acids at even

Table 11.8 Fatty acid composition (as g/100 g total lipid) and cholesterol (as g/100 g cholesterol esters (CE)) of goat, cow, and human milks. Ranges are in parentheses. Tr = trace amount.

	Total lipid (g/100 g fat)			Cholesterol (g/100 g CE)	
Fatty acid	Goat[42,43]	Cow[44,45]	Human[46]	Goat[45]	Cow[46]
4:0	2.6 (3.3–4.8)	3.3 (2.5–6.2)	—		
6:0	2.9 (1.7–3.0)	1.6 (1.5–3.8)	Tr		
8:0	2.7 (1.5–3.6)	1.3 (1.0–1.9)	Tr		
10:0	8.4 (6.4–11.1)	3.0 (2.1–4.0)	1.3	5.2	2.9
10:1	Tr	Tr	Tr	Tr	0.3
12:0	3.3 (2.5–5.0)	3.1 (2.3–4.7)	3.1	4.2	4.1
12:1	Tr	Tr	Tr	1.0	0.2
13:0	Tr	Tr	Tr	Tr	Tr
13:1	Tr	Tr	Tr	0.9	11.0
14:0	10.3 (8.5–11.2)	9.5 (8.5–12.8)	5.1	9.2	6.9
14:1	Tr	Tr (0–1.5)	Tr	1.4	0.5
15:0	Tr	Tr	Tr	1.3	2.1
15:1	Tr	Tr	Tr	Tr	2.6
16:0	24.6 (25.1–38.4)	26.5 (24.0–33.3)	20.2	39.3	26.9
16:1	2.2 (0.7–1.7)	2.3 (1.3–2.8)	5.7	Tr	11.9
17:0	Tr	Tr	Tr	Tr	Tr
18:0	12.5 (5.9–14.9)	14.6 (6.2–13.6)	6.0	9.0	6.7
18:1	28.5 (15.6–28.2)	29.8 (19.7–31.2)	46.4	26.5	13.7
18:2	2.2 (1.8–4.0)	2.5 (1.5–5.2)	13.0	2.1	10.1
18:3	Tr	1.8	1.4		

higher ratios (18:1 and 18:2) than goat and cow milks, which appears species specific in milk fat composition.

A positive characteristic of ruminant lipids (milk and adipose tissue) is a relatively high level of conjugated linoleic acid (CLA), which is considered to have anti-oxidative and anti-carcinogenic properties. Milk and cheese are important sources of CLA, which can be increased by increasing the level of linoleic acid in the animal's diet.[9,26]

Table 11.9 compares the cholesterol and fatty acid contents of different species' milks published by the Royal Society of Chemistry.[48] The respective cholesterol levels of normal fluid goat, cow, sheep, and human milks are 10, 14, 11, and 16 mg/100 g, indicating that goat milk has the lowest cholesterol content among these four species' milks.[47] There is some similarity in the cholesterol values reported in the USDA Agricultural Handbook[34] and others.[10,42] Human milk has the highest cholesterol among those milks, and its colostrum has 31 mg/100 g. Cow milk powder has substantially higher cholesterol (120 mg/100 g) since it is a dried and concentrated product. The cholesterol contents of cheeses in cow, goat, and sheep milks are expected to have a similar trend of compositions to that of original species milk, although some variations in the cholesterol levels of cheeses may occur due to different breeds within a species milk.

Table 11.9 Fatty acid and cholesterol compositions of different milks.[48] Chol. = cholesterol, Sat. = saturated fat, Monoun. = monounsaturated fat, Polyun. = polyunsaturated fat, Tr = trace.

	Fatty acids (g/100 g)			
Species	Sat.	Monoun.	Polyun.	Chol. (mg/100 g)
Cow milk				
Whole	2.4	1.1	0.1	14
Skim	0.1	Tr	Tr	2
Dried whole	16.5	7.6	0.8	120
Goat milk	2.3	0.8	0.1	10
Sheep milk	3.8	1.5	0.3	11
Human milk				
Colostrum	1.1	1.1	0.3	31
Mature	1.8	1.6	0.5	16
Soya milk	0.3	0.4	1.1	0

With respect to fatty acid composition, normal cow, goat, sheep, and human milks contain 2.4, 2.3, 3.8, and 1.8 g/100 g saturated fatty acids, respectively, suggesting that human milk has lower saturated fat than the 3 major dairy species' milks (Table 11.9). The opposite phenomenon was observed for the mono- and polyunsaturated fatty acid levels in human milk compared to the other three species' milks (Table 11.9). The significantly higher unsaturated fatty acids and cholesterol in human milk compared to the three animal species' milk appear to be interesting and may be related to certain health issues such as coronary heart diseases. The lower levels of monounsaturated fatty acids in goat milk than those in cow and sheep milks can be reflected in the respective species' milk cheeses, which may need further investigations in future studies.

Numerous studies have been reported on the nutritive values of milk and cheeses in relation to feeding control and nutrient supplemented diets in milking animals, especially in compositions of fat, fatty acids, volatiles, cholesterol, and other lipid moieties. As an example, Luna *et al.*[49] studied the FA composition (% of total fatty acid methyl esters, FAME) of milk fat from ewes fed a control diet (CD), and supplement enriched in linseed (SEL) diet, and cheeses at 1, 2, and 4 months of ripening made from the milk of animal fed the SEL. The FA profile of ewes' milk fat was altered by the feed, while the development of the major FA content was similar for both control and SEL herds during the testing period (Table 11.10). The proportions of saturated FA from 8:0 to 16:0 decreased with the SEL diet, whereas 4:0 and 6:0 hardly varied. Lipid supplementation with α-linolenic acid increased the 18:0 and 18:3 percentages as well as most of the 18:1 and 18:2 and isomers (Table 11.10). Ewes' milk fat contained very small amounts of C20 and C22 polyunsaturated fatty acids (PUFA), identified by using GC–MS.

Researchers found that the most prominent PUFA present in ewe's milk cheeses were 20:4, 20:5 (eicosapentaenoic acid, EPA), and 22:5, whereas 22:6 (docosahexaenoic acid, DHA) was absent (Table 11.10). There was a

Table 11.10 Fatty acid composition (% of total fatty acid methyl esters) of milk fat from ewes fed the control diet (CD), a supplement enriched in linseed (LIN), and cheeses at 1, 2, and 4 months of ripening made from the milk of animal fed a supplement enriched in linseed.[49]

	Milk		Cheese		
Fatty acid	CD	LIN	1 month	2 months	4 months
4:0	3.76 ± 0.38	3.68 ± 0.53	3.74 ± 0.42	3.72 ± 0.56	3.70 ± 0.47
6:0	3.13 ± 0.13	2.78 ± 0.14	2.76 ± 0.18	2.81 ± 0.16	2.70 ± 0.12
8:0	2.62 ± 0.04	2.22 ± 0.10	2.20 ± 0.11	2.24 ± 0.14	2.22 ± 0.30
9:0	0.06 ± 0.01	0.05 ± 0.01	0.05 ± 0.01	0.05 ± 0.01	0.06 ± 0.01
10:0	7.42 ± 0.11	5.95 ± 0.20	5.94 ± 0.19	5.98 ± 0.25	5.97 ± 0.32
11:0	0.29 ± 0.01	0.21 ± 0.01	0.21 ± 0.01	0.21 ± 0.01	0.21 ± 0.01
12:0	3.73 ± 0.05	3.02 ± 0.06	3.02 ± 0.05	3.03 ± 0.07	2.99 ± 0.07
13:0	0.12 ± 0.01	0.09 ± 0.01	0.09 ± 0.01	0.10 ± 0.01	0.09 ± 0.01
14:0 iso	0.06 ± 0.01	0.06 ± 0.01	0.06 ± 0.01	0.07 ± 0.01	0.06 ± 0.01
14:0	9.30 ± 0.19	8.07 ± 0.15	8.09 ± 0.16	8.05 ± 0.15	8.03 ± 0.21
15:0 iso	0.15 ± 0.01	0.15 ± 0.01	0.16 ± 0.01	0.16 ± 0.01	0.15 ± 0.01
15:0 anteiso	0.28 ± 0.01	0.31 ± 0.01	0.31 ± 0.01	0.31 ± 0.01	0.31 ± 0.01
14:1	0.16 ± 0.01	0.10 ± 0.01	0.10 ± 0.01	0.10 ± 0.01	0.10 ± 0.01
15:0	0.78 ± 0.02	0.72 ± 0.02	0.72 ± 0.02	0.71 ± 0.02	0.71 ± 0.02
15:1	0.08 ± 0.01	0.07 ± 0.01	0.07 ± 0.01	0.07 ± 0.01	0.07 ± 0.01
16:0 iso	0.25 ± 0.01	0.25 ± 0.01	0.25 ± 0.01	0.25 ± 0.01	0.24 ± 0.01
16:0	28.34 ± 0.33	24.95 ± 0.26	25.01 ± 0.34	24.79 ± 0.24	24.88 ± 0.54
17:0 iso	0.45 ± 0.01	0.57 ± 0.03	0.57 ± 0.03	0.57 ± 0.03	0.55 ± 0.02
17:0 anteiso	0.23 ± 0.01	0.28 ± 0.01	0.29 ± 0.03	0.28 ± 0.01	0.27 ± 0.01
16:1	1.25 ± 0.02	1.08 ± 0.02	1.08 ± 0.02	1.09 ± 0.03	1.07 ± 0.04
17:0	0.58 ± 0.02	0.59 ± 0.01	0.59 ± 0.01	0.58 ± 0.02	0.57 ± 0.01
17:1	0.15 ± 0.01	0.14 ± 0.01	0.14 ± 0.01	0.14 ± 0.01	0.14 ± 0.01
18:0	8.12 ± 0.13	10.50 ± 0.18	10.58 ± 0.10	10.37 ± 0.13	10.55 ± 0.39
18:1 (6–8 *trans*)	0.25 ± 0.02	0.29 ± 0.02	0.30 ± 0.03	0.30 ± 0.02	0.31 ± 0.04
18:1 (9 *trans*)	0.33 ± 0.04	0.37 ± 0.04	0.35 ± 0.02	0.34 ± 0.04	0.34 ± 0.05
18:1 (10 *trans*)	1.17 ± 0.03	1.60 ± 0.05	1.65 ± 0.07	1.74 ± 0.04	1.60 ± 0.09
18:1 (11 *trans*)	2.39 ± 0.06	3.62 ± 0.10	3.64 ± 0.10	3.54 ± 0.07	3.71 ± 0.13
18:1 (9–11 *cis*) + (12–15 *trans*)	17.65 ± 0.26	20.12 ± 0.43	19.87 ± 0.15	20.20 ± 0.26	20.07 ± 0.81
18:1 (12–13 *cis*)	0.46 ± 0.01	0.65 ± 0.01	0.66 ± 0.01	0.66 ± 0.01	0.65 ± 0.02
18:1 (16 *trans* + 14 *cis*)	0.25 ± 0.01	0.42 ± 0.01	0.42 ± 0.01	0.42 ± 0.01	0.43 ± 0.01
18:1 (15 *cis*) + 19:0	0.16 ± 0.01	0.29 ± 0.01	0.29 ± 0.01	0.29 ± 0.01	0.29 ± 0.01
18:2 (9 *trans*, 12 *trans*) + (9 *cis*, 13 *trans*)	0.28 ± 0.01	0.44 ± 0.01	0.44 ± 0.01	0.44 ± 0.01	0.43 ± 0.01
18:2 (8 *trans*, 13 *cis* + 9 *cis*, 12 *trans*)	0.15 ± 0.01	0.20 ± 0.01	0.20 ± 0.01	0.20 ± 0.01	0.20 ± 0.01
18:2 (11 *trans*, 15 *cis*)	0.03 ± 0.01	0.21 ± 0.02	0.21 ± 0.02	0.21 ± 0.02	0.20 ± 0.02
18:2 (9 *cis*, 12 *cis*)	3.28 ± 0.13	2.84 ± 0.10	2.81 ± 0.08	2.84 ± 0.10	2.79 ± 0.07
20:0	0.17 ± 0.01	0.18 ± 0.01	0.18 ± 0.01	0.17 ± 0.01	0.18 ± 0.01
18:3	0.49 ± 0.01	1.05 ± 0.01	1.04 ± 0.01	1.05 ± 0.01	1.06 ± 0.03
CLA	0.80 ± 0.01	1.02 ± 0.01	1.01 ± 0.01	1.02 ± 0.01	1.02 ± 0.03

Table 11.10 (*Continued*)

	Milk		Cheese		
Fatty acid	CD	LIN	1 month	2 months	4 months
22:0	0.07 ± 0.01	0.07 ± 0.01	0.07 ± 0.01	0.06 ± 0.01	0.07 ± 0.01
20:4	0.18 ± 0.01	0.15 ± 0.01	0.15 ± 0.01	0.15 ± 0.01	0.15 ± 0.01
20:5	0.05 ± 0.01	0.06 ± 0.01	0.05 ± 0.01	0.06 ± 0.01	0.05 ± 0.01
22:5	0.06 ± 0.01	0.07 ± 0.01	0.07 ± 0.01	0.07 ± 0.01	0.04 ± 0.03
Others	0.47	0.51	0.56	0.56	0.76

remarkable increase in the *trans*-11, *cis*-15 18:2 level in the milk and cheese fat of ewes fed SEL. In previous studies, *trans*-11, *cis*-15 18:2 was first recognized as the main intermediate in the ruminal hydrogenation of α-linolenic acid *in vitro*, which was also confirmed *in vivo* as the major 18:2 isomer produced during hydrogenation of α-linolenic acid in the rumen.[49,50] This isomerization pathway involves three steps: an initial isomerization to a conjugated triene (*cis*-9, *trans*-11, *cis*-15 18:3), followed by reductions in the double bonds at carbons 9, 15, and 11 to yield *trans*-11, *cis*-15 18:2, *trans*-11 18:1, then 18:0, respectively.[49] In modern human nutrition, the efforts of enhancing the n-3 FA content in milk fat has become important. Two of the most important n-3 PUFA (EPA and DHA) usually have a very low level in traditional dairy diets (<0.1% of total FAME).[49] The amount of DHA is present only in trace quantities in milk fat unless the ruminant diet is supplemented with marine products. Thus, very low levels of n-3 PUFA in ewe milk and cheeses were observed in their study.[49]

A linear relationship was also observed between *trans*-vaccenic acid (*trans*-11-octadecenoic acid; *trans*-11 18:1) and 9-*cis*,11-*trans* CLA in ewes' milk fat (Table 11.10). With regard to the CLA isomer profile, increases in the 11,13- and 12,14-18:2 positional isomers were considerable when linseed was included in the diet. The organoleptic characteristics of cheeses made with CLA-enriched milk did not substantially differ from those made with non-supplemented ewes' milk. During cheese ripening, the total CLA content and isomer profile were not changed. The inclusion of long-chain FA in the experimental diet reduced the medium-chain saturated FA level in the milk mainly by inhibiting mammary *de novo* FA synthesis. They postulated the increases in the proportion of 18:0, 18:1, and 18:2 isomers might be associated with biohydrogenation activity in the rumen, which is stimulated by feeding PUFA-enriched lipids. However, the process of biohydrogenation in the rumen appeared to be limited because noticeable amounts of α-linolenic acid contents escaped hydrogenation and increased the amounts of these FA in the milk fat in the SEL fed animals.

In a very recent study, Balthazar *et al.*[51] identified 46 volatile organic compounds in control and probiotic Minas fresh cheese samples, which were 3 ketones, 13 esters, 17 acids, 6 alcohols, and 7 other compounds (Table 11.11). The authors observed a total of 21 compounds in both cheese types. These aroma compounds included acidic (acetic acid), sweet

Table 11.11 Volatile organic profiles of control and probiotic Minas fresh cheeses.[51] Compounds present in the formulation are indicated by × and compounds absent by —. RI = retention index, LRI = linear retention index.

Compound	RI (min)	LRI	Control	Probiotic
Ketones				
2-Heptanone	10.925	1162	×	×
2-Nonanone	15.911	1368	×	×
2-Undecanone	20.998	1579	×	—
Esters				
Butanoic acid, ethyl ester	7.661	1019	×	—
Butanoic acid, 2-methylpropyl ester	10.231	1132	×	—
Hexanoic acid, ethyl ester	11.994	1207	×	×
Propanoic acid, 2-methyl-, 3-methylbutyl ester	12.759	1239	×	—
Heptanoic acid, ethyl ester	14.445	1308	×	—
Acetic acid, 2-ethylhexyl ester	15.703	1360	×	—
Octanoic acid, methyl ester	15.852	1366	×	—
Octanoic acid, ethyl ester	16.979	1412	×	×
Isopentyl hexanoate	17.573	1437	×	—
n-Caprylic acid, isobutyl ester	19.864	1531	×	—
Decanoic acid, methyl ester	20.879	1574	×	—
Decanoic acid, ethyl ester	21.906	1618		—
Hexanedioic acid, bis(2-ethylhexyl) ester	44.739	2796	×	×
Acids				
Acetic acid	17.478	1433	×	×
Butanoic acid	21.52	1601	×	×
Butanoic acid, 3-methyl-	22.523	1645	×	×
Pentanoic acid	24.054	1711	×	×
Hexanoic acid	26.399	1817	×	×
Hexanoic acid, 4-methyl-	28.091	1897		×
Heptanoic acid	28.619	1922	×	×
Octanoic acid	30.791	2029	×	×
Nonanoic acid	32.797	2133	×	×
n-Decanoic acid	34.756	2238	×	×
9-Decenoic acid	35.812	2296	×	×
Benzoic acid	37.587	2398	×	—
Dodecanoic acid	38.406	2447	×	×
Tetradecanoic acid	41.872	2652	×	×
Pentadecanoic acid	43.902	2756		×
n-Hexadecanoic acid	46.341	2860	×	×
Octadecanoic acid	53.209	3108		×
Alcohols				
1-Butanol, 3-methyl-	11.78	1198	×	—
3-Pentanol	14.967	1330	×	—
1-Hexanol, 2-ethyl-	15.686	1359		×
2,3-Butanediol	20.268	1548	×	—
Phenylethyl alcohol	27.794	1883	×	—
Phenol	29.687	1974	×	×
Other				
Furfural	13.448	1267		×
Acetoin	13.739	1279	×	—
Benzaldehyde	16.072	1375		×
Indene	18.256	1465	×	×
Camphor	19.152	1502	×	×
Limonene	11.003	1165	×	×

(hexanoic acid), fatty (heptanoic acid), fruity (octanoic acid and 2-nonanone), sour (nonanoic acid and *n*-decanoic acid), rancid (butanoic acid), cheese aroma (2-heptanone), and sweaty (3-methylbutanoic acid).[52] They suggested these compounds would be typical of Minas Frescal cheeses, whereby these aroma compounds may be used for quality control programs in the cheese industry.

Table 11.11 also specifically reveals that the control cheese contained 17 compounds that did not exist in the probiotic cheese: 2-undecanone (floral), 1-butanol, 3-methyl (fresh cheese), and acetoin (creamy).[52] From the six alcohols identified in the cheeses, four were present only in the control cheese (1-butanol, 3-methyl-3-pentanol, 2,3-butanediol, and phenylethyl alcohol). The alcohol compounds in Table 11.11 may have originated from the breakdown products of fatty acids and from the reduction of methyl ketones.[53] Many esters were also present in control cheese including butanoic acid, ethyl ester; butanoic acid, 2-methyl propyl ester; propanoic acid, 2-methyl-, 3-methylbutyl ester; acetic acid, 2-ethylhexyl ester; octanoic acid, methyl ester; *n*-caprylic acid, isobutyl ester; decanoic acid, methyl ester; and decanoic acid, ethyl ester.[51] It has been also found that esters in cheese are primarily generated by the esterification of short-chain free fatty acids with alcohols, and the higher concentration of fatty acids observed in the NMR in the control cheese may have resulted in a higher concentration of alcohols and esters. Furthermore, they can provide fruity and floral notes to cheeses, minimizing the sharpness of fatty acids and the bitterness of some compounds.[54]

The authors also found that probiotic culture addition resulted in the development of seven compounds that were not present in the control formulation: benzaldehyde, furfural, 2-ethyl-1-hexanol, octadecanoic acid, pentadecanoic acid, 4-methyl-hexanoic acid, and hexanedioic acid, bis(2-ethylhexyl) ester. Among the 17 volatile acids identified in the cheeses, three were found only in the probiotic cheese (4-methylhexanoic acid, pentadecanoic acid, and octadecanoic acid). These compounds are precursors of other compounds such as aldehydes, ketones, and esters and can contribute to the flavour profile of cheeses.[55] These acid compounds may be produced by lipolysis, degradation of amino acids and lactose, or oxidation of esters, ketones, and aldehydes.[56] The formation of these acids may originate from proteolysis observed in the NMR in the probiotic cheese.

Balthazar *et al.*[51] also identified the presence of benzoic acid only in the control cheese, while benzaldehyde was found in the probiotic cheese, corroborating the NMR data. This indicates that the auto-oxidation of benzaldehyde to benzoic acid probably did not occur in the probiotic cheese. The addition of probiotic culture altered the aromatic profile of the cheeses, which may require a sensory analysis of the products.

With regard to free fatty acid (FFA) concentrations in different species cheeses, the qualitative and quantitative profiles of most branched chain FFA were similar in cow, sheep, and goat milks, except that 4-ethyloctanoic acid was found in cow milk cheese.[57] Table 11.12 displays the levels of volatile FFA (VFFA) and volatile total FA (VTFA) in goat and sheep

Table 11.12 Concentrations ($\mu g\,g^{-1}$ cheese) of volatile free fatty acids (VFFA) and volatile total fatty acids (VTFA) in cheeses from goat and sheep milks.[57]

	Goat milk cheese				Sheep milk cheese	
	VFFA			VTFA	VFFA	VTFA
Fatty acids	A	B	C	A	A	A
Butanoic	3.50	0.72	31.8	7030	32	18.18
2-Methylbutanoic	ND[a]	ND	2.26	1.18	2.48	3.20
3-Methylbutanoic	0.05	ND	ND	20.4	20.4	22.6
2-Ethylbutanoic	0.95	ND	ND	4.86	ND	ND
Pentanoic	0.02	ND	0.31	5.33	0.08	2.16
3-Methylpentanoic	ND	ND	ND	2.71	ND	0.66
4-Methylpentanoic	ND	ND	ND	0.45	ND	0.78
Hexanoic	11.3	0.76	61.2	6000	40.7	7230
2-Ethylhexanoic	0.18	0.18	ND	0.60	0.18	0.75
4-Methylpentanoic	0.05	0.05	3.23	ND	0.13	1.19
Heptanoic	0.33	0.03	0.90	21.6	0.51	10.9
2,4-Dimethylheptanoic	0.04	0.21	ND	1.43	ND	ND
A methylheptanoic[b]	0.03	0.01	ND	0.45	0.06	0.65
An ethylheptanoic[b]	0.03	0.02	ND	0.83	ND	ND
An ethylheptenoic[b]	1.82	1.88	0.39	6.59	ND	ND
Octanoic	30.9	3.69	70.3	6006	38.3	7577
4-Ethylheptanoic	0.11	0.06	ND	0.97	0.28	0.31
4-Methyloctanoic	0.09	0.02	0.26	9.70	0.08	2.79
A dimethyloctanoic[b]	ND	ND	ND	ND	0.09	0.22
6-Methyloctanoic	0.04	0.02	ND	0.12	0.08	0.28
Nonanoic	0.38	0.06	1.30	19.6	0.64	13.5
4-Ethyloctanoic	0.01	0.01	0.05	2.84	0.13	0.19
4-Methylnonanoic	0.05	0.01	0.11	1.80	0.07	1.17
8-Methylnononoic	0.41	ND	ND	0.63	ND	ND
Decanoic	88.2	21.0	183	21 410	88.4	23 320
A methyldecanoic[b]	ND	0.09	0.13	3.15	0.08	0.72
2-Ethyldecanoic	ND	ND	0.07	2.90	0.03	1.59
9-Decenoic	2.30	0.51	4.01	63.5	2.56	53.3

[a] Not detected.
[b] Tentative identification.

milk cheeses. The milk fat of cows contained low concentrations of 4-methyloctanoic acid, while those of sheep and goat contained significant amounts of both 4-methyloctanoic and 4-ethyloctanoic acids, whereby these two FA contributed mutton-like and goat-like flavours.[57] Goat milk cheeses have higher levels of C6:0, C8:0 and C10:0 which endow a strong goaty flavour,[10,20,58] whereas cow milk cheese contains fewer short chain FA.

Elevation of FFA levels in a cheese indicates rancid or off-flavour development. The higher the FFA, the stronger the flavour development. An example is that flavour intensity increased in Italian cheese as short chain FFA concentrations increased. The FFA profiles of goat, sheep, and cow cheeses were similar, with the exception of 4-ethyloctanoic acid which was present only in goat and sheep cheeses, but not in the cow cheese.[57] The levels of 4-methyloctanoic acid in goat Cheddar cheese were significantly increased from day 1 to week 12 of the aging period.[59] At concentrations below 100 ppb, 4-methyloctanoic acid exhibited a mutton-like aroma, whereas 4-methyloctanoic acid blended easily with the goaty aroma of 4-ethyloctanoic acid to produce distinctive goaty flavour.[57,59] It was found that the threshold concentration of 4-ethyloctanoic acid for goaty aroma was 1.8 ppb[60] and 6.0 ppb[61] in dilute citric acid solution at pH 2.0. The FFA content of cheese increased during refrigeration (4 °C) storage, where the initial FFA consisted of short chain acids, but the 16:0 and 18:0 contents increased after 10 days.[62]

Fat degradation in cheeses changes the lipid composition of the products. Lipolysis in goat milk increases during storage at room temperature for 4 h and 12 h and further extended storage.[1,63,64] Chilliard *et al.*[65] found that goat milk had a significant correlation between spontaneous lipolysis and lipoprotein lipase activity, while no correlation was found in cow milk. This is because goat milk has higher sensitivity to spontaneous lipolysis than cow milk owing to the difference in lipase distribution. Jin and Park[66] demonstrated that the acid degree value (ADV) of goat milk cheeses steadily increased as the aging period advanced, where ADV is a measure of lipolysis or the degree of formation of FFA in milk and dairy products.

11.6 Protein

Much of the varietal difference among cheeses is attributed to the nature of physical and chemical changes during ripening, which are influenced by the cultures, chemicals, or flavour ingredients added to the curd during manufacturing.[4,5] However, hard type cheeses can be manufactured easily from cow milk than goat milk due to milk protein composition. Cow milk contains a significantly higher amount of α_{s1}-casein (α_{s1}-CN) than goat milk (Table 11.12). This lack of α_{s1}-CN in goat milk compared to cow milk presents a natural problem of coagulation in soft curd formation, whereby most goat milk cheeses are soft curd body resulting in soft cheeses.[1,4]

As shown in Table 11.7, cheeses contain 4–40% high-quality protein.[9] For most cheese varieties (rennet- or acid-coagulated), the vast majority of the protein is casein, which is deficient in sulphur-containing amino acids, resulting in a nutritional value of 91–97 compared to 100 for total milk protein.[37] As in heat-acid coagulated varieties and some cheeses manufactured from UF retentate, the incorporation of whey proteins in cheese increases the nutritional value of its proteins. Cheese proteins are highly digestible and better digestible than milk protein which is attributed to proteolysis of cheese proteins during ripening.[9]

Most goat cheeses are soft cheese varieties, which are consumed fresh. They are set by biological acidification or by an acid (hydrochloric, lactic, vinegar, lemon, lime, *etc.*) coagulation process, aided usually by low amounts of rennet, *i.e.*, lactic-type cheeses. There are also varieties consumed after ripening, which are generally manufactured by the enzyme rennet (chymosin) setting process. Lactic-type cheeses emerged as a result of the predominance of low levels of α_{s1}-CN alleles in the milk of high-productivity goats.[1]

Semi-hard and hard type cheeses contain significantly higher protein contents than soft type cheeses. As shown in Table 11.7, hard type cheeses, such as Parmesan, reduced-fat Cheddar, Emmental, Edam, and Gruyere, contain substantially higher protein content than the other types of cheeses, indicating cheeses are highly nutritious and rich in animal protein. The caseins contain several bioactive peptides encrypted within the sequences of the caseins, some of which are released during ripening, especially angiotensin-converting enzyme inhibitors, casomorphins and phosphopeptides.[67] Phosphopeptides in cheeses are postulated to serve as good carriers of nutritionally important metals for food supplementation, which may require further confirmation.

Cheese is considered important for teeth development, both by providing Ca and P for initial development and by reducing cariogenesis. These good cariostatic effects of cheese are mainly by hardening tooth enamel damaged by acids in some foods or produced in plaque.[9] Extensive decarboxylation can occur in surface smear-ripened varieties, and some of the resulting amines, especially histamine and tyramine, are bioactive but cause problems only in patients using monoamine oxidase inhibitors.[9]

The percent compositions of different protein fractions in cow and goat milks are summarized in Table 11.13. The protein composition can influence the coagulation and rheological properties of cheese products. The percentages of α_{s1}-CN and α_{s2}-casein (α_{s2}-CN) in goat milk are markedly different

Table 11.13 Caseins, minor proteins, and enzyme contents of goat milk in comparison with those of cow and human milks.[26,68–70]

Proteins	Goat	Cow	Human
Protein (%)	3.5	3.3	1.2
Total casein (g/100 mL)	2.11	2.70	0.40
α_{s1} (% of total casein)	5.6	38.0	—
α_{s2} (% of total casein)	19.2	12.0	—
β (% of total casein)	54.8	36.0	60–70.0
κ (% of total casein)	20.4	14.0	7.0
Whey protein (%) (albumin and globulin)	0.6	0.6	0.7
Nonprotein N (%)	0.4	0.2	0.5
Lactoferrin ($\mu g\, mL^{-1}$)	20–200	20–200	<2000
Transferrin ($\mu g\, mL^{-1}$)	20–200	20–200	<50
Prolactin ($\mu g\, mL^{-1}$)	44	50	40–160
Folate-binding protein ($\mu g\, mL^{-1}$)	12	8	—
Immunoglobulin			
IgA (milk: $\mu g\, mL^{-1}$)	30–80	140	1000
IgA (colostrum: $mg\, mL^{-1}$)	0.9–2.4	3.9	17.35
IgM (milk: $\mu g\, mL^{-1}$)	10–40	50	100
IgM (colostrum: $mg\, mL^{-1}$)	1.6–5.2	4.2	1.59
IgG (milk: $\mu g\, mL^{-1}$)	100–400	590	40
IgG (colostrum: $mg\, mL^{-1}$)	50–60	47.6	0.43
Lysozyme (μg/100 mL)	25	10–35	4–40
Ribonuclease (μg/100 mL)	425	1000–2000	10–20
Xanthine oxidase (μl O_2 per h per mL)	19–113	120	—

from those in cow milk, where goat milk has much lower α_{s1} and higher α_{s2}-CN than cow milk.[69] The higher α_{s1}-CN in cow milk causes firmer curd formation during cheese making. Electrophoretic mobility under standard conditions shows that β-casein (β-CN) is the major component of the casein fraction in goat milk, whereas α_{s1}-CN is the major casein in cow milk.[1] The total casein content of goat milk is slightly lower than that of cow milk, which causes the yield of goat cheese to be lower than that of cow cheese.

Because of very low levels of α_{s1}-CN present in goat milk, most goat milk cheeses are soft type,[7] which affect the texture and flavour as well as ripening characteristics compared to bovine milk cheeses. Goat milk contains five principal proteins, β-lactoglobulin (β-Lg), α-lactalbumin (α-La), κ-casein (κ-CN), β-CN, and α_{s2}-CN, which were named after their corresponding proteins of cow milk due to their homologous nature in composition and properties.[71] The casein composition in goat milk is influenced by genetic polymorphism on the casein loci.[72] Polymorphism of α_{s1}-CN controls the level of α_{s1}-CN excretion in milk, and more than 18 alleles have been identified in goat milk.[19] These alleles are distributed among seven different classes of protein variants (α_{s1}-casein A-G) and associated with 4 levels of α_{s1}-CN expression.[19,73]

As far as minor proteins and enzymes of milk are concerned, the lactoferrin, transferrin, and prolactin contents of goat milk are comparable to those of cow milk (Table 11.13). Human milk contains more than 2 mg lactoferrin per mL, which is 10–100 times higher than that in goat milk. Cow and goat milks contain transferrin levels of 20–200 $\mu g\,mL^{-1}$, while human milk contains <50 $\mu g\,mL^{-1}$. Goat milk also contains immunoglobulins IgG, IgA, and IgM. Cow milk has similar ranges of immunoglobulins to those of goat and sheep milks and colostrums. Distribution of enzymes in goat milk is quite different from that in cow milk.[69,74] The ribonuclease level in cow milk is much greater than that in goat milk, where this enzyme is identical to bovine pancreatic ribonuclease.[44] The lysozyme concentrations of goat and cow milks are comparable, which are low in milks of artiodactyl (even-toed ungulates). Goat milk contains on average 25 μg lysozyme, 425 μg ribonuclease, and 36 μM per min lipase per 100 mL.[74] The effects of these minor proteins and enzymes on cheese formation, maturation and nutritive values are not known and require further investigations.

Fortification of nutrients in diets of milking animals or direct addition of nutrients to the milk can change the nutrient compositions of milk as well as those of manufactured products. For example, CLA, vitamins, and minerals can be added to animal diets or directly fortified in cheese milk. These nutrient supplementations can alter the composition of cheese nutrients. In a study of regular ferrous sulphate (RFS) and large microencapsulated ferrous sulphate (LMFS) fortification in goat Cheddar cheeses, iron added cheeses revealed higher levels of fat and protein than those of non-fortified control cheeses.[75] The addition of iron salt in the goat cheese might have formed further chemical structural bridges within casein networks, which can result in more complex and rigid textural formation. Thus, this situation may lead to better fat entrapment into the cheese with higher fat content in iron fortified cheeses as compared to the control cheeses.

Fe fortification in caprine Cheddar cheeses has also resulted in the elevation of protein contents, where caseins contain phosphorylated serine residues that serve as major binding sites for minerals including added Fe molecules.[75,76] The clusters of phosphoserine residues are responsible for the high metal binding capacity of caseins and develop multi-dentate ligand properties in which Fe forms coordinate bonds with the O of the phosphorylated residue.[76,77] The structure of the protein matrix of cheeses begins with the rearrangement of casein micelles as demonstrated and analysed by Boutrou *et al.*[78]

With regard to the pore numbers of microstructures in goat Cheddar cheeses (Table 11.14), LMFS added cheeses had significantly higher ($P<0.05$) pore numbers than those of control and RFS counterparts, suggesting that LMFS salt caused greater numbers of pores the control and RFS cheeses. In contrast, opposite trends were observed for other microstructural parameters, such as the pore area, pore perimeter and iron crystal size, where the average values of these parameters of LMFS cheeses were significantly lower ($P<0.05$) than those of control and RFS fortified cheeses.

Changes in the protein matrix structure, protein degradation, and protein microstructure can occur through proteolysis of cheese proteins during storage.[64,79] It was observed that ripening of full fat Cheddar cheese altered the pore size of the cheese microstructure during 33 days of storage at 20 °C, and these structural changes were associated with increased proteolysis through the solubilisation of the protein network.[79] However, no significant

Table 11.14 Comparison of mean differences in microstructural parameters between control and two Fe fortified experimental goat milk Cheddar cheeses after 0, 2, and 4 months of storage.[a] [75]

		Month 0	Month 2	Month 4
Parameter	Type	Mean ± SD	Mean ± SD	Mean ± SD
Pore number (n)	CC	163.32^{ay} ± 0.96	134.56^{by} ± 1.71	113.52^{cz} ± 1.77
	RFS	177.67^{ax} ± 0.12	155.82^{bxy} ± 0.85	146.65^{cy} ± 1.84
	LMFS	179.43^{ax} ± 0.45	157.60^{bx} ± 0.76	149.94^{cx} ± 1.98
Pore area (μm^2)	CC	323.45^{cy} ± 2.55	362.67^{bz} ± 2.40	590.67^{ay} ± 1.64
	RFS	343.76^{cy} ± 2.35	553.65^{bx} ± 2.08	686.56^{ax} ± 0.55
	LMFS	416.34^{cx} ± 2.56	449.89^{by} ± 2.11	691.78^{ax} ± 1.77
Size of Fe crystals (μm)	CC	0.00^{ay} ± 0.00	0.00^{az} ± 0.00	0.00^{az} ± 0.00
	RFS	41.23^{cx} ± 2.66	58.40^{bx} ± 1.91	81.31^{ax} ± 1.39
	LMFS	40.05^{cx} ± 5.09	52.05^{by} ± 1.90	78.42^{ay} ± 1.88
Pore perimeter (μm)	CC	42.06^{cy} ± 0.12	49.66^{by} ± 0.05	57.61^{az} ± 0.15
	RFS	51.87^{cxy} ± 0.14	67.66^{bx} ± 0.05	71.64^{ax} ± 0.03
	LMFS	55.64^{cx} ± 0.24	65.69^{bx} ± 0.13	69.17^{ay} ± 0.07

[a] a,b,c Means with different superscripts within a row between storage periods are significantly different ($P<0.05$). x,y,z Means with different superscripts within a column between cheese types are significantly different ($P<0.05$). CC = control cheese (not fortified with Fe); RFS = cheese fortified with regular ferrous sulphate; LMFS = cheese fortified with large microencapsulated ferrous sulphate.

differences were observed in the microstructure in the goatskin bag cheese Bouhezza during 6 weeks of storage at −20 °C, as observed by scanning electron microscopy and confocal scanning electron microscopy.[80]

11.7 Carbohydrates

The major carbohydrate in milk is lactose (β-galactose-1,4-glucose), which is a disaccharide that makes up around 2–8% of milk by weight. Lactase deficiency causes inefficient hydrolysis of lactose (lactose maldigestion), and individuals are said to have lactose intolerance if physiological symptoms are present.[81] Rennet coagulated fresh-curd cheese contains about 1% lactose, which is catabolised during aging. Cheeses with washed acid curd, such as Cottage, contain no lactose. Thus, lactose-intolerant people can consume Cottage cheese or some hard or semi-hard cheeses which contain negligible amounts of lactose. During ripening, lactose is fermented to lactic acid, which relieves lactose load in the cheese products.[82]

Lactose is present only in mammalian milk, about 7.2 g/100 mL in mature human milk and 4.7 g/100 mL in cow milk.[26,83] Low amounts of lactose are present in cheeses in the fresh stage, and its level further decreases as ripening progresses. Lactose is the main source of energy during the first year of human life, and it provides almost half of the total energy requirement of infants.[84] The derivatives of lactose are lactulose, lactitol, and galacto-oligosaccharides, which can be applied in foods and pharmaceutical preparations as prebiotics to promote gut health. Similarly, compounds such as tagatose and lactobionic acid have potential applications as bioactive ingredients in foods.[85] Since the glycemic index (GI) of lactose is relatively low, the minimal level of lactose in cheeses can be beneficial for those who are sensitive to hyperglycemia.[85]

11.8 Organic Acids

Lactic acid and acetic acid are the degradation products of lactose in cheeses and cultured dairy products. Although lactic acid is the major organic acid in ripened dairy products, citric acid is the predominant organic acid in milk. During storage of milk and cheeses, citric acid disappears rapidly because of the action of bacteria.[1,86] Orotic acid is an intermediary product in the biosynthesis of pyrimidine nucleotides, which occurs specifically in milk products.[86] Orotic acid, as well as total creatinine and uric acid, are suitable indicators for the determination of the proportion of milk in foods.[86] The levels of orotic acid, creatinine and uric acid in milk vary among different species. The concentration of orotic acid in buffalo milk is low (19.6 μg/100 mL) as compared to cow milk (52.6 μg/100 mL), while creatinine is higher (246 mg/100 mL *versus* 167 mg/100 mL) and uric acid is the same (0.26 mg/100 mL).[1,4,86]

Acidity can be developed in raw milk that is produced under normal conditions, and it has been known that high acid milk does not putrefy. Thus, the other milk constituents can be naturally preserved if milk is allowed to develop acidity.[86] Fresh bovine milk is particularly suitable as a fermentation substrate for most microorganisms since it contains 5% lactose and 3.3% protein, and has a water activity near 1.0 and a pH of 6.7. Even though other species contain different compositions of protein and lactose, they are also suitable for microbial growth.[4,86]

A number of acids are produced from sugar fermentation by anaerobic bacteria depending on the microorganism involved and the end products in milk.[86] The lactic acid fermentation is the most important one in milk and is central to many processes. For example, if only streptococci and lactobacilli are present, the end product will be lactic acid.[87] Lactic acid bacteria are saccharolytic and fermentative, and therefore they are ideally suited for growth in milk. They will generally out-compete other microorganisms for lactose and will produce an unfriendly environment for competitors due to acidification.[86]

Propionic fermentation is a mixed-acid fermentation and is used in the manufacture of Swiss cheese varieties.[86] Propionibacteria ferment lactose, which produces lactic acid, propionic acid, acetic acid, and carbon dioxide. Yeasts, such as *Candida* and *Torula*, will produce ethanol and carbon dioxide. Coliform bacteria cause the undesirable spoilage fermentation, and large numbers of coliform bacteria in milk indicate poor hygiene. Coliform fermentation can also occur with the production of lactic acid, acetic acid, ethanol, carbon dioxide, and hydrogen.[86] The coliform fermentation disrupts lactic acid fermentation resulting in spoilage of cheese. Different bacteria may be used for fermentation, providing cheeses or other products with special flavour and aroma, and with several potential health beneficial metabolites.[88]

Organic acids contain one or more carboxylic acid groups, which may be covalently linked in groups such as amides, esters, and peptides.[1,4,86] Organic acids play an important role in fermented milks and cheese, and they can be altered due to different processing and storage options. The organic acid composition of several types of goat cheese were studied by Park *et al.*[10] Differences were found in tartaric, formic and uric acid contents between fresh and frozen/thawed plain soft cheese. Freezing caused an increase in formic and uric acids and a decrease in tartaric acid (Table 11.15). In contrast, Califano and Bevilacqua[89] did not find a significant effect of freezing on the variations in the organic acid contents of cow milk Mozzarella cheese. Park *et al.*[10] observed that refrigerated storage of plain soft goat cheese at 4 °C for 28 days promoted changes in only three acids (orotic, malic, and butyric). In Monterey Jack cheese, frozen and thawed cheese contained higher acetic, butyric, citric, malic, propionic, and pyruvic acids compared to the unfrozen control cheese (Table 11.15). In addition, aging time elevated the levels of acetic, butyric, malic, and orotic acids in the experimental caprine cheeses.

11.9 Minerals

Rennet-coagulated cheeses are rich sources of Ca and phosphate, while acid-coagulated varieties contain much lower levels.[9] Cheese provides a significant

Table 11.15 Comparison of organic acid contents (mg g^{-1} cheese) among nonfrozen control (NFC), 0 month frozen stored (0 MF), 3 months frozen stored (3 MF), and 6 months frozen stored (6 MF) plain soft (PS) and Monterey Jack (MJ) goat milk cheeses aged at 4 °C for 4 weeks. Several unknown acids were also detected but are not listed.[10]

	NFC			0 MF			3 MF			6 MF		
Acid	0 days	14 days	28 days	0 days	14 days	28 days	0 days	14 days	28 days	0 days	14 days	28 days
Tartaric (TA)												
PS (iso-TA)	0.86	0.93	0.93	0.75	0.67	0.70	1.53	1.49	1.42	1.62	1.79	1.92
MJ (TA)	ND[a]	ND	0.38	0.34	0.40	0.49	ND	0.16	0.57	0.21	0.19	0.48
Formic												
PS	2.32	2.23	2.21	2.66	2.50	2.59	1.63	1.74	1.81	2.31	2.37	2.61
MJ	2.59	2.62	2.62	2.63	2.40	2.44	2.72	2.41	2.10	3.00	2.73	2.18
Orotic[b]												
PS	0.042	0.011	0.012	0.043	0.014	0.013	0.011	0.037	0.036	ND	ND	ND
MJ	0.012	0.010	0.009	0.004	0.008	0.013	0.010	0.014	0.028	0.012	0.018	ND
Malic												
PS	1.13	1.32	1.44	1.22	1.40	1.21	0.42	0.44	0.30	0.03	0.03	0.02
MJ	1.26	1.00	1.24	1.91	1.96	2.34	ND	ND	ND	0.70	0.65	0.28
Lactic												
PS	10.04	10.22	10.37	8.96	10.15	9.83	11.19	10.97	10.70	12.47	12.30	12.91
MJ	10.89	9.56	9.16	9.72	10.00	10.24	7.02	10.29	9.12	13.05	12.14	11.05
Acetic												
PS	2.86	4.20	4.34	5.01	4.03	3.69	3.24	3.32	2.65	3.04	3.41	4.41
MJ	1.82	1.86	2.43	2.00	2.19	2.82	2.47	2.94	2.61	3.51	4.93	4.84
Citric												
PS	0.69	0.08	0.87	0.88	0.89	0.81	2.12	1.73	1.58	0.58	0.92	1.42
MJ	1.30	0.85	0.77	1.05	1.31	1.32	1.20	1.22	0.01	1.63	1.27	0.93
Uric[b]												
PS	0.017	0.020	0.015	0.029	0.038	0.034	0.037	0.043	0.040	0.083	0.085	0.084
MJ	0.062	0.072	0.064	0.064	0.065	0.063	0.008	0.020	0.013	0.013	0.016	ND
Propionic												
PS	0.71	0.79	0.69	1.28	0.60	0.83	1.53	2.13	2.15	5.36	4.25	4.84
MJ	0.66	0.36	5.96	6.31	5.53	7.16	2.34	3.34	3.62	4.88	4.71	3.65
Pyruvic												
PS	ND	ND	ND	ND	ND	ND	ND	ND	ND	ND	ND	ND
MJ	1.58	1.29	1.11	2.04	1.80	1.96	ND	ND	ND	ND	ND	1.57
Butyric[bc]												
PS	1.07	1.20	1.29	1.01	1.21	1.62	1.16	1.53	1.83	2.76	2.93	4.93
MJ	9.56	10.58	11.94	10.52	13.04	14.04	28.01	37.57	38.56	35.36	48.91	32.28

[a] ND = not detectable.
[b] Differences between cheese types are significant ($P < 0.05$).
[c] Differences among storage treatments are significant ($P < 0.05$).

amount (up to 50%) of the dietary Ca intake in Western countries. Ca is essential for the development and maintenance of bone and teeth, and a high intake of milk and dairy products during childhood and early adulthood has proven to decrease the risk of osteoporosis in later stages of life.[9,10]

On the other hand, as shown in Table 11.16, milk and cheese contain very low levels of Fe, where all varieties of cheeses and cultured milk products including yogurt, sour cream, and processed cheese foods are poor sources of dietary Fe. However, Fe is a potent pro-oxidant, so that fortification of milk with Fe leads to the development of lipid oxidation (oxidative rancidity). Nevertheless, Zhang and Mahony[90] reported that Cheddar and processed cheese can be fortified with Fe without ill-effects, possibly because cheese has a low redox potential (−200 mV).[9]

Most cheeses are salted during processing which has important effects on cheese ripening and quality. Pasteurized process cheese contains the highest Na due to the high amount of NaCl added for processing, and Blue and Parmesan cheeses have the next highest contents of Na among the listed products in Table 11.16. Processed foods are major contributors of dietary Na, but substantial quantities may be added during cooking and by individual consumers at the table. Although cheese is not a major contributor of dietary Na (12% of the RDA), high Na intake is strongly related to hypertension. Therefore, reduction of Na levels in cheeses is desired and its impact on cheese quality has been studied by many researchers.

Park[7] studied the major and trace mineral concentrations of commercial goat milk cheeses produced in the US (Table 11.17). There were wide

Table 11.16 The mineral contents (in mg/100 g) of cheese and fermented milk foods.[91]

Cheese type	Ca	P	Fe	Na	K
Blue	528	387	0.31	1146	256
Brick	674	451	0.43	560	136
Camembert	388	187	0.33	842	187
Cheddar	710	455	0.14	653	76
Cottage, creamed	53	113	0.16	344	90
Cottage, uncreamed	86	190	0.15	372	137
Cream	97	107	0.11	314	132
Limburger	497	128	0.13	800	128
Mozzarella, low moisture part skim	697	548	0.22	666	188
Parmesan	1187	694	0.82	1376	92
Swiss	890	574	0.13	187	72
Pasteurized process					
American	682	438	0.26	1284	255
Swiss	723	526	0.60	1552	284
Cheese spread	562	875	0.33	1625	242
Other fermented milk products					
Buttermilk	115	85	0.03	105	135
Sour cream	141	71	0.07	83	212
Yogurt, plain	183	144	0.08	70	234

Table 11.17 Nutrient profiles of selected varieties of commercial goat milk cheeses in the US.[7]

Variety	H_2O	Fat	Protein	Ash	S	P	K	Mg	Ca	Na	Cl	Fe	Al	Mn	Cu	Zn
	g/100 g				mg/100 g							μg g^{-1}				
Fresh soft, plain	60	23	19.8	1.7	3.5	275	26	15	172	416	293	18	15	1.0	7.4	9.1
Fresh soft, garlic	64	18	16.7	1.3	4.1	247	15	17	117	331	145	13	38	0.9	8.1	14
Fresh soft, pepper	57	23	21.6	1.3	4.1	236	28	18	107	298	201	29	45	1.2	6.5	8.8
Fresh soft, herb	59	22	17.3	1.6	3.0	225	35	15	112	336	165	18	9	1.1	6.7	7.8
Feta	52	25	25.1	4.3	3.5	544	62	18	639	916	1260	6.1	15	1.0	6.5	16
Camembert capri	47	29	29.1	1.7	2.4	264	7	22	229	279	301	8.9	13	0.9	5.4	16
Blue	26	32	20.2	3.3	8.1	575	89	35	841	924	398	15	36	0.9	5.6	6.9
Cheddar	42	27	30.3	3.6	7.8	526	11	42	599	361	1030	7.7	15	0.9	4.9	6.5
Shepherd's hard	32	33	28.9	3.9	4.1	737	54	62	1035	285	114	27	12	4.0	6.6	19
Montasio's hard	26	37	31.9	3.5	4.9	720	42	55	990	260	96	11	16	1.1	6.0	41[a]
Ancho chile	42	30	22.2	4.1	3.2	688	53	71	939	464	162	25	16	1.1	6.4	9.1

[a] Extremely high value.

variations in the concentrations of minerals among and within the varieties of 30 different goat cheeses. The spiced cheeses had significantly lower macrominerals than plain soft cheeses. The differences in trace minerals among six types of cheeses were minimal, except for Zn. The high Fe and Al contents in semi-soft and soft cheeses might be due to the possible contamination of the two minerals from processing utensils during manufacturing of farmstead cheeses.[7] Similar observations were reported in goat milk yogurt.[92]

Among the major minerals, the K content displayed marked differences between different cheeses and other goat milk products. The levels of K in plain goat milk yogurt and plain soft cheese were lower than those in fluid milk. Furthermore, the level of K in goat milk Cheddar cheese was significantly lower than that in fluid milk, which is due to the loss of K in the whey. The other manufactured products having higher total solid contents explicitly displayed higher levels of all other mineral contents.[7,92]

The S contents of cheeses, especially hard cheeses such as Blue and Cheddar, were considerably greater than those of the other varieties of cheeses (Table 11.17). Reports indicate that fermented dairy products are more nutritious than the original milk from which they are made,[93] implying that certain nutrients such as S-containing proteins may be synthesized by the bacteria during the fermentation processes of cheeses and goat yogurts.

Goat milk contains more Ca, P, K, Mg, and Cl, and less Na and S contents than cow milk.[94,95] Trace mineral contents of goat milk are affected by diet, breed, animals, and stages of lactation.[95] The mean levels of Mn, Cu, and Fe in Alpine goat milk were 0.33, 5.0, and 1.7 $mg\,L^{-1}$, while Nubian goat milk contained significantly higher levels of Cu (1.36 *vs.* 1.69 $mg\,L^{-1}$) and Zn (7.9 *vs.* 11.9 $mg\,L^{-1}$).[95] Among the trace minerals, Zn was present at the greatest amount, and goat and cow milks had more Zn than human milk.[95] Sheep milk has around 0.9% total minerals or ash compared to 0.7% in cow milk. The levels of Ca, P, Mg, Zn, Fe, and Cu are higher in sheep than those in cow milk, while the opposite appears to be the case for K, Na, and Mn.[26] The mineral composition of milk can influence the texture and aging properties of cheese. Trace minerals such as Fe and Cu may have a greater impact on lipolysis and proteolysis during storage due to accelerated ripening. Significant changes can occur in the composition of fatty acids, free fatty acids, peptides and amino acids in ripened cheeses.

11.10 Vitamins

Cheese completely retains the fat-soluble vitamins from the cheese milk, whereas most of the water-soluble vitamins in the milk are lost into the whey. Hence all cheeses are poor sources of these water-soluble vitamins, although some biosynthesis of these vitamins is accomplished by GI bacteria during ripening.[1,9,97] Table 11.18 displays the vitamin concentrations of milk and dairy products of different species.

Table 11.18 The vitamin contents (in µg/100 g) of cheese and fermented milk foods.[91] Thiamine = vitamin B_1, riboflavin = B_2, niacin = B_3, pantothenic acid = B_5, folate = B_7, NM = not measured.

Cheese type	A	B_1	B_2	B_3	B_5	B_6	B_9	B_{12}	C	D	E	K
Blue	198	29	382	1016	1729	166	36	1.22	0.0	0.5	250	2.4
Brick	292	14	351	118	288	65	20	1.26	0.0	0.5	260	2.5
Camembert	241	28	488	630	1364	227	62	1.30	0.0	0.4	210	2.0
Cheddar	330	29	428	59	410	66	27	1.10	0.0	0.6	710	2.4
Cottage, creamed	37	27	163	99	557	46	12	0.43	0.0	0.1	80	0.0
Cottage, uncreamed	2	23	226	144	446	16	9	0.46	0.0	0.0	10	0.0
Cream	308	23	230	91	517	56	9	0.22	0.0	0.0	860	2.1
Limburger	340	80	503	158	1177	86	58	1.04	0.0	0.5	230	2.3
Mozzarella, low moisture part skim	223	24	353	111	429	100	27	1.68	0.0	0.4	500	1.3
Parmesan	207	39	332	271	453	91	7	1.20	0.0	0.5	220	1.7
Swiss	288	11	302	64	353	71	10	3.06	0.0	0.0	600	1.4
Pasteurized process												
American	201	35	360	155	855	102	7	1.33	0.0	2.5	630	3.4
Swiss	198	14	276	38	260	36	6	1.23	0.0	0.5	340	2.2
Cheese spread	173	48	431	131	686	117	7	0.40	0.0	0.4	190	1.8
Other fermented milk products												
Buttermilk	47	47	172	90	380	36	5	0.46	0.0	1.3	70	0.3
Sour cream	90	40	120	70	NM	20	11	0.42	0.9	0.2	300	0.5
Yogurt, plain	14	44	214	114	591	49	11	0.56	0.8	0.0	30	0.2

Goat and sheep milks generally have higher amounts of vitamin A than cow milk with some variations between breeds within a species. Because goats convert all β-carotene (which imparts a bright yellow-orange colour to cheese) into vitamin A in the milk, caprine milk is whiter than bovine milk.[10,42] Goat milk contains adequate amounts of vitamin A and niacin, and excess amounts of thiamine, riboflavin and pantothenate for a human infant.[4,42,97] Vitamin B levels in goat and cow milks are a result of rumen synthesis and are somewhat independent of diet.[27,96,97]

Compared to cow milk, goat milk is significantly deficient in folic acid and vitamin B_{12}, which causes "goat milk anemia".[42,97] The levels of folate and vitamin B_{12} in cow milk are five times higher than those in goat milk, where folate is necessary for the synthesis of haemoglobin.[4,98] Megaloblastic anemia has been mainly attributed to the folate deficiency in goat milk. Both goat and cow milks are equally deficient in vitamins B_6, C, and D, where these vitamins must be supplemented from dietary food sources other than cheeses and milk products.[99]

In pasteurization of goat milk, high temperature and short-time (HTST) heat treatment is the best method for preserving various vitamins as well as extending the shelf-life of the milk.[100] Lavigne *et al.* also found that losses of thiamine, riboflavin, and vitamin C were reduced if the milk was processed by HTST, flash and UHT processes than by LTLT (low temperature long time) and autoclave treatment methods. The vitamin levels of cheese products are undoubtedly influenced by those of cheese milk. Vitamin contents in sheep milk are mostly higher than those in cow milk, except for carotene. The vitamin levels of sheep milk are even higher than those of goat milk, although there is a paucity of research data on the vitamin values of sheep milk.[97]

Studies on the vitamin contents of milk and dairy products are limited. The Composition of Foods published by the Royal Society of Chemistry[48] and the USDA National Nutrient Database for Standard Reference[91] provide a significant amount of information on the vitamin and other nutrient compositions of cheeses and dairy products. Using these published data, the vitamin compositions of major species' milk products were compiled in several reports[1,4] (Table 11.18). The reported data revealed that the milk and dairy product samples showed considerable variations in compositions due to the differences in the sources of milk, such as raw and pasteurized, breed of dairy species, diet, stages of lactation, location, feeding management, *etc.*[1]

Some loss of vitamins is inevitable when milk and dairy products are processed and stored, while some vitamin contents can be increased in the products by fortification such as vitamin A and D addition to the pasteurized fluid milk.[4] Some vitamins also increased in cultured dairy foods such as yogurt and cheese products due to fermentation by bacterial cultures. Moatsou and Park[1] reported that vitamin contents in dairy products generally depend on the following conditions: (a) the type of dairy product, *e.g.*, cheese is enriched with fat-soluble vitamins, since milk fat is concentrated in the curd; (b) the conditions of heat treatment, *i.e.* combination of temperature/time; (c) packaging, *i.e.* exposure to light and oxygen permeability; and (d) the conditions of storage, *i.e.* temperature and duration. Milk processing can change riboflavin (vitamin B2) and vitamin C, which in turn can decrease the levels of other vitamins as well as the off-flavours of milk and cheeses.[101,102] With regard to the effect of heat treatment on thiamine, riboflavin, and vitamin C, HTST pasteurization (74 °C, 16 s) under vacuum was considered the best process for goat milk and cheese processing.[1]

Lucas *et al.*[103] confirmed that the cheese compositional variability in β-carotene, xanthophylls, and vitamin E depended mainly on the composition of the original milk, whereas vitamin A was partially influenced by both the original milk composition and the cheese-making process. They suggested that a portion of folates interacted with the cheese dry matter and/or microbial synthesis occurred during the cheese manufacturing. The authors concluded that animal species as well as the feeding scheme affected the vitamins in cheese, in particular the basic fodder ration.[104] In addition, the cheese made from the milk of grazed goats contained higher amounts of α-tocopherol compared to cheeses from indoor-fed animals.[105]

Table 11.18 clearly indicates the wide range of vitamin contents in cheese and other fermented products. The differences are due to the composition of the milk, production techniques, and the vitamins that the microorganisms consume and produce. Milk contains low levels of vitamin D and no vitamin C, and none is generated in cheese.

11.11 Summary

The worldwide increases in the production and consumption of cheese are strong indicators that people consider cheese to be a highly nutritious and palatable food. The fat, proteins, and carbohydrates break down during ripening to form various volatile compounds responsible for flavour, and the minerals and vitamins contribute to the nutrition of cheese, which can be part of a healthy diet.

References

1. G. Moatsou and Y. W. Park, in *Handbook of Milk of Non-Bovine Mammals*, ed. Y. W. Park, G. F. W. Haenlein and W. L. Wendorff, Wiley-Blackwell, Oxford, UK, 2nd edn, 2017, p. 84.
2. P. Walstra, J. T. M. Wouters and T. J. Geurts, in *Dairy Science and Technology*, CRC Press, Boca Raton, FL, 2nd edn, 2006, p. 641.
3. F. V. Kosikowski, in *Cheese and Fermented Milk Foods*, Kosikowski and Assoc., Brooktondale, NY, 2nd edn, 1982.
4. Y. W. Park and M. Guo, in *Handbook of Milk of Non-Bovine Mammals*, ed. Y. W. Park, G. F. W. Haenlein and W. L. Wendorff, Wiley-Blackwell, Oxford, UK, 2006.
5. M. Loewenstein, S. J. Speck, H. M. Barnhart and J. H. Frank, *J. Dairy Sci.*, 1980, **63**, 1631.
6. H. Lame and M. Hekmati, *Lait*, 1975, **55**, 418.
7. Y. W. Park, *J. Dairy Sci.*, 1990, **73**, 3059.
8. FAO/WHO Food Standards Programme, *Codex Alimentarius*, 2nd edn, 2013, p. 276.
9. P. F. Fox and T. P. Guinee, in *Milk and Dairy Products in Human Nutrition*, Wiley-Blackwell, Oxford, UK, 2013, p. 357.
10. Y. W. Park, in *Handbook of Milk of Non-Bovine Mammals*, ed. Y. W. Park and G. F. W. Haenlein, Wiley-Blackwell, Oxford, UK, 1st edn, 2006, p. 34.
11. USDA, Agricultural Research Service, Office of Communications, photo gallery, image number K7227-2.
12. Y. Chilliard and A. Ferlay, *Reprod., Nutr., Dev.*, 2004, **44**, 467.
13. Y. Chilliard, J. Rouel, A. Ferlay, L. Bernard, P. Gaborit, K. Raynal-Ljutovac and A. Lauret, *Int. Dairy Fed.*, 2005, **5**, 297.
14. USDA, Foreign Agricultural Service, *Dairy: World Markets and Trade, Dairy Production and Trade Developments,* 2021, https://www.fas.usda.gov/data/dairy-world-markets-and-trade.
15. USDA, Economic Research Service, *Annual Consumption of Cheese Worldwide by selected countries*, 2021, https://www.ers.usda.gov/data-products/dairy-data/.
16. M. Shahbandeh, Global cheese production *2015–2020*, 2021, https://www.statista.com/statistics/1120911/cheese-production-worldwide/.
17. P. J. M. Reis and F. X. Malcata, *Small Rumin. Res.*, 2011, **101**, 122.

18. B. Vlahović, A. Popović-Vranješ and I. Mugoša, *Econ. Insights Trends Challenges*, 2014, **3**, 35.
19. P. Mikkelsen, *World Cheese Production: Special Focus on Asia and the Middle East*, PM Food & Dairy Consulting, Højbjerg, Denmark, 2021.
20. Eurostat, Milk and Milk Product Statistics, 2017, http://ec.europa.eu/eurostat/statisticsxplained/index.php/Milk_and_milk_product_statistics.
21. International Dairy Federation, *The World Dairy Situation*, Bull. 470, Brussels, Belgium, 2013, p. 7.
22. National Dairy Development Board, *Statistic of Milk Production in India*, 2014, http://www.nddb.org/English/Statistics/Pages/MilkProduction.aspx.
23. B. A. Underwood, *16th Meeting of the Pan American Health Organization Advisory Committee on Medical Research*, 1977, https://iris.paho.org/bitstream/handle/10665.2/35040/ACMR16_26.pdf?sequence=1.
24. Y. W. Park and G. F. W. Haenlein, in *Handbook of Food Products Manufacturing*, ed. Y. H. Hui, John Wiley & Sons, New York, NY, 2007, p. 447.
25. E. I. El-Agamy, in *Handbook of Milk of Non-Bovine Mammals*, ed. Y. W. Park, G. F. W. Haenlein and W. L. Wendorff, Wiley-Blackwell, Oxford, UK, 2nd edn, 2017, p. 409.
26. Y. W. Park, in *Handbook of Milk of Non-Bovine Mammals*, ed. Y. W. Park, G. F. W. Haenlein and W. L. Wendorff, Wiley-Blackwell, Oxford, UK, 2nd edn, 2017, p. 42.
27. G. F. W. Haenlein and R. Caccese, in *Extension Goat Handbook*, ed. G. F. W. Haenlein and D. L. Ace, USDA, Washington, DC, 1984, p. 1.
28. A. Lainé, C. Bastin, C. Grelet, H. Hammami, F. G. Colinet, L. M. Dale, A. Gillon, J. Vandenplas, F. Dehareng and N. Gengler, *J. Dairy Sci.*, 2017, **100**, 2863.
29. R. R. Peters, *Dairy Goat J.*, 1990, **68**, 223.
30. G. F. W. Haenlein, *National Symposium on Goat Meat Production and Marketing*, 1992, p. 112.
31. M. R. Guo, P. H. Dixon, Y. W. Park, J. A. Gilmore and P. S. Kindstedt, *J. Dairy Sci.*, 2001, **84**, E79.
32. G. H. Schmidt, in *Biology of Lactation*, WH Freeman, San Francisco, CA, 1971, p.178.
33. S. Carpino, S. Mallia, S. La Terra, C. Melilli, G. Licitra, T. E. Acree, D. M. Barbano and P. J. Van Soest, *J. Dairy Sci.*, 2004, **87**, 816.
34. L. P. Posati and L. Orr, *Composition of Foods. Dairy and Egg Products, Raw, Processed, Prepared*, Agricultural Handbook No. 8-1, Agricultural Research Service, USDA, Washington, DC, 1976.
35. T. P. Guinee and K. N. Kilcawley, in *Cheese: Chemistry, Physics and Microbiology*, ed. P. F. Fox, P. L. H. McSweeney, T. M. Cogan and T. P. Guinee, Elsevier, Oxford, UK, 3rd edn, 2004, vol. 1, p. 393.
36. M. A. Fenelon and T. P. Guinee, *J. Dairy Sci.*, 1999, **82**, 2287.
37. E. Renner, in *Cheese: Chemistry, Physics and Microbiology*, ed. P. F. Fox, Chapman and Hall, London, UK, 2nd edn, 1993, vol. 1, p. 557.

38. M. E. Johnson, R. Kapoor, D. J. McMahon, D. R. McCoy and R. G. Narasimmon, *Compr. Rev. Food Sci. Food Saf.*, 2009, **8**, 252.
39. D. L. Van Hekken, Y. W. Park and M. H. Tunick, *Small Rumin. Res.*, 2013, **110**, 46.
40. J. L. Iverson and A. J. Sheppard, *J. Dairy Sci.*, 1989, **72**, 1707.
41. V. Palo and K. Simkova, *Polnohospodarstvo*, 1981, **27**, 155.
42. R. Jenness, *J. Dairy Sci.*, 1980, **63**, 1605.
43. S. Gönç, R. Schmidt and E. Renner, *Milchwissenschaft*, 1979, **34**, 684.
44. M. Juàrez and M. Ramos, *Intl. Dairy Bull.*, 1986, **202**, 54.
45. I. Martinez Castro, M. Suárez and P. J. Álvarez, *Milchwissenschaft*, 1979, **34**, 207.
46. R. G. Jensen, A. M. Ferris, C. J. Lammi-Keefe and R. A. Henderson, *J. Dairy Sci.*, 1990, **73**, 223.
47. T. W. Keenan and S. Patton, *Lipids*, 1970, **5**, 42.
48. B. Holland, I. D. Unwin, D. H. Buss and R. A. McCance, in *The Composition of Foods*, Royal Society of Chemistry, Cambridge, UK, 1989.
49. P. Luna, J. Fontecha, M. Juárez and M. A. de la Fuent, *Lipids*, 2005, **40**, 445.
50. IDF, *Sensory Evaluation of Dairy Products*, International Dairy Federation Std. 99A, Brussels, Belgium, 1987.
51. C. F. Balthazar, J. T. Guimarães, R. Silva, E. G. A. Filho, E. S. Brito, T. C. Pimentel and S. Rodrigues, *et al.*, *J. Dairy Sci.*, 2021, **104**, 5133.
52. C. S. Ranadheera, C. A. Evans, S. K. Baines, C. F. Balthazar, A. G. Cruz, E. A. Esmerino and M. Q. Freitas, *et al.*, *Compr. Rev. Food Sci. Food Saf.*, 2019, **18**, 867.
53. I. V. Wolf, M. C. Perotti, S. M. Bernal and C. A. Zalazar, *Food Res. Int.*, 2010, **43**, 1204.
54. T. G. Bontinis, H. Mallatou, E. C. Pappa, T. Massouras and E. Alichanidis, *Small Rumin. Res.*, 2012, **105**, 193.
55. J. Hong-Xin, S. Mi-Ya and G. Guang-Yu, *CyTA–J. Food*, 2015, **13**, 464.
56. E. Vítová, B. Loupancova, J. Zemanova, H. Štoudková, P. Březina and L. Babak, *Czech J. Food Sci.*, 2006, **24**, 268.
57. J. K. Ha and R. C. Lindsay, *J. Dairy Sci.*, 1991, **74**, 3267.
58. A. H. S. Woo, S. Kollodge and R. C. Lindsay, *J. Dairy Sci.*, 1984, **67**, 874.
59. R. Attaie and R. L. Richter, *J. Dairy Sci.*, 1996, **79**, 717.
60. H. Boelens, H. G. Haring and D. de Reijke, *Perfum. Flavor.*, 1983, **8**, 71.
61. C. P. Brennand, J. K. Ha and R. C. Lindsay, *J. Sens. Stud.*, 1989, **4**, 105.
62. P. Bas, P. Morand-Fehr, A. Rouzeau, *XX International Dairy Congress Paris, E*, 1978, p. 301.
63. S. P. Singh and M. P. Gupta, *Asian J. Dairy Res.*, 1985, **4**, 15.
64. Y. W. Park, *J. Dairy Sci.*, 2001, **84**, E84.
65. Y. Chilliard, G. Selselet-Attou, P. Bas and P. Morand-Fehr, *J. Dairy Sci.*, 1984, **67**, 2216.
66. Y. K. Jin and Y. W. Park, *J. Dairy Sci.*, 1995, **78**, 2598.
67. R. J. FitzGerald and H. Meisel, in *Advanced Dairy Chemistry*, ed. P. F. Fox and P. L. H. McSweeney, Kluwer Academic, New York, NY, 3rd edn, 2003, vol. 1, p. 675.

68. R. C. Chandan, R. M. Parry and K. M. Shahani, *J. Dairy Sci.*, 1968, **51**, 606.
69. F. Remeuf and J. Lenoir, *Intl. Dairy Bull.*, 1986, **202**, 68.
70. E. Renner, G. Schaafsma and K. J. Scott, in *Micronutrients in Milk and Milk-Based Products*, ed. E. Renner, Elsevier, New York, 1989, p. 1.
71. R. M. Whitney, J. R. Brunner, K. E. Ebner, H. M. Farrell, R. V. Josephson, C. V. Morr and H. E. Swaisgood, *J. Dairy Sci.*, 1976, **59**, 795.
72. A. Tziboula-Clarke, in *Encyclopedia of Dairy Sciences*, ed. H. Roginski, J. Fuquay and P. F. Fox, Academic Press, Cambridge, MA, 2003, p. 1270.
73. C. Carles, *J. Dairy Res.*, 1986, **53**, 35.
74. R. C. Chandan, R. Attaie, K. M. Shahani, *Proceedings of the V International Conference on Goats*, 1992, vol. 2, p. 399.
75. A. Siddique and Y. W. Park, *J. Dairy Sci.*, 2019, **102**, 2890.
76. J. Hegenauer, P. Saltman, D. Ludwig, L. Ripley and A. Ley, *J. Agric. Food Chem.*, 1979, **27**, 1294.
77. E. Bernos, J. M. Girardet, G. Humbert and G. Linden, *Biochim. Biophys. Acta, Protein Struct. Mol. Enzymol.*, 1997, **1337**, 149.
78. R. Boutrou, M. H. Famelart, F. Gaucheron, Y. Le Graet, J. Y. Gassi, M. Piot and J. Leonil, *J. Dairy Res.*, 2002, **69**, 60.
79. K. Soodam, L. Ong, I. B. Powell, S. E. Kentish and S. L. Gras, *Food Struct.*, 2017, **14**, 816.
80. O. Aissaoui, S. Carpino, N. Fucà and M. L. Mansour, *3rd International Multidisciplinary Microscopy and Microanalysis Congress*, 2017, vol. 3, p. 175.
81. L. Harrington and J. Mayberry, *Int. J. Clin. Pract.*, 2008, **62**, 1541.
82. S. A. Ibrahim and R. Gyawali, in *Milk and Dairy Products in Human Nutrition*, ed. Y. W. Park and G. F. W. Haenlein, Wiley-Blackwell, Oxford, UK, 2013, p. 246.
83. M. Lomer, G. Parkes and J. Sanderson, *Aliment. Pharmacol. Ther.*, 2008, **27**, 93.
84. E. van Griethuysen-Dilber, E. Flaschel and A. Renken, *Process Biochem.*, 1988, **23**, 55.
85. G. Schaafsma, *Int. Dairy J.*, 2008, **18**, 458.
86. L. G. Rodrigues, in *Milk and Dairy Products in Human Nutrition*, ed. Y. W. Park and G. F. W. Haenlein, Wiley-Blackwell, Oxford, UK, 2013, p. 237.
87. W. J. Mullin and D. B. Emmons, *Food Res. Int.*, 1997, **30**, 147.
88. E. Rossland, T. Langsrud, P. E. Granum and T. Sorhaug, *Int. J. Food Microbiol.*, 2005, **98**, 193.
89. A. N. Califano and A. E. Bevilacqua, *Food Chem.*, 1999, **64**, 193.
90. D. Zhang and A. W. Mahoney, *J. Dairy Sci.*, 1991, **74**, 353.
91. USDA, Agricultural Research Service, *USDA National Nutrient Database for Standard Reference*, Release 28, May 2016.
92. Y. W. Park, *Small Rumin. Res.*, 1994, **13**, 63.
93. H. Lee, B. A. Friend and K. M. Shahani, *J. Dairy Sci.*, 1988, **71**, 3203.
94. Y. W. Park and H. I. Chukwu, *Small Rumin. Res.*, 1988, **1**, 157.

95. Y. W. Park and H. I. Chukwu, *J. Food Compos. Anal.*, 1989, **2**, 161.
96. E. J. Mann, *Dairy Ind. Int.*, 1988, **53**, 23.
97. Y. W. Park, M. Juárez, M. Ramos and G. F. W. Haenlein, *Small Rumin. Res.*, 2007, **68**, 88.
98. G. P. Davidson and R. R. W. Townley, *J. Pediatr.*, 1977, **90**, 590.
99. D. T. McClenathan and W. A. Walker, *Post Grad. Med.*, 1982, **72**, 233.
100. C. Lavigne, J. A. Zee, R. E. Simard and B. Béliveau, *J. Food Sci.*, 1989, **54**, 30.
101. P. Walstra, J. T. M. Wouters and T. J. Geurts, *Dairy Science and Technology*, CRC Press, Boca Raton, FL, 2nd edn, 2006.
102. G. Moatsou, in *Milk and Dairy Products in Human Nutrition*, ed. Y. W. Park and G. F. W. Haenlein, Wiley-Blackwell, Oxford, UK, 2013, p. 288.
103. A. Lucas, E. Rock, J.-F. Chamba, I. Isabelle Verdier-Metz, P. Brachet and J.-B. Coulon, *Lait*, 2006, **86**, 21.
104. A. Lucas, J. B. Coulon, C. Agabriel, Y. Chilliard and E. Rock, *Small Rumin. Res.*, 2008, **74**, 91.
105. M. A. Galina, F. Osnaya, H. M. Cuchillo and G. F. W. Haenlein, *Small Rumin. Res.*, 2004, **71**, 264.

CHAPTER 12

Non-dairy Cheese

MICHAEL H. TUNICK

Department of Food & Hospitality Management, College of Nursing & Health Professions, Drexel University, Philadelphia, PA 19104, USA
Email: mht39@drexel.edu

12.1 Introduction

The purpose of this brief final chapter is to review the growing field of cheeses made without milk. The products are usually labelled as analogues, alternatives, replacements, or substitutes, with the adjectives non-dairy, dairy-free, vegan, or plant-based. Non-dairy cheese products are not a new phenomenon. Tofu, which was developed in China some 2000 years ago, is prepared in a manner analogous to cheesemaking. Soybeans are soaked and ground in water to form a slurry which is then boiled and filtered to obtain soy milk. The soy protein, mostly glycinin and β-conglycinin, is coagulated with a salt (traditionally calcium sulphate), an acid (such as citric or lactic), or an enzyme (such as papain). The resulting curd is pressed to obtain firm or extra-firm tofu, also known as soy paneer.[1] Non-dairy cheese thus has a history stretching across two millennia.

12.1.1 Rationale

Non-dairy cheese in the 21st century has been described as a trend and not a fad, and is being purchased by consumers for various reasons, including expense, sustainability, dietary restrictions, and nutritional concerns.[2] Many consumers are concerned with greenhouse gas emissions from livestock production, which have been identified as important contributors

Food Chemistry, Function and Analysis No. 40
Handbook of Cheese Chemistry
Edited by Michael H. Tunick

Published by the Royal Society of Chemistry, www.rsc.org

to climate change and animal welfare.[3] Some medical professionals recommend that meat consumption in developed countries should be halved to reduce diet-related diseases. "De-animalisation" will be the result of these trends.[3]

12.1.2 Market

Cheese alternatives are an expanding market, especially with vegans and lactose intolerant people, but duplicating the flavour, mouthfeel, and meltability of dairy cheese has been described by executives in the industry as being "difficult" and even "literally impossible".[2] In 2020, non-dairy cheese sales in the US reached $270 million, a 42.5% increase over 2019, making plant-based cheese alternatives one of the fastest growing categories of plant-based foods.[4] The market for vegan cheese has been predicted to increase by almost 12% from 2021 through 2028.[5] The market research publisher Packaged Facts predicts that better technology and processes will eventually be developed that will make plant-based cheese products more dairy-like.[6] They also project sales of these products will grow faster during 2024–2029 than during 2019–2024, and that plant-based cheese will become the second largest category in the plant-based dairy space by 2029.[6]

Some of the research on cheese alternatives is performed by companies for their own use and is proprietary and unpublished. Other manufacturers apply for patents, which can be read by anyone, and a few universities and institutes are investigating dairy substitutes in cheese and reporting their results. What follows here is a summary of the publicly available work in this area.

12.2 Plants Used

Grains, legumes (especially soybeans), nuts (especially coconut), seeds, and tubers have been used to manufacture non-dairy cheese. A recent survey of 245 plant-based cheeses in the western US showed that 43% were made from a mixture of coconut oil, potato, tapioca, and corn starch; another 25% were based on soy and coconut oil; 14% were based on cashews; and the remaining 18% was manufactured from almonds, oats, soy, and various oils and starches, either alone or in combination.[5] Eighty percent of plant-based cheeses in the UK are based on coconut or palm oil.[7] Table 12.1 lists the advantages and disadvantages of some common plants used in non-dairy cheeses.

Cheese has a complex structure, which makes the manufacture of non-dairy alternatives problematic. High-quality plant-based ingredients with the required functional qualities are hard to come by, and the processing technologies for transforming these ingredients into desirable products are still being developed.[8]

Table 12.1 Advantages and disadvantages of selected plants used in non-dairy cheese compared with cow's milk. Data reproduced from ref. 13 and 14, https://doi.org/10.3390/pr812154 and https://doi.org/10.1016/j.fufo.2021.100098, under the terms of CC BY 4.0 license, http://creativecommons.org/licenses/by/4.0/.

Plant	Advantages	Disadvantages
Almond	Contains high levels of protein and monosaturated fatty acids. Good source of Mn and vitamins A and E. Contains some bioactive compounds.	Contains potentially allergenic proteins. Susceptible to rancidity.
Coconut	Contains a high level of saturated fat. Good source of vitamin E. Good availability of Cu, Fe, and Mn. Not allergenic.	Low levels of monosaturated and polyunsaturated fatty acids.
Oat	High levels of carbohydrate, fibre, and lipids. Gluten-free and not allergenic.	Contains trypsin inhibitors and phytates. Low Ca level. Susceptible to rancidity.
Quinoa	Good source of cysteine, lysine, methionine, tocopherol, Ca, Cu, Fe, K, Mg, and Mn. Gluten-free. High levels of starch and protein digestibility.	Uncommon plant compared with the others in this list. Bitter taste.
Rice	High levels of carbohydrate, starch, K, Mg, and P. Good source of B vitamins and vitamin E. Contains phytosterols. Gluten-free and not allergenic.	Low levels of monounsaturated and polyunsaturated fatty acids. High sugar content. Protein is at a low level and poorly digestible. Contains trypsin inhibitors and phytates.
Soy	High protein digestibility and availability of Cu, Fe, and Mg. Contains bioactive compounds, phytochemicals, and polyunsaturated fatty acids.	Has beany flavour and contains potentially allergenic proteins, and trypsin inhibitors.

12.3 Manufacture

12.3.1 Procedures

Non-dairy cheeses are protein-based (and usually fermented with lactic acid bacteria) or oil- and starch-based (not fermented). Emulsions based on oil and starch have a continuous phase containing oil droplets and soluble polysaccharides such as modified starch, inulin, or xanthan.[9] The starch helps with stretching of the product and the oil provides meltability.[10]

Plant proteins for use in non-dairy cheese often undergo a combination of soaking, roasting, dehulling, and milling prior to fractionation or tissue disruption,[11] resulting in colloidal dispersions consisting of a viscoelastic polysaccharide and/or protein network with lipid droplets embedded within it.[8,11] Emulsions based on polysaccharides and plant proteins have continuous phases similar to those based on oil and starch, but they also contain insoluble fibre.[9] Homogenization in water with emulsifiers, flavours, stabilizers, and other ingredients leads to colloids in the 5–20 μm range;[8] in comparison, bovine milk fat globules are 2–15 μm in size and casein micelles are <0.5 μm.[12] Lactic acid bacterium cultures are then added to ferment the mixture until the desired texture is achieved.[13] Non-dairy cheese frequently undergoes pasteurization for safety, but this process thermally degrades some bioactive compounds and changes the sensory properties. Some innovative technologies, including high pressure processing, ohmic heating, pulsed electric field, ultrasound, and ultraviolet radiation, have been used in place of pasteurization.[14]

The use of acids, enzymes, and heat may be required to obtain the desired texture and meltability of non-dairy cheese,[15] leaving room for much experimentation with various combinations. The required firmness may be produced by adding partially hydrogenated vegetable oils.[13] Alternatively, gel strength may be increased by adding cross-linking agents such as transglutaminase, which form covalent cross-links between protein molecules, or calcium or other cations, which cause electrostatic bridging of anionic proteins.[8] Coconut oil is often preferred because it contains a high percentage of saturated fatty acids, giving the product firmness along with a creamy texture.[15]

12.3.2 Patented Processes

A number of patents and patent applications for non-dairy cheeses have been published. To give three examples: in a patent for pea flour cheese, water and yellow pea flour are mixed at 60 °C, blended with canola oil, heated to 90 °C, combined with $CaSO_4$, and cooled.[16] Vegetable and microbial rennets are used in a patent application for preparing cheese from coconut, avocado, and durian milks. The milk is heated to 84 °C, then cooled to 35 °C, the rennets are added, and the curd is drained over a period of 12–48 h before pressing.[17] In another patent application, safflower and/or sunflower seeds are soaked, rinsed with heat, blended with sea salt, oil, and water, and boiled. After draining and cooling, $CaCO_3$, sugars, microbes, and flavouring agents are added, and the curd is dehydrated or incubated to accelerate fermentation.[18]

12.4 Properties

The properties that make cheese desirable include flavour, texture, and meltability. Flavour in cheese originates from the volatile compounds present, texture is related to hardness and fracture characteristics, and meltability depends on the fat globules in the casein matrix.[19] A recent comparison of commercial plant-based cheese blocks with Cheddar and processed cheeses showed that the plant-based products were significantly different in composition, microstructure, and some physicochemical properties, which were strongly influenced by their formulation. The protein content of the non-dairy cheese was significantly lower than that of the dairy-based products and microscopy revealed that the non-dairy cheese lacked the continuous protein network present in regular cheese.[20] However, zein (maize protein) has a relatively open structure compared with other plant proteins, reportedly resulting in a product with melting, stretching, rheology, and texture similar to Cheddar.[10]

12.5 Nutrition

Fermentation has been used for centuries to preserve food by outcompeting pathogens and spoilage microbes and generating antimicrobial compounds, including acid. The bioavailability of nutrients in soy cheese is increased by fermentation while reducing antinutritive components.[21] Extra vitamins and minerals may be added as well using more than one plant, which would provide a wider spectrum of nutrients.[22]

Craig *et al.* tabulated the calories and nutrients contained in the products in their survey and found that only 3% of the products contained 5 g of protein per serving, <20% were fortified with calcium, 6% had added vitamin B_{12}, and only 1% had vitamin D added.[5] Fresán and Rippin recently compared 40 plant-based cheeses in Spanish supermarkets and found that majority contained high levels of calories, fat, saturated fat, and salt, and were low in fibre and protein.[23] Non-dairy cheeses based on cashew and tofu provided better nutritional profiles, but these products were scarce.[23]

12.6 Sensory Attributes

Soy proteins are large oligomers, which result in gritty textures.[11] Soy is also noted for having a beany flavour that comes from pentanal and hexanal, which are breakdown products of unsaturated fatty acids.[21] This undesirable flavour was greatly reduced by adding a fungus, *Geotrichum candidum*, to induce proteolysis and lipolysis of soy protein and fat.[24] Tofu, Mozzarella, and Cheddar blends in various ratios led to processed cream cheeses with good acceptability scores.[25] The proteolytic pathways of many lactic acid bacteria are known, and selected cultures will result in breakdown products that should enhance flavour.[26] Mixing dairy milk with soy milk often improves the sensory properties of the final product.[21] Soaking soybeans overnight in a sodium bicarbonate solution before undergoing fermentation and coagulation with lactic acid bacteria and glucono-delta-lactone significantly improved likability, though dairy cheese scored better.[27] Oyeyinka *et al.* studied beske, a traditional Nigerian soy cheese, and found that the overall acceptability of 100% soy milk beske was not significantly different from that of a 60% soy milk–40% cashew milk blend, with both being rated as a 7.0 out of a possible 9.0.[28]

Cheeses made from cow, coconut, and soy milks were prepared by coagulation with an acid or rennet and their composition, microbiology, and sensory properties were measured. The coconut and soy cheese compared favourably with cow cheese in appearance, aroma, taste, and texture.[29] Experiments with blends of soy and coconut milks showed that a 50 : 50 blend performed best.[30]

Many protein-based non-dairy cheeses are based on cashew nuts, apparently because of consumer aversion and allergies to soy.[15] However, the literature covering the analysis of cashew-based cheese products is scanty.[15]

12.7 Consumer Perceptions

Studies show that about one-third of US consumers believe that plant-based cheese tastes as good as regular cheese, about one-third disagree, and the remaining third are neutral.[22] Strategies toward improving consumer perceptions toward non-dairy cheese and other plant-based foods have been investigated in recent years, and a summary was published by Short *et al.*[22] They selected 12 papers dealing with sensory analysis of plant-based cheese substitutes, 11 of which were published since 2013. Differences in methodology, including the number of panellists and their level of training, limit the usefulness of the results. No study directly measured the beany and gritty sensory attributes of soy-based cheese, and four of the papers did not include statistical analysis.[22] It is clear that more standardized sensory testing and correlations with instrumental methods are needed to provide useful comparisons among plant-based cheeses and with dairy cheese.

12.8 Conclusion

The science behind the production of non-dairy cheeses is still developing. Various plants have been used to manufacture these products, others are being tested, and improvements in nutritional and sensory properties are being investigated. Substituting plant milk for dairy milk is a daunting challenge since the structure of genuine cheese is different than that of non-dairy cheese, so duplicating milk-based cheese using plants will be quite challenging.

References

1. L. Zheng, J. M. Regenstein, F. Teng and Y. Li, *Compr. Rev. Food Sci.*, 2020, **19**, 3683.
2. A. Boisseau, *New Nondairy Offerings Are More Than Alternatives*, 2021, https://www.dairyfoods.com/articles/95212-new-nondairy-offerings-are-more-than-alternatives.
3. C. Morris, J. Mylan and E. Beech, *Int. J. Soc. Agric. Food*, 2018, **25**, 42.

4. Plant-based Foods Association, *Retail Sales Data*, 2021, https://www.plantbasedfoods.org/retail-sales-data/.
5. W. J. Craig, A. R. Mangels and C. J. Brothers, *Nutrients*, 2022, **14**, 1247.
6. B. Garrison, *Dairy Foods Plant-Based Alternatives*, 2021, https://digitaledition.preparedfoods.com/march-2021/dairy-foods-plant-based/.
7. M. Saraco and J. Blaxland, *Br. Food J.*, 2020, **122**, 3727.
8. D. J. McClements and L. Grossmann, *Compr. Rev. Food Sci.*, 2021, **20**, 4049.
9. R. Kamath, S. Basak and J. Gokhale, *LWT*, 2021, **17**, 112991.
10. K. D. Mattice and A. G. Marangoni, *Curr. Res. Food Sci.*, 2020, **3**, 59.
11. A. Pua, V. C. Tang, R. M. Goh, J. Sun, B. Lassabliere and S. Q. Liu, *Foods*, 2022, **11**, 875.
12. A. Logan, L. Day, A. Pin, M. Auldist, A. Leis, A. Puvanenthiran and M. A. Augustin, *Food Bioprocess Technol.*, 2014, **7**, 3175.
13. W. Fu and H. Yano, *Processes*, 2020, **8**, 1541.
14. R. Bocker and E. K. Silva, *Future Foods*, 2022, **5**, 100098.
15. A. R. Harper, R. C. Dobson, V. K. Morris and G. J. Moggré, *Microb. Biotechnol.*, 2022, **15**, 1404.
16. W. C. Thresher, *US Pat.*, 6777016B2, 2021.
17. E. E. Iruthayathasan and B. Lahousse, *US Pat.*, US2021/0106018A1, 2021.
18. P. Randolph, *US Pat.*, US2022/0030900A1, 2022.
19. M. H. Tunick, *The Science of Cheese*, Oxford University Press, New York, 2014.
20. N. Grasso, Y. H. Roos, S. V. Crowley, E. K. Arendt and J. A. O'Mahony, *Future Foods*, 2021, **4**, 100048.
21. R. K. Jeewanthi and H. D. Paik, *J. Food Sci. Technol.*, 2018, **55**, 4384.
22. E. C. Short, A. J. Kinchla and A. A. Nolden, *Foods*, 2021, **30**, 725.
23. U. Fresán and H. Rippin, *Nutrients*, 2021, **21**, 3291.
24. Y. Li, X. Zhang, J. J. Yang, X. Y. Ma, X. D. Jia, P. Du and A. L. Li, *J. Food Process. Preserv.*, 2020, **44**, e14823.
25. S. K. Verma, S. Upadhyay, R. Chandra and A. Paul, *Int. J. Food Sci. Nutr.*, 2013, **2**, 19.
26. N. Ahmad, L. Li, X. Q. Yang, Z. X. Ning and M. A. Randhawa, *Food Technol. Biotechnol.*, 2008, **46**, 252.
27. Q. Li, Y. Xia, L. Zhou and J. Xie, *Food Bioprod. Process.*, 2013, **91**, 429.
28. A. T. Oyeyinka, J. O. Odukoya and Y. S. Adebayo, *J. Food Process. Preserv.*, 2019, **43**, e14285.
29. A. A. Ayodeji, D. Ahure, E. E. Efiong and I. O. Acham, *Eur. J. Nutr. Food Saf.*, 2020, **12**, 1.
30. V. S. Kadbhane, G. N. Shelke and S. L. Thorat, *Pharma Innovation*, 2019, **8**, 56.

Subject Index